An Expanding World
Volume 14

Metals and Monies in an Emerging Global Economy

AN EXPANDING WORLD

The European Impact on World History, 1450–1800

General Editor: A.J.R. Russell-Wood

Please note titles may change prior to publication

An Expanding World
The European Impact on World History 1450–1800

Volume 14

Metals and Monies in an Emerging Global Economy

edited by
Dennis O. Flynn and
Arturo Giráldez

VARIORUM
1997

Published by VARIORUM
Ashgate Publishing Limited
Gower House, Croft Road
Aldershot, Hampshire GU11 3HR
Great Britain

Ashgate Publishing Company
Old Post Road
Brookfield, Vermont 05036
USA

ISBN 0-86078-531-9

British Library CIP data

Metals and Monies in an Emerging Global Economy.
(An Expanding World: The European Impact on World History, 1450–1800: Vol. 14).
1. Money–History. 2. Precious metal industries–History. 3. Economic history.
I. Flynn, Dennis O. II. Giráldez, Arturo.
332. 4' 9

US Library of Congress CIP data

Metals and Monies in an Emerging Global Economy/edited by Dennis O. Flynn and Arturo Giráldez.
p. cm. – (An Expanding World: Vol. 14).
Text in English, French and Spanish. A collection of articles previously published between 1926 and 1995. Includes bibliographical references and index. 1. Precious metals—History. 2. Money–History. 3. Money Supply–History.
I. Flynn, Dennis Owen, 1945– . II. Giráldez, Arturo. III. Series
HG265.M47 1997 96–46512
332.4–dc21 CIP

This book is printed on acid free paper.

Printed in Great Britain by Galliard (Printers) Ltd, Great Yarmouth

AN EXPANDING WORLD 14

Contents

Acknowledgements

The chapters in this volume are taken from the sources listed below, for which the editor and publishers wish to thank their authors, original publishers or other copyright holders for permission to use their material as follows:

Chapter 1: John U. Nef, 'Silver Production in Central Europe, 1450–1618', *Journal of Political Economy* XLIX (Chicago, 1941), pp. 575–591. Copyright © 1941 by the University of Chicago Press.

Chapter 2: Zacarías Moutoukias, 'Una forma de oposición: el contabando', *Governare il mondo. L'impero spagnolo dal XV al XIX secolo*, ed. Massimo Ganci and Ruggiero Romano (Palermo, 1991), pp. 333–368. Copyright © 1991 by the Società Siciliana per la Storia Patria.

Chapter 3: A. Kobata, 'The Production and Uses of Gold and Silver in Sixteenth and Seventeenth-Century Japan', *Economic History Review* XLIII, no. 2 (Oxford, 1965), pp. 245–266. Copyright 1965 by A. Kobata.

Chapter 4: Tashiro Kazui, 'Exports of Japan's Silver to China via Korea and Changes in the Tokugawa Monetary System During the Seventeenth and Eighteenth Centuries', in ed. Eddy H.G. Van Cauwenberghe, *Precious Metals, Coinage and the Changes of Monetary Structures in Latin America, Europe and Asia* (Leuven, 1989), pp. 99–116. Copyright © 1989 by Leuven University Press.

Chapter 5: Kristof Glamann, 'The Dutch East India Company's Trade in Japanese Copper, 1645–1736', *Scandinavian Economic History Review* I (Lund, 1953), pp. 41–79. Copyright © 1953 by Kristof Glamann.

Chapter 6: Richard Bean, 'A Note on the Relative Importance of Slaves and Gold in West African Exports', *Journal of African History* XV, no. 3 (Cambridge, 1974), pp. 351–356. Copyright © 1974 by Cambridge University Press.

Chapter 7: William S. Atwell, 'International Bullion Flows and the Chinese Economy, *circa* 1530–1650', *Past and Present* XC (Oxford, 1982), pp. 68–90. Copyright © 1982 by the Past and Present Society.

Chapter 8: Brian Moloughney and Xia Weizhong, 'Silver and the Fall of the Ming: A Reassessment', *Papers on Far Eastern History* XL (Canberra, 1989), pp. 51–78. Copyright © 1989 by *Papers on Far Eastern History*, Department of Far Eastern History, Australian National University.

Chapter 9: Marion Johnson, 'The Cowrie Currencies of West Africa', *Journal of African History* XI, nos. 1 and 3 (Cambridge, 1970), pp. 17–48 and 331–353. Copyright © 1970 by Cambridge University Press.

Chapter 10: Frank Perlin, 'A History of Money in Asian Perspective', *The Journal of Peasant Studies* VII, no. 2 (Ilford, Essex, 1980), pp. 235–244. Copyright © 1980 by Frank Cass and Co. Ltd.

Chapter 11: Dennis O. Flynn and Arturo Giráldez, 'Born with a 'Silver Spoon': the Origin of World Trade in 1571', *Journal of World History* VI, no. 2 (Hawaii, 1995), pp. 201–221. Copyright © 1995 by the University of Hawaii Press.

Chapter 12: Chuan Hang-Sheng, 'Trade Between China, the Philippines and the Americas During the Sixteenth and Seventeenth Centuries', *Proceedings of the International Conference of Sinology: Section on History and Archaeology* (Taipei: Academia Sinica, 1981), pp. 849–854. Copyright © 1981 by the Academia Sinica, Taiwan.

Chapter 13: J.-G. Van Dillen, 'Amsterdam, marché mondial des métaux précieux au XVIIe et XVIIIe siècles', *Revue Historique* CLII (Paris, 1926–1927), pp. 194–201. Copyright © 1926 by Presses Universitaires de France.

Chapter 14: K.N. Chaudhuri, 'World Silver Flows and Monetary Factors as a Force of International Economic Integration, 1658–1758', in eds. W. Fisher, R. McInnis and J. Schneider, *The Emergence of a World Economy, 1500–1914: Papers of the IX International Congress of Economic History* (Stuttgart, 1986), pp. 61–82. Copyright © by K.N. Chaudhuri.

Chapter 15: Pamuk Şevket, 'The Disintegration of the Ottoman Monetary System During the Seventeenth Century', *Princeton Papers in Near Eastern Studies* II (Princeton, 1993), pp. 67–81. Copyright © 1993 by the Darwin Press, Inc.

Chapter 16: V.B. Gupta, 'Imports of Treasure and Surat's Trade in the Seventeenth Century', in ed. Eddy H.G. Van Cauwenberghe, *Money, Coins, and Commerce: Essays in the Monetary History of Asia and Europe* (Leuven, 1991), pp. 455–471. Copyright © 1991 by Leuven University Press.

General Editor's Preface

A.J.R. Russell-Wood

An Expanding World: The European Impact on World History, 1450–1800 is designed to meet two objectives: first, each volume covers a specific aspect of the European initiative and reaction across time and space; second, the series represents a superb overview and compendium of knowledge and is an invaluable reference source on the European presence beyond Europe in the early modern period, interaction with non-Europeans, and experiences of peoples of other continents, religions, and races in relation to Europe and Europeans. The series reflects revisionist interpretations and new approaches to what has been called 'the expansion of Europe' and whose historiography traditionally bore the hallmarks of a narrowly Eurocentric perspective, focus on the achievements of individual nations, and characterization of the European presence as one of dominance, conquest, and control. Fragmentation characterized much of this literature: fragmentation by national groups, by geography, and by chronology.

The volumes of *An Expanding World* seek to transcend nationalist histories and to examine on the global stage rather than in discrete regions important selected facets of the European presence overseas. One result has been to bring to the fore the multicontinental, multi-oceanic and multinational dimension of the European activities. A further outcome is compensatory in the emphasis placed on the cross-cultural context of European activities and on how collaboration and cooperation between peoples transcended real or perceived boundaries of religion, nationality, race, and language and were no less important aspects of the European experience in Africa, Asia, the Americas, and Australia than the highly publicized confrontational, bellicose, and exploitative dimensions. Recent scholarship has not only led to greater understanding of peoples, cultures, and institutions of Africa, Asia, the Americas, and Australasia with whom Europeans interacted and the complexity of such interactions and transactions, but also of relations between Europeans of different nationalities and religious persuasions.

The initial five volumes reflect the changing historiography and set the stage for volumes encompassing the broad themes of technology and science, trade and commerce, exploitation as reflected in agriculture and the extractive industries and through systems of forced and coerced labour, government of empire, and society and culture in European colonies and settlements overseas. Final volumes examine the image of Europe and Europeans as 'the other' and the impact of the wider world on European *mentalités* and mores.

An international team of editors was selected to reflect a diversity of educational backgrounds, nationalities, and scholars at different stages of their professional careers. Few would claim to be 'world historians', but each is a recognized authority in his or her field and has the demonstrated capacity to ask the significant questions and provide a conceptual framework for the selection of articles which combine analysis with interpretation. Editors were exhorted to place their specific subjects within a global context and over the *longue durée*. I have been delighted by the enthusiasm with which they took up this intellectual challenge, their courage in venturing beyond their immediate research fields to look over the fences into the gardens of their academic neighbours, and the collegiality which has led to a generous informal exchange of information. Editors were posed the daunting task of surveying a rich historical literature and selecting those essays which they regarded as significant contributions to an understanding of the specific field or representative of the historiography. They were asked to give priority to articles in scholarly journals; essays from conference volumes and *Festschriften* were acceptable; excluded (with some few exceptions) were excerpts from recent monographs or paperback volumes. After much discussion and agonizing, the decision was taken to incorporate essays only in English, French, and Spanish. This has led to the exclusion of the extensive scholarly literature in Danish, Dutch, German and Portuguese. The ramifications of these decisions and how these have had an impact on the representative quality of selections of articles have varied, depending on the theme, and have been addressed by editors in their introductions.

The introduction to each volume enables readers to assess the importance of the topic *per se* and place this in the broader context of European activities overseas. It acquaints readers with broad trends in the historiography and alerts them to controversies and conflicting interpretations. Editors clarify the conceptual framework for each volume and explain the rationale for the selection of articles and how they relate to each other. Introductions permit volume editors to assess the impact on their treatments of discrete topics of constraints of language, format, and chronology, assess the completeness of the journal literature, and address *lacunae*. A further charge to editors was to describe and evaluate the importance of change over time, explain differences attributable to differing geographical, cultural, institutional, and economic circumstances and suggest the potential for cross-cultural, comparative, and interdisciplinary approaches. The addition of notes and bibliographies enhances the scholarly value of the introductions and suggests avenues for further enquiry.

I should like to express my thanks to the volume editors for their willing participation, enthusiasm, sage counsel, invaluable suggestions, and good judgment. Evidence of the timeliness and importance of the series was illustrated by the decision, based on extensive consultation with the scholarly community, to expand a series, which had originally been projected not to exceed eight volumes, to more than thirty volumes. It was John Smedley's initiative which gave rise to discussions as to the viability and need for such a series and he has overseen the publishing, publicity, and marketing of *An Expanding World*. As

General Editor, my task was greatly facilitated by the assistance of Dr Mark Steele who was initially responsible for the 'operations' component of the series as it got under way; latterly this assistance has been provided by staff at Variorum.

The Department of History,
The Johns Hopkins University

Introduction

Dennis O. Flynn and Arturo Giráldez

I. Organizational comments

Both the selection of articles for this volume, as well as the scheme for organizing them, are unconventional. First, most monetary histories of the early-modern period focus excessively on Europe. This volume aspires to offer a more even geographical balance, one which we hope encourages others to view monetary events from a global perspective. Europeans played a significant role in the movement of monies around the world, to be sure, but they were not necessarily prime movers. In many important respects, European merchants played the role of middlemen.

Second, articles chosen for inclusion tend to focus on demand and supply forces affecting the specific markets for gold, silver, copper, and cowries during particular time intervals. It is essential to keep in mind that the history of the supply of, and demand for, each of these substances is unique (interrelationships among unique markets notwithstanding). Since Europe, Spanish America, and Japan were the world's major supply-side producers of silver during our period, for example, we have grouped together articles which discuss these silver suppliers under 'Source Areas' (see Section I: Source Areas) along with essays on gold and copper suppliers. This scheme facilitates a view of globally-integrating market forces, in contrast to the market-shrouding alternative of fragmenting integrated markets into their 'Asian', 'American', 'African', 'European', and 'Middle Eastern' components. Monies and metals impacted all of these geographical areas simultaneously; it is essential that we break the habit of thinking in 'regional-history' terms when discussing globally-traded monies and metals.

Early-modern monetary history tends to focus too heavily upon European issues, but we feel that this characteristic has less to do with chauvinistic attitudes than with distinctions imbedded within modern monetary theory itself. Specifically, macroeconomic monetary theory instructs that various coined substances should be added up to help form a unit-of-analysis, 'money'. This convention of aggregating diverse monetary items into an abstract category 'money' has led to confusion in the literature on the early-modern period. Dispute over the cause of Euro-Asian trade in the sixteenth and seventeenth centuries illustrates the nature of this confusion about how to best conceptualize monies. At issue is whether it is best to perceive precious metals as cause or as effect.

The great French historian Fernand Braudel maintained that it 'was the economic development of Europe and the demand thus created which stimulated and dictated the labour of the gold panners and the Indians in the silver mines...';

that is, European demand can be considered cause, while mine activity in America was an effect (responding to European demand).[1] We will take issue with Braudel's conclusion regarding end-market demand in a moment, but his underlying intuition points to a key question: Why would entrepreneurs bother to produce any item in the absence of purchasers on the demand-side of the market? It makes no sense to produce a commercial item, whether it is stamped into money later or not, unless there are plenty of customers willing to purchase it at a price which yields profit. Elsewhere in the same volume, Braudel portrays a more complex picture, one which acknowledges that Spanish-American silver merely passed through – not to – Europe on its mercantile voyage to Asia:

> So, during the sixteenth and seventeenth centuries there circulated throughout the vast Asian continent, source of spices, drugs, and silk, the precious gold and above all silver coins minted at Venice, Genoa, or Florence, and later the famous Spanish pieces of eight. Away to the east flowed these currencies, out of the Mediterranean circuit into which it had often required so much patience to introduce them. The Mediterranean...horded them only to lose them all to India, China, and the East Indies. The great discoveries may have revolutionized routes and prices, but they did not alter this fundamental situation, no doubt because it was still a major advantage to westerners to have access to the precious merchandise of the East...; no doubt also because in the sixteenth century, as in the past, the purchasing power of precious metals rose above that of Christian countries as soon as one crossed the border into the Orient.[2]

Let us take a closer look at the European-demand explanation for precious metal flows to Asia, one reiterated repeatedly in textbooks on the subject. According to this line of reasoning, an enormous European demand for spices, silks, porcelains, and other Asian products is juxtaposed with a meager Asian demand for European products. For some reason, Asian people – often depicted as traditional, inward-looking, or insular in character – had not yet developed an appreciation for European products and therefore purchased few of them. Thus, huge European imports from Asia, in conjunction with meager European exports to Asia, created a European balance-of-trade deficit with the East. Precious metals flowed to Asia, according to this scenario, as balancing items. Demand-led dynamism within Europe caused treasure to migrate to Asia.

The problem is, this European balance-of-trade hypothesis contradicts the historical record. 'Precious metals' did not flow eastward between 1550 and 1650 – *silver* did. Moreover, one of the products flowing in the opposite direction was

---

[1] Fernand Braudel, *The Mediterranean and the Mediterranean World in the Age of Philip II* (New York, Evanston, San Francisco, London, 1972), vol.I, p.522.

[2] Ibid., p.464.

gold:[3] '[In the sixteenth century, the] balance of payments was definitely favorable to Portugal. In Goa, commodities were partly paid for in bills of exchange, and partly in gold sent to Lisbon. This gold might in turn go to Hamburg, to be exchanged for copper and silver'.[4] Supple devotes an entire chapter to discussion of whether England's silver exports (1600–1642) were equal in value to its imports of gold, focusing on the question of whether England's money supply may thereby have been stable.[5] In a detailed study of the English East India Company, K.N. Chaudhuri concludes:

> The East India Company was aware of the true reason for the export of English silver coins, and the Deputy Governor's reference to the import of gold for East India commodities sold in Europe is another example which bears out Dr. Supple's conclusion that the outflow of silver was matched by an inflow of gold.[6]

Furthermore, this century-long silver-drain should not be viewed in East-West terms. American silver was not exported to abstract 'Asia'; its predominant destination was China, a country simultaneously exporting significant amounts of gold. Japan was the world's second largest producer of silver during this period, again exporting its silver to China in exchange for silk, gold, and other items. Once more, if precious metals were exported by Europeans to Asia for the purpose of balancing a European trade deficit, then why did silver dominate the flow to China, while gold sometimes flowed in the opposite direction to Europe? Both were monetized metals, but each flowed independently in response to its own specific market conditions.

Throughout the seventeenth century, Pacific galleons carried two million pesos in silver annually (i.e. more than 50 tons) from Acapulco to Manila, where it was quickly transhipped by merchants to China.[7] Why has no one argued that the flow of American treasure to China was the result of Mexico's balance-of-trade deficit with China? That is, why does no one maintain that this drain of silver was caused by dynamic Mexican demand for Asian products (in the face of stagnant Asian demand for Spanish-American output)? There is no theoretical

---

[3] Our analysis follows in the tradition of Chaudhuri, whose work is based on the reasoning of Classical economics, especially that of David Ricardo. K.N. Chaudhuri, *The Trading World of Asia and the English East India Company, 1660–1760* (Cambridge, 1978), p.156.

[4] Pierre Vilar, *A History of Gold and Money, 1450–1920.* (London, 1976), p.100.

[5] B.E. Supple, *Commercial Crisis and Change in England 1600–1642* (Cambridge, 1964), pp.163–194.

[6] K.N. Chaudhuri, *The English East India Company* (London, 1965), p.132.

[7] Hang-Sheng Chuan, 'The Inflow of American Silver into China from the Late Ming to the Mid-Ch'ing Period', *The Journal of the Institute of Chinese Studies of the Chinese University of Hong Kong*, no.2, p.79; D.O. Flynn and Arturo Giráldez, 'China and the Manila Galleons', in eds. A.J.H Latham and Heita Kawakatsu, *Japanese Industrialization and the Asian Economy* (London, 1994), p.82.

justification for treatment of Latin America differently than Europe, yet their exports of silver are not portrayed similarly in the literature.

During the second half of the seventeenth century, the historical record also contradicts the balance-of-payments hypothesis. Japan became the world's largest copper producer (and a significant source of gold as well) after Japanese silver production declined later in the seventeenth century. This time, it was copper which was shipped to Europe from Asia in exchange for both silver and gold in a counter-flow out of Europe:

> the importance of Europe as a buyer of copper in the 1650's and 1670's was considerably greater than in later years. As will be mentioned in another connection, there is reason to assume that the great European demand from the middle of the 1650's gave an impetus to the [Dutch East India] company's copper exports from Japan (see chapter 5 below).[8]

But this sea change in metals trade means that two monetized metals (gold and silver) were now exported out of Europe, in exchange for one monetized metal (copper) imported to Europe from Japan. Again, the practice of aggregating these three metals, each with a unique history, into a unit of analysis called 'money' distracts attention from the underlying forces driving these events.

In sum, prodigious quantities of silver flowed to China from America and Japan between 1550–1650, while China exported substantial quantities of gold to Japan; lesser amounts of Asian gold were shipped to Europe because Africa and America were Europe's main sources of gold. During the second half of the 17th century, Japan exported copper and gold to China (as well as elsewhere in Asia) and copper to Europe, while gold and silver flowed out of Europe into Asian market-places. The 'dynamic-European-trade-deficit' hypothesis conflicts with all of these historical trends (as would a 'dynamic-Mexican-trade-deficit' argument, if anyone were to offer a parallel argument). If the trade-deficit argument were true, then various types of monies would be expected to have gravitated toward Asia in the early-modern period. This is not what happened. The specific metal, silver, headed to China from all corners of the earth. Discovery of why China was such a giant vacuum cleaner for silver – *bomba aspirante* (suction pump) in Vitorino Magalhães Godinho's colourful terms – requires conceptual disaggregation of silver from other monetized substances.[9] Aggregation of all of these items into a unit of analysis called 'money' precludes proper understanding of the economic forces driving intercontinental trade. Each substance must be subjected to individual supply and demand analysis in its own right. Section I:

---

[8] Kristof Glamann, 'The Dutch East India Company's Trade in Japanese Copper, 1645–1736', *Scandinavian Economic History Review* (1953), p.61.

[9] Vitorino Magalhães Godinho, *Os Descobrimentos e a Economia Mundial*, 2nd. edn. IV (Lisboa, 1984), pp. 432–65.

Source Areas supports this disaggregation scheme by offering essays which sketch most of the important areas of gold, silver, and copper production.

Section II: End Markets contains articles which focus primarily on the demand side for particular monies, and Section III: Intermediary Trade Routes deals with avenues of trade which connected source areas with end markets. As Braudel's citation above suggests, it does not make sense to produce commercial products unless there are end-market customers. The purpose of Section II is to identify market forces on the demand-side for the four monetary substances discussed in this volume: Gold, silver, copper, and cowries. Each one of these money substances will be discussed further in a moment, but we should first complete the story of the 'silverization' of China, perhaps one of the most important and neglected aspects of the 'Age of Expansion'.

Our 'Born with a Silver Spoon: The Origin of World Trade in 1571' (see chapter 11 below) offers an overview of the 'silverization' of China and its impact on world trade patterns. China's paper money system had fallen apart and merchants, especially in maritime regions of China, converted to the use of silver as a monetary base.[10] As this trend toward conversion to silver money expanded, local and regional government entities increasingly converted tax payments to silver as well. Although the Ming Dynasty had long opposed the silverization movement which had emerged 'from the bottom up', Ming acquiescence to this irresistible trend was eventually institutionalized in the form of the 'Single-Whip Tax Reform' in the 1570s; a multitude of taxes were consolidated into a single empire-wide tax, payable exclusively in silver (even by peasants). Since China contained a quarter of the world's population and huge cities (e.g. 1 million in Nanjing and 660,000 in Beijing), the conversion of China's monetary and fiscal systems to a silver base was of enormous importance. Moreover, because of China's extensive tributary system, domestic silverization created a ripple effect reaching far beyond Chinese borders. Takeshi Hamashita summarizes this relationship:

> The entire tribute and interregional trade zone had its own structural rules which exercised a systematic control through silver circulation and with the Chinese tribute trade at the center. This system, encompassing East and South-East Asia was articulated with neighboring trade zones such as India, the Islamic region and Europe.[11]

The Chinese economy was so enormous that its conversion to silver caused the market value of the white metal to soar far higher in China than in America, Japan, Europe, and elsewhere in the world. Regionally-divergent bimetallic ratios

---

[10] For a long-term monetary history of China, see Richard von Glahn's landmark treatise. Richard Von Glahn, *Fountain of Fortune: Money and Monetary Policy in China, 1000–1700* (Berkeley, 1996a).

[11] Takeshi Hamashita, 'The Tribute System of Modern Asia', in eds. A.J.H. Latham and Heita Kawakatsu, *Japanese Industrialization and the Asian Economy* (London, 1994), p.97.

provide a clear indicator of the elevated value of silver in China *vis-à-vis* the rest of the world. In the early sixteenth century, the gold/silver ratio in China stood at 1:6, while in 'contrast the gold/silver ratio hovered around 1:12 in Europe, 1:10 in Persia, and 1:8 in India'.[12] No wonder that the specific metal, silver, flooded into China for centuries; the Chinese simply offered the best price for this product. Rather than depicting the flow of silver to China as a passive *effect* of disequilibrium in non-monetary trade, therefore, it is better to recognize that disequilibrium within the silver market itself was an active *cause* of global trade. The worldwide silver trade is best understood in the context of domestic developments within China (in conjunction with cost-reducing factors in silver mines in Europe, America, and Japan). There is no doubt that Europeans played an important role as intermediaries, facilitating the movement of tens of thousands of tons of silver around the globe, but the most critical element of dynamism in this case should be attributed to end-customer China.

Each essay included in this volume was selected on the basis of its contribution to a global perspective on where, when, how, and why the monetary substances – gold, silver, copper, and cowries – were respectively produced and traded. Space restrictions preclude reproduction of many relevant and important selections we would like to have included. Many landmark articles and books are merely suggested for further reading because they are already well known and easily accessible. We decided to avoid altogether the important literature on price history, especially the Price Revolution, for fear that its size and complexity would render it unwieldy, and would potentially distract attention away from underlying supply-and-demand factors which accounted for the production and intercontinental migrations of specific metals and monetary substances.[13]

II. Production and exports of metals and monies

John Nef's 'Silver Production in Central Europe, 1450–1618' (see chapter 1 below) documents an important mid-fifteenth century surge in Central European silver production, one propelled by investments in new technology.

> The year following… 1451, marked the Duke of Saxony's grant of the right to use a new invention for separating silver from rich argentiferous copper ores with the help of lead. More than any other invention, this accounts for the great prosperity of mining and metallurgy during the century which followed.[14]

---

[12] Richard Von Glahn, (1996a), p.127.

[13] For a statement of our view that silver production caused the Price Revolution around the world, see Kerry W. Doherty and D.O. Flynn, 'A Microeconomic Quantity Theory of Money and the Price Revolution', in Eddy H.G. Van Cauwenberghe *Precious Metals, Coinage and the Changes of Monetary Structures in Latin America, Europe and Asia* (Leuven, 1989), pp.185–208.

[14] John U. Nef, 'Silver Production in Central Europe, 1450–1618', *Journal of Political Economy* XLI (1941), p.576 (see chapter 1 below).

Central European silver production may seem modest in retrospect (peaking at 100 tons of silver mined during the decade 1526–35, equal to about 4 million pesos), compared with the spectacular American mine output of the sixteenth and seventeenth centuries, but two aspects of Central European silver production deserve mention here.[15] First, some have argued that the Price Revolution began in the second half of the fifteenth century, prior to the arrival of metals from America.[16] But this is consistent with surging silver production in Central Europe. One would not be surprised to learn that a rise in European production may have reduced silver's value toward its new, lower European cost of production; price inflation (in silver-content terms) would result.[17] Secondly, Nef reported that although Central European mine production peaked in the 1530s – at a rate of output not reattained until the eighteenth century – it was the flood of silver from America which caused European production to fall off drastically soon thereafter.[18] The cost of producing silver in America was much lower, of course, so the gradual decline in silver's market value caused the closure of less-efficient European mines first. Because of the continued decline in silver's market value, however, even the world's most prolific silver mines at Potosí, fifteen thousand feet into the Andes and source of an estimated 60 per cent of total world silver output, were themselves facing a profit squeeze a few decades later.[19] Potosí's subsequent mining boom was a consequence of reforms introduced by the Viceroy Francisco de Toledo after 1570, including the mercury-amalgamation process, construction of water mills, and creation of the *mita*, an indigenous system of forced labour: 'Supplemented with gold from Colombia, the mines of Mexico and Peru yielded quantities of bullion greater than the world had ever seen'.[20] Thus, Central European and American silver producers can be viewed as competitors in an emerging world economy; both struggled with steadily declining silver prices, although the tremendous surge in demand emanating from China

[15] Nef, (1941), p.585.

[16] e.g. F. Braudel and F. Spooner, 'Prices in Europe from 1450–1750', in E.E. Rich and C.H.Wilson *The Economy of Expanding Europe in the Sixteenth and Seventeenth Centuries*. Vol. IV of *The Cambridge Economic History of Europe* (Cambridge, 1967), pp. 401–2.

[17] Van der Wee provides a telling example of a surge in silver mintage in the Low Countries as a result of enhanced German silver production. Herman Van der Wee, *The Low Countries in the Early Modern Period* (Aldershot, pp.179–80). When American silver later flooded into Europe, maintenance of a bimetallic system became impossible and 'the silver standard remained characteristic for the monetary system of Brabant and the surrounding area until the end of the Ancien Regime'. (Van der Wee, ibid.), p.182.

[18] Nef, (1941), p.589.

[19] Harry E. Cross, 'South American Bullion Production and Export, 1550–1750', in John F. Richards, *Precious Metals in the Later Medieval and Early Modern Worlds* (Durham, North Carolina, 1983), p.403.

[20] Cross, ibid., p.401.

certainly slowed the rate at which silver lost value. More than anywhere else, customers in China were responsible for the demand-side impetus for the silver booms in Europe, Japan, and America.

Since Peter Bakewell's *Mines of Silver and Gold*, a 1997 volume in this *Expanding World* series, already summarizes the state of knowledge on Spanish American mining, no essays dealing directly with American mine production are included here. Useful for our purposes, however, is an attempt to outline the trade routes of American metals after their production. Zacarias Moutoukias' work (see chapter 2 below) on American contraband trade is enlightening on this score. There has long been vigorous debate, of course, over the fraction of overall silver trade comprised by smuggling activity. Moutoukias' study suggests that contraband trade was far more significant than is normally acknowledged.

Anecdotal evidence about illegal practices is abundant. In the case of a shipwreck between Cádiz and Gibraltar in 1555, for example, 150,000 pesos in treasure were officially registered, but the quantities actually recuperated from the shipwreck proved that unregistered treasure was double the declared amount. Spanish authorities were well aware of this situation, as attested in a letter from Emperor Charles V to his daughter Juana in March of 1557, where he complained about a ship carrying a large amount of undeclared treasure with the cooperation of officials at the *Casa de Contratación*. Prior to arrival in Seville or Cádiz, fleets or single ships could and did stop over in the Azorean islands or Portuguese ports under some pretext to drop off quantities of silver.[21] The Canary Islands were likewise involved in illegal trade. Nearly a century later, in 1650 Alonso Merlo de la Fuente complained that 'all the silver and gold that comes unregistered, usually eight times the registered... goes from the merchant ships to the foreign ships that are waiting for them in the ports of Cádiz and Sanlucar... without paying any taxes'; this observation is consistent with Van Dillen's remarks about Dutch prominence in Cádiz.[22] According to Moutoukias, undeclared precious metals between 1630–1640 were double the amount legally registered; smuggled treasure represented between 40 and 90 per cent of total treasure transported between 1650–1659; and, pertaining to the Río de la Plata (River Plate), the *Navios de Registro* indicate that skippers declared only 10 per cent to 20 per cent of the precious metals they transported to Spain.[23]

---

[21] Ramón Carande, *Carlos V y sus banqueros. 1: La vida económica en Castilla*, 4 edn. (Barcelona, 1990) p.238.

[22] Quoted by Charles R. Boxer, 'Plata es Sangre: Sidelights on the Drain of Spanish-American Silver in the Far East, 1550–1700', *Philippine Studies* XVIII (1970), n.3, p.470. J.G. Van Dillen, 'Amsterdam, marché mondial des métaux précieux aux XVII et XVIII siècles', *Revue Historique* CLIII (1926–27), pp.196–8.

[23] Zacarías Moutakias, 'Una forma de oposición: el contrabando', in eds. Massimo Ganci e Ruggiero Romano *Governare il Mondo. L'Impero spagnolo dal XV al XIX secolo* (Palermo: 1991), p.339.

The two Spanish fleets which annually plied the Atlantic were the main carriers of silver in the sixteenth and seventeenth century, but American silver also found outlets outside the galleons of the Spanish crown. European powers were engaged in active contraband trade with Spanish America during the seventeenth century, all of which implied flows of silver. One of the most important routes of illegal trade was the commercial axis between the Spanish enclave of Rio de la Plata, the Portuguese colony of Sacramento (across the same estuary), and the viceroyalty of Peru. Trade between Potosí and Buenos Aires began almost immediately upon the arrival of Spaniards. Local products – sugar from Brazil, European manufactures, and slaves introduced by the Portuguese since 1595 – were all exchanged for Peruvian silver.[24] This was the so-called 'Back Door' for the viceroyalty of Peru, one of the routes through which enormous quantities of silver escaped the control of Spanish kings: 'During its most successful years, no less than 1–2 million pesos (roughly 25,000 to 50,000 kg) flowed illegally from the mines of Peru out through the port of Buenos Aires. These totals equalled from 15 percent to 30 percent of the silver output of Potosí'.[25] According to Moutoukias, several European powers participated actively along the Rio de la Plata:

> From 1648 to 1702, about 200 ships entered the Plata River without permits – of which, 129 could be verified – amounting to more than two-thirds of total Atlantic traffic (with and without authorization). Fifty percent of them were Dutch; in order of importance, the rest were Portuguese, English, and French.[26]

Fraud was also customary within Perú. Maria Helmer mentions the practice of transporting silver to poor districts offering lower taxes; also, contraband in mercury from Huancavelica was very active. Helmer concludes that, due to tax-avoidance behavior, official mercury figures cannot be used to accurately estimate mine production: 'There is no way of determining the silver production that was spread over the world during the great days of Potosí'.[27] In addition, production of silver was sometimes stopped at an intermediate stage (piña) so as to avoid paying the *quinto* tax, but piñas nonetheless circulated as currency within the viceroyalty.

Fraud along the Pacific side of the silver trade was rampant. To discourage contraband via the Acapulco-Manila galleon route, Philip IV sent a *visitador* in

---

[24] Four thousand cows and sixty thousand mules were exported from Tucuman, Argentina, to Peru in 1610; such figures were common in prosperous years. (Maria Helmer, 'Comércio y Contrabando entre Bahia e Potosi no Século XVI', *Revista de Historia* IV (1953), p.206. Thirty thousand slaves were shipped to the Rio de la Plata during the first half of the seventeenth century. (Moutoukias, (1991), p.353)

[25] Cross, (1983), p.414.

[26] Moutakias, (1991), p.341.

[27] Helmer, (1953), pp.204–5.

1635, Pedro de Quiroga, to put teeth into regulations applicable to the trade. For two years, merchants in Manila refused to send any more merchandise until customary practices were resumed (i.e. officials looked the other way). Evidently, partial tax payments were judged superior to none at all, since the Crown yielded to pressure and removed Quiroga, so that business could resume on a normal basis. Solid evidence of fraud has been documented in the case of the galleon San Francisco Javier, sunk in the vicinity of Manila bay in 1654. This ship was found to have carried 1,180,865 pesos, 228,000 belonging to the *situado*, and 11,049 to help with construction of a cathedral; only 418,323 pesos were registered, however, so nearly two-thirds (64.58 per cent) of the treasure was unregistered.[28]

Official estimates of overall trade activity via the Pacific are unreliable because of the prevalence of contraband: 'That the Philippines siphoned off large sums of silver from the New World cannot be denied, but measuring that flow is virtually impossible'.[29] Admittedly-conservative estimates for annual silver exports from Acapulco vary from about 1/2 million pesos to 2/3 million pesos.[30] Based mainly on Pierre Chaunu's study of *almojarifazgo* tax receipts, the conventional belief is that Acapulco silver exports were very large in the early-seventeenth century, followed by a slight decline after 1620, and a dramatic drop off after 1640.[31] Not all scholars are willing to accept Chaunu's pessimistic conclusions, however, because his tax receipts pertained only to non-smuggled silver. Cross estimates that an average of 2 to 3 million pesos (53,000 kg to 79,000 kg) in silver were sent annually from Peru to Mexico between 1580 and 1610; according to Borah, inclusion of the contraband trade raises this total to 4 million pesos (100,000+ kg).[32] But there was no reason to send silver from Peru to Mexico except to purchase Asian products, so how could 2–4 million pesos have been shipped up Mexico's Pacific Coast yearly, while fewer than one million pesos are said to have been forwarded annually to China (via Acapulco and Manila)? If South Americans alone were exporting 2–4 million pesos of silver in exchange for Asian products – and we have not yet mentioned Mexican silver

[28] Ana M. Prieto Lucena, *Filipinas durante el gobierno de Manrique de Lara 1653–1663* (Sevilla, 1984), p.50.

[29] John TePaske, 'New World Silver, castile and the Philippines, 1590–1800', in ed. John F. Richards, *Precious Metals in the Later Medieval and Early Modern Worlds* (Durham, North Carolina, 1983), p.437.

[30] Ibid., p.437; Ward Barrett, 'World Bullion Flows, 1450–1800', in ed. James D. Tracy *The Rise of Merchant Empires Long Distance Trade in the Early Modern World, 1350–1750* Cambridge, 1990), p.236–49.

[31] Pierre Chaunu, *Les Philippines et le Pacifique des Ibériques, XVIe, XVIIe, XVIIIe, siècles* (Paris, 1960), p.250.

[32] Cross, (1983), p.412; W. Borah, *Early Colonial Trade and Navigation between Mexico and Peru* (Berkeley & Los Angeles, 1954), pp. 88, 123.

production, which surged in the seventeenth century – then the amount of silver exported from Acapulco must have been in the millions of pesos per year.

Chuan has provided figures indicating that 2 million pesos annually crossed the Pacific via the Manila Galleons throughout the entire seventeenth century.[33] Chuan's numbers are consistent with other indicators. Manila was an entrepôt with no economic purpose other than the swap of American silver for Chinese silk. In the middle of the seventeenth century the city contained 42,000 inhabitants, approximately the same population as Barcelona, Valladolid, Danzig, Vienna, Marseille, and other major European trading centers.[34] If there were a seventeenth-century collapse in trade via Manila, as suggested by Chaunu's official *almojarifazgo* tax receipts, then why did this city maintain vitality through the century? Its survival depended entirely upon the silk-for-silver trade, so evidence of a powerful and vital city seems inconsistent with the contention that the silver trade shriveled.

Since most of the seventeenth-century silver exported from Acapulco to China (via Manila) was smuggled, it is obviously difficult to document the value of the Acapulco-Manila trade. Flynn and Giráldez approach this problem indirectly by noting that 2,000,000 pesos in Chinese silk crossed the Pacific annually via Manila; again, all available evidence suggests that the silk trade flourished throughout the seventeenth century.[35] Since the Acapulco-Manila trade involved a straightforward silk-for-silver swap, the high and sustained value of silk imports into Acapulco provide strong evidence in favor of an equal value of silver exports from Mexico. This view is also consistent with conclusions reached by Cross:

> Quantities of silver left the New World through the ports of Buenos Aires and Sacramento and through the Manila Galleons. At the peak of these activities, perhaps as much as 6 million pesos per year (159,000 kg.), or half the output of Peru, was diverted to these channels from the Seville trade.[36]

Barrett points to a glaring discrepancy between Spanish American production figures and estimated exports to Europe – 5.5 million pesos per year.[37] There is no compelling reason to suspect that the relatively small, depopulated American economy could have retained such vast amounts of silver in the face of demand-side competition from the world's largest economies.

---

[33] Chuan, (1969), p.79.

[34] Eric Wolf, *Europe and the People Without History* (Berkeley, Los Angeles, London, 1982), p.153.; Jan De Vries, *European Urbanization, 1500–1800* (Cambridge, Mass.,) Appendix 1; Roger Mols, 'Population in Europe, 1500–1700', in ed. C.M. Cipolla *The Fontana Economic History of Europe*, II: *The Sixteenth and Seventeenth Centuries* (Glasgow, 1977), pp.42–3.

[35] D.O. Flynn and Arturo Giraldéz, 'Silk for Silver: Manila-Macao Trade in the 17th Century', *Philippine Studies* XLIV (1996), First Quarter, pp.52–68.

[36] Cross, (1983), p.420.

[37] Barrett, (1990), p.236.

Kobata's (see chapter 3 below) work on Japanese silver, gold and copper production supports this volume's theme that these metals must be viewed from a global perspective:

> Japan's production of precious metals is therefore closely related with the history of colonization and the Far Eastern trade of Europeans. It is probable that the bringing of huge amounts of silver from Mexico for the China trade, the similar import into China of silver from India, and the carrying of gold from South-East Asia to Europe were also basic factors in the development of relationships between the values of precious metals in the sphere of Far Eastern trade.[38]

Although Kobata admits that calculation of actual quantities of gold and silver production in Japan is impossible, he offers 'a bold conjecture that the export of Japanese silver at the beginning of the seventeenth century would very likely have amounted to 200,000 kg. annually'.[39] Roughly equivalent to 8 million pesos per year, Japan would thereby clearly have been a major supply-side competitor for Spanish American mines during the 'Silver Century' (mid-16th to mid-17th century). Once again, the composition of the Sino-Japanese and the Sino-Mexican trades was similar. Kobata reports that in the 1580s Chinese silk and silk textiles were Japan's most important imports via Macao, followed by gold: 'In 1590... 2,000 pieces of gold bullion...were exported from China to Japan'.[40] During the second half of the seventeenth century copper replaced silver as Japan's pre-eminent export, as mentioned earlier, and again the most important market for Japanese copper exports was China.

Tashiro's work (see chapter 4 below) on Japanese silver complements that of Kobata by tracing silver's precise trade routes through Tsushima and Korea, and on to China. The Tsushima-Korea trade, unaffected by the so-called 'closure' of Japanese trade in the late 1630s, was five times larger than the Nagasaki trade by 1686. It took two forms: one, official trade stipulated by diplomatic provisions, and two, private merchant trade, held six times per month and subject to no restrictions.[41] Not surprisingly, silver was smuggled along the Sino-Korean border, as was the case elsewhere throughout the world. The Korean government dispatched 3 or 4 official missions per year during the seventeenth century, thereby affording Tashiro the opportunity to document monthly shipments of silver

---

[38] A. Kobata, 'The Production and Uses of Gold and Silver in Sixteenth and Seventeenth-Century Japan', *Economic History Review* XLIII (1965), n.2, p.247.

[39] Kobata, (1965), p.248.

[40] Kobata, (1965), p.248.

[41] Kazui Tashiro, 'Exports of Japan's Silver to China via Korea and Changes in the Tokugawa System During the 17th and 18th Centuries', in ed. Eddy H.G. Van Cauwenberghe, *Precious Metals, Coinage and the Changes of Monetary Structures in Latin America, Europe and Asia* (Leuven, 1989), pp.101–2.

as they were required by the rhythms of the tribute-trade system focused on China.

Although silk was by far China's dominant export to Japan and Mexico, there is also clear evidence of significant exports of gold during the 'Silver Century'. The Chinese-European trade picture is extremely convoluted by comparison. Europe's primary supplier of silk during this period was Persia, not China.[42] Moreover, while there is evidence of Asian exports of gold to Europe (already cited), nobody argues that Chinese gold flowed to Europe on the scale that it flowed to Japan. It seems that Africa and America were the sources of most of Europe's gold imports, not Asia. Although only able to offer an 'educated guess' about African gold exports during the early-modern period, Bean (see chapter 6 below) surmises that 200,000 pounds sterling is a reasonable estimate for annual gold exports to Europe over the interval 1500 to 1700.[43] Comparing this figure with slave exports from the Gold Coast, Bean concludes that 'it was only for a ten-year span in the 18th century that slaves clearly earned more foreign exchange annually for the Gold Coast than gold exports had earned per year in the 17th century'.[44] Nigeria and Angola produced no gold at all while exporting many slaves, on the other hand, so this does not mean that the value gold exports were greater than the value of slave exports overall. Bean's point is simply that African exports of gold to Europe were extremely significant to both source and recipient areas.

We would like to suggest that both gold and slave exports from Africa were tied to the dominant global silver trade outlined above. In discussing the Rio de la Plata trade, mention has already been made of the fact that Portuguese slaves were essentially bartered in exchange for Peruvian silver via America's 'Back Door'. Imagine if this silver would have been exported directly to West Africa – which it was not – rather than to Europe. Under this counter-factual scenario, West Africa would then have re-exported the silver, we contend, and this silver would have joined with other silver flows eventually winding up in China. What happened instead was that silver passed through Europe on its way to China; European cities may have simply been more convenient transhipment centers. Nonetheless, African slaves were bartered for American silver and this slave-silver barter activity depended upon the silverization of China.

How did African gold tie in at the global level? It too was exchanged for American silver, in a manner suggested in a passage from Braudel:

---

[42] Chaudhuri, pp. 203–6; Niels Steensgaard, *The Asian Trade Revolution of the Seventeenth Century: The East India Companies and the Decline of the Caravan Trade* (Chicago, 1974), Ch.9; Kristof Glamann, *Dutch-Asiatic Trade, 1620–1740*, 2nd. edn. (The Hague, 1981), Ch. 6.

[43] Richard Bean, 'A Note of the Relative Importance of Slaves and Gold in West African Exports', *Journal of African History* XV (1974), n.3, p.353.

[44] Bean, (1974), p.355.

> So Italy stood at the crossroads where the south-north axis maintained by Spanish policy and the Genoese *asientos* met the east-west axis running to the Levant and the Far East, where the golden road from Genoa to Antwerp met the silver road to the east.[45]

The Spanish Crown needed European financiers to transfer gold to soldiers in Flanders, who always demanded at least part of their pay in gold.[46] American silver was therefore exchanged for African gold in Italy, whereupon American silver headed off to China and African gold was forwarded to the Low Counties. African gold and slaves were both exchanged for American silver. Silver was a crucial commodity connecting various trades because of powerful market forces emanating out of China.

Cowrie shells were among the most significant products imported into West Africa, where they had been used as currency since at least the eleventh century. (see chapter 9 below).[47] According to Hogendorn and Johnson, in the 1350s Ibn Battuta observed enormous profits in this cowrie trade: 'The figures he [Battuta] quotes show at the very least a West African cowrie price about 350 times greater than the going price in the Maldives'.[48] Although the Maldives' own cowrie currency served as their sole money in Ibn Battuta's time, in subsequent centuries it came to be supplemented, then supplanted by the silver *larin*.[49] Strong external demand for cowries, in conjunction with importation of silver from abroad, caused Maldivian cowrie prices to rise during the seventeenth century, 'perhaps as much as four to seven times according to Heimann's calculations, which include some depreciation of silver during that century'.[50] This situation is somewhat reminiscent of early seventeenth-century Spain – a country on a copper-money standard while simultaneously producing prodigious quantities of silver money for use in the rest of the world – as the Maldives harvested enormous volumes of cowrie shells, but ceased using them domestically because of conversion to a silver standard.

Cowries from the Maldive Islands have been used as money for at least 1,000 years and continued in limited use in West Africa and Bengal even in recent times.[51] They circulated widely in Asia, the Middle East, and Europe. Cowries were never used as currency in Europe, however, as they were re-exported from

---

[45] Braudel, (1972), vol.I, p.499.

[46] Braudel and Spooner, (1967), p.385.

[47] Marion Johnson, 'The Cowrie Currencies of West Africa', *Journal of African History* XI (1970), n.1, p.33.

[48] Jan Hogendorn and Marion Johnson, *The Shell Money of the Slave Trade* (Cambridge, 1986), p.25.

[49] Hogendorn and Johnson, (1986), p.26.

[50] Hogendorn and Johnson, (1986), p.40.

[51] Johnson, (1970), p.18.

Europe to Africa. The King of Portugal licensed a Portuguese merchant to bring cowries as ballast in 1515, without charge for freight, whereupon the shells were re-exported to Guinea, Benin, and the Congo.[52] After the Dutch took over Ceylon from the Portuguese (1658), cowries began showing up in Rotterdam, where they were packed in barrels for the Guinea trade. English, French and Germans were also active in the cowrie trade. By 1720, Europeans were importing and re-exporting around a million pounds weight in cowries per year, equal to approximately one-third of the value of the 20,000 African slaves exported annually at that time.[53] By the end of the eighteenth century, the English alone were shipping some 100 tons of cowries per year to Africa. The cowrie trade experienced crisis when the slave trade was abolished in 1806; within five or ten years, European shipments dropped to about 6 tons per year.[54]

Cowries were also imported in exchange for African gold. Indeed, cowries originally served as fractional change for gold currency in the Niger Bend country. Johnson speculates that one would expect that the 'large imports of cowries in the sixteenth century would tend to lead to a depreciation of cowries (as is known to have happened in the Congo)' – relative to gold – if a system of pegging cowries to gold were to have broken down. She says that cowries did in fact go off the 'gold standard' in the second half of the eighteenth century, which would make it 'interesting to see if there is evidence of such a breakdown in the sixteenth or early seventeenth centuries'.[55] In any case, one prominent Asian product, cowries, provides a partial answer to the puzzle of exactly what products were exchanged – both directly and indirectly – for the vast quantity of American silver flowing to China. Another important African import was copper, used as a medium of exchange from at least the eleventh century: '[T]here can be little doubt that the bulk of West Africa's copper has come from beyond her borders: from northwest Mauritania, southern Morocco, the Byzantine Empire (Malfante's 'Romania') and Central Europe'.[56]

So, hundreds of tons of silver flowed annually to the Far East – including to the Maldive Islands themselves – while hundreds of tons of cowries simultaneously flowed from South Asia. Africa's exports of slaves and gold were connected with imports of cowries and copper. The tie between Africa's slave trade and smuggled silver via the Rio de la Plata is easy to see on the American side; more difficult to detect are the gold-cowrie-slave triangular trade connections with silver as the white metal passed through Europe on its voyage to China.

---

52 Johnson, (1970), p.19.

53 Johnson, (1970), p.21.

54 Johnson, (1970), p.22.

55 Johnson, (1970), vol.XI, n.3, p.33.

56 Eugenia W. Herbert, 'Aspects of the Use of Copper in Pre-Colonial West Africa', *Journal of African History* XIV (1973), n.2, p.180.

Japan's role as a huge silver source through two-thirds of the seventeenth century has already been noted, as has the less-dramatic role of Japanese gold exports. Japanese copper became increasingly prominent in intercontinental trade during the early-modern period, sometimes for use as 'petty monies' (with a role similar to that of cowries) and sometimes as plate. Sweden was Europe's most important source of copper, but Japanese output was twice as large.[57] Although 'the principal part of the Japanese copper remained in the East' (see chapter 5 below), it offered serious competition in Europe as well. Exported as early as the 1620s, Japanese copper 'in 1656 was stimulated by vigorous demand from Europe...Accordingly 39,140 lbs. were sent to Holland in 1657 and in 1658 the vast quantity of 1,007,160 lbs'.[58] Since the seventeenth century is sometimes referred to as Europe's 'copper century' in terms of monetary history, it is perhaps understandable that copper flowed to Europe from Asia. In monetary terms, the silverization of China occurred against a backdrop of a de-silverization of Europe through much of the early-modern period. Citing Spooner (1972), Pamuk (see chapter 15 below) summarizes the trend toward copper in Europe:

> By the middle of the [17th] century, copper coinage was being used heavily in parts of Europe. In Spain, Sweden, and Russia, the great majority of transactions were being undertaken with copper coinage.[59]

The point is that particular metals flowed globally in response to changes in their respective stock demands and costs of production. Since copper substituted for silver (e.g. in Europe) while silver was substituting for paper money (e.g. in China), it is not surprising that merchants reacted by shipping both silver to Chinese customers and copper to European buyers. These areas simply furnished the most profitable markets for particular products during specific time intervals. This is the way markets function, irrespective of whether the items traded had been monetized or not.

Strong demand for copper within Europe notwithstanding, by the late 1670s and early 1680s no copper was carried to Europe, even though Japanese copper exports peaked during these years. According to Directors of the Dutch Company in 1640, 'the Japanese and Chinese trade [w]as the principal trade in the entire Indies', however, the export of Japanese copper by Chinese (nearly equal to those of the Dutch by 1679) was 'partly carried to Taiwan, partly to Siam, where it was sold to the English, the Danes and others, and carried on to Coromandel

---

[57] Glamann, (1981), p.174.

[58] Kristof Glamann, 'The Dutch East India Company's Trade in Japanese Copper, 1645–1736', *Scandinavian Economic History Review*, I (1953), pp.47, 56.

[59] Şevket Pamuk, 'The Disintegration of the Ottoman Monetary System During the Seventeenth Century', *Princeton Papers in Near Eastern Studies*, n.2, p.73, who cites Frank Spooner, (1972), *The International Economy and Monetary Movements in France* (Cambridge, Mass., 1972)

and Bengal...; the Chinese – who exported not only copper bars but also scrap copper and copper plates – seem to have taken the lead, superseding the Dutch, at the close of the 1680s'.[60] The Dutch Company retaliated by trying to corner the Japanese copper market in 1683–84, but they were unable to establish control.

Glamann's emphasis on the cost of transporting copper, a relatively bulky item, is noteworthy. Copper was shipped to Europe as ballast (like cowries), which explains why the market for this relatively-bulky item was linked over such vast distances.[61] Saltpetre was usually the dominant Dutch ballast commodity for shipments to Europe, so 'it is remarkable that the company's importation of copper was large exactly in those years when the prices of saltpetre in Europe showed a falling tendency'.[62] It is clear that monetized commodities were locked in vigorous global competition with one, other monetized commodities, as well as two, non-monetized items in geographically-dispersed markets throughout the early-modern period.

III. Transhipment of metals and monies

Considering the prominent role of the Ottoman Empire in transhipping silver to Asia, surprisingly little scholarly attention has been devoted to study of Middle Eastern trade linkages.[63] It was mainly American silver which was reexported through the Middle East, of course, as is made clear in this passage from Pamuk (see chapter 15 below):

> The Middle East, however, was often a transit zone for these intercontinental flows. While the Ottoman government welcomed the arrival of large amounts of silver and silver coinage from the West, it could not prevent their outflow towards Iran and India, as the Empire continued to run deficits towards the East.[64]

'Pragmatism' is the term used by Pamuk to describe Ottoman monetary policy. Policy was pragmatic in the sense that Ottomans did not insist upon monopolization of domestic money supplies, an important issue because 'contrary to the prevailing view, the use of money was not limited to segments of the urban population but was quite widely spread during the sixteenth century'.[65] Foreign

---

[60] Glamann, (1953), pp.45, 77–88.

[61] Porcelain was an extremely important ballast item in Chinese shipping. Souza lists three main Chinese export items: (1) gold, (2) silk and cotton, and (3) 'ballast' items (including porcelain). George B. Souza, *The Survival of the Empire Portuguese Trade and Society in China and the South China Sea, 1630–1754* (Cambridge, 1986), p.5.

[62] Glamann, (1953), p.62.

[63] While not focusing on precious metals flows *per se*, an interesting account of the importance of Islamic trading networks can be found in Stephen F. Dale, *Indian Merchants and Eurasian Trade, 1600–1750* (Cambridge, 1994).

[64] Pamuk, (1993), p.73.

[65] Pamuk, (1993), p.70.

coins were allowed to circulate freely throughout the region under Ottoman control:

> Foreign coins in fact became the leading forms of actual money in the akçe region, from the Balkans and Istanbul to Anatolia and Syria... the Dutch thaler... Even more important were the Spanish 'pieces of eight'... The Ottoman government did not attempt to restrict the circulation of these coins. In fact, the government regularly accepted, and sometimes even demanded, payment in European coinage.[66]

This Ottoman pragmatism is seen as both a healthy trend (a pro-market philosophy) as well as negative, in the sense that it provided 'strong evidence regarding the limitations of its [the Empire's] power'.[67] Below, we will briefly revisit this issue of the role of silver in the decline of Ottoman (and Ming) financial power.

East of the Ottoman Empire, from the Yemen and Persia to the Indonesian archipelago, coinage and media of exchange were silver-based: 'From the Red Sea to the Pacific wherever the Company's servants went (with the exception of southern India of course) they met with coins and moneys of account whose value was fixed in silver'.[68] In Mughal India the silver rupee was used widely for trade and tax purposes; the gold coins, 'muhr' or 'ashrafi', enjoyed limited circulation, and copper coinage, based on the 'dam', was employed for small transactions.[69] Chaudhuri states that 'the currency standard in areas of Mughal rule can be said to have been unambiguously based on silver'.[70] In contrast, the gold 'pagoda' prevailed along the Malabar coast and in southern India.[71]

No one disputes the fact that India was one of the big poles attracting silver during the sixteenth and seventeenth centuries. Conversion of the tax system in the Mughal empire to silver, which spread southward through the seventeenth century, indicates widespread silverization in the Indian economy.[72] Substantial

---

[66] Pamuk, (1993), p.75.

[67] Pamuk, (1993), p.76.

[68] K.N. Chaudhuri, 'The Economic and Monetary Problems of European trade with Asia during the 17th and 18th Centuries', *The Journal of European Economic History* IV (1975), n.2, p.348.

[69] Irfan Habib, 'Monetary System and Prices', in eds. Raychaudhuri, Tapan and Irfan Habib *The Cambridge Economic History of India, Volume I: c.1200–c.1700* (Cambridge, 1982), p.360.

[70] Chaudhuri, (1975), p.347.

[71] The division of India into gold and silver zones is an interesting phenomenon, but not unique. Japan was also divided into gold and silver zones (east-west). As we have seen, China experienced no such dichotomy, having emerged as a united silver zone.

[72] K.N. Chaudhuri, 'World Silver Flows and Monetary Factors as a Force of International Economic Integration, 1685–1758', in eds. W. Fisher, R.E. McInnis, and J. Schneider *The Emergence of a World Economy, 1500–1914: Papers of the IX International Congress of Economic History* I (1986), p.74; also see Frank Perlin, *The Invisible City: Monetary, Administrative and Popular Infrastructures in Asia and Europe, 1500–1900* (Aldershot, 1993), p.154.

imports of silver into India have led many historians to claim that perhaps India absorbed the white metal nearly to the extent of China: 'India, which heavily imported silver and hardly sent any out, must have had its supply of silver augmented to a much greater degree than other areas of the world'.[73] While India no doubt absorbed silver, what is doubtful is the contention that India 'hardly sent any [silver] out'. There is evidence of trading of Indian rupees in Tibet; mints were established in Katmandu and in Cooch Bahar already by the 1560s. Indian silver was carried to Lhasa during the 1680s, and from there to Xining. Silver also could have left India via routes linking Yunnan Province with the Bay of Bengal and Burma, 'though most likely not in significant quantities' (see chapter 8 below).[74]

Perhaps the most compelling indirect evidence that a substantial fraction of Indian silver must have continued on to China rests on the observed divergence of India's bimetallic ratio from that of China. During the late sixteenth century, for example, the gold/silver ratio ranged from 1:5.5 to 1:7.5 in Canton while simultaneously standing at 1:9 in India.[75] Since the high value of silver in China is what prompted entrepreneurs to mine and ship silver from all corners of the earth for centuries, it only stands to reason that this mechanism had to have been in operation for two enormous neighboring economies. Basing his conclusion upon both archival and theoretical considerations, Chaudhuri (see chapter 14 below) is emphatic that 'India was the penultimate destination of the American silver. For beyond South Asia there were still the vast areas of Imperial China...'.[76] India certainly absorbed significant amounts of silver, but both direct and indirect evidence supports Chaudhuri's contention that China was most dominant on the demand side of the silver equation.

Perlin's review essay (see chapter 10 below) constitutes a call for more balanced and thoughtful treatments of Third World histories in general and of India in particular. He justifiably criticizes Vilar (1976) for 'Eurocentric bias', while acknowledging Vilar's sound analysis in other respects:

> Precisely as a zone of attraction, Asia may, ironically, have had a far from passive role in Europe's growth... It can hardly be disputed that trade in precious metals was a fundamental lever of European trading ventures in Asia, and that the rapid growth in the organization of overseas trade thereby depended on the zones of reception provided by Asia's economies for coinable metals.[77]

---

73 Habib, (1982), p.368–9.

74 Brian Moloughney, and Xia Weizhong, 'Silver and the Fall of the Ming: A Reassessment', *Papers on Far Eastern History* (1989), n.40, p.65–6.

75 Habib, (1982), p.367; Von Glahn, (1996a), pp.127–8.

76 Chaudhuri, (1986), p.324.

77 Frank Perlin, 'A History of Money in Asian Perspective', *The Journal of Peasant Studies*, VII (1980), n.2, p.239.

But Vilar's shortcomings in emphasis are misdemeanors compared with the position advanced in *Spenders and Hoarders*, where Kindleberger (1989) attributes Asia's importation of precious metals to a fundamental psychological flaw, namely that Asians are notorious 'hoarders' (a term which he never defines). For Kindleberger, hoarding of precious metals is a pejorative term correlating with backwardness.[78] But the studies collected in this volume make it clear that monies of various types simply migrated around the world in search of their best market. Chinese did not hoard silver, any more than Europeans hoarded copper and bills of exchange, Africans hoarded cowries and copper, Japanese and Indians hoarded gold and silver, or Maldivians hoarded silver *larins*. There is no logical *a priori* reason to associate economic advance or decline with any particular medium of exchange. Each situation must be evaluated according to its particular historical circumstances.

Perlin prefers the terminology 'monetary media' over 'precious metals' because all monies – not just gold and silver – were interrelated. 'Lesser' monies were extremely important and were structurally linked with the circulation of silver and gold; since Indian 'imports of copper and *cowries* were complementary with importation with the precious metals, their study then must assume major significance'.[79] Perlin's argument is buttressed by Marion Johnson's similar rejection of what she considers 'patronizing' characterizations of 'primitive monies':

> West African cowrie currencies, though they have some unusual qualities, are in no sense a 'primitive' money, but a sophisticated form of currency capable of adaptation to the particular needs of West African trade. The currency is naturally best remembered in its latest and most debased stage...but it had behind it centuries of useful existence before its breakdown.[80]

We should keep an open mind regarding the roles of monetary substances, including everyday monies, if we hope to more fully comprehend the nature of global trade linkages during early-modern times.

IV. Monies, power and empires

Probably no one would dispute that the emergence of intercontinental trade in the early-modern period implied prodigious profits for successful participants.

---

[78] For criticism of the hoarding thesis, see Chaudhuri, (1986); Flynn, (1990), '[Review of] Charles Kindleberger's *Spenders and Hoarders: The World Distribution of Spanish American Silver, 1550–1750*', *Journal of Economic History* no.3 (1990), pp.332–59; Perlin, (1980), p.242; Prakash, (1986), 'Precious Metal flows in Asia and World Economic Integration in the 17th century,' in eds. R. Wolfram Fischer, Marvin McInnis and Jurgen Schneider *The Emergence of a World Economy, 1500–1914. Papers of the IX International Congress of Economic History* (Wiesbaden, 1986); among others.

[79] Perlin, (1993), p.137.

[80] Johnson, (1970), p.17.

Production of key products yielded profit, as did processing, shipping, wholesale and retail sale. Governments profited directly through ownership/control and indirectly through taxation (including seigniorage) of products connected with long-distance trade. While we do not possess the knowledge required to assess the causes of the rise and decline of myriad entities discussed throughout this volume, we can offer a brief review of arguments concerning the role of silver (our specialty) in the rise and decline of empires around the world. It is possible that these arguments could be of use to scholars specializing in the study of other commodities or different periods.

Over the past two decades, the most persistent proponent for a strong connection between silver imports and the economy of China is unquestionably William Atwell (see chapter 7 below): 'Japanese and Spanish-American silver may well have been the most significant factor in the vigorous economic expansion which occurred in China during the period in question'.[81] Based on a variety of sources which seemed at the time to indicate that Chinese imports of silver declined in the late Ming period (the Qing overthrow occurred in 1644), Atwell reasoned that declining silver imports created a crisis which ultimately led to the decline of the Ming Empire. Keeping in mind that exports from China's huge silk industry depended largely upon the ability to swap silk for foreign silver, this did not seem like an unreasonable argument. Moloughney and Xia (see chapter 8 below), Goldstone, and von Glahn, however, have subsequently attacked Atwell's argument on empirical grounds.[82] Recent information seems to indicate that Chinese silver imports peaked, rather than collapsed, precisely during the late Ming period.

While serious differences remain, we contend that these authors' core arguments are compatible.[83] The key is to view silver as a commodity produced for profit, as would be the case for a normal, non-monetary item. The continued accumulation of silver stocks within China is what caused its market value to decline there. In this context, Ming policy of fixing tax payments in terms of certain *quantities* of silver payments turned out to be a fatal mistake. In modern

---

81 William S. Atwell, 'Notes on Silver, Foreign Trade, and the Late Ming Economy', *Ch'ing-shih wen-t'i* III, (1977), no.8, pp.1–33; Atwell, (1982), 'International Bullion Flows and the Chinese Economy *circa* 1530–1650', *Past and Present* no.95 (1982), p.68–90; Atwell, (1986), 'Some Observations on the 'Seventeenth Century Crisis' in China and Japan', *Journal of Asian Studies* XLV, no.2 (1986), pp.223–44.; Atwell, (1988), 'Ming Observations of Ming Decline: Some Chinese Views on the 'Seventeenth-Century Crisis' in Comparative Perspective', *Journal of the Royal Asiatic Society* no.2 (1988), pp.316–48. Atwell, (1982), p.5.

82 Brian Moloughney and Xia Weizhong, (1989); Von Glahn (1996); Jack Goldstone, *Revolution and Rebellion in the Early Modern World* (Berkeley & Los Angeles, 1991), p.371.

83 D.O. Flynn, and Arturo Giráldez, 'China and the Manila Galleons', in eds. A.J.H. Latham and Heita Kawakatsu *Japanese Industrialization and the Asian Economy* (London, 1994), pp. 437–40; Flynn and Giráldez (1995), 'Born with a 'Silver Spoon': The Origin of World Trade in 1571', *Journal of World History*, VI, no.2 (1965), pp.213–14.

terminology, the Ming did not 'index' their taxes (e.g. a tax of US$20 fifty years ago was certainly more beneficial to a government agency than would be a US$20 tax today). Collection of a fixed quantity of anything results in a loss of purchasing power for the taxing authority, when the value of the item collected steadily erodes in value. Based on Geiss, Goldstone therefore argues that it was an inflation-based fiscal crisis which led to decline in Ming power; we simply fortify this argument by pointing out that it was the abundance of silver within China which led to a loss in its purchasing power.[84] The same explanation can be extended to declining revenues in the Ottoman Empire, another jurisdiction which made the mistake of fixing taxes in terms of specific quantities.

This argument fits into the global pattern of silver trade outlined earlier. High demand for silver, in conjunction with low costs of production in Japan and America, created unusual profits for entities controlling the mines. Conspicuous in this regard were the Spanish Crown and the Tokugawa Shogunate, both of which were financed by profits from silver mines. When the market value of silver finally fell to its cost of production, super-normal mine profits eroded; the fiscal foundation of both the Spanish Empire and the Shogun was undermined.[85]

If asked to choose a critical period for the world silver trade, we would choose the years around 1640, by which time bimetallic ratios had converged globally. After a century-long boom, the price of silver had finally fallen to its American and Japanese costs of production. Spain had lost European hegemony; its bankruptcy destroyed powerful financiers throughout Europe. The Ottomans had been experiencing silver-related crisis from the late sixteenth century into the seventeenth century.[86] The Ming were overthrown by the Qing in 1644. The end of the Price Revolution coincided with the close of an era which might be described as the 'Silver Century'.

Select Bibliography

Attman, Artur, *The Bullion Flow between Europe and the East, 1000–1750* (Göteborg, 1981).

Attman, Artur, *Dutch Enterprise in the World Bullion Trade, 1500–1800* (Göteborg, 1983).

[84] James Geiss, 'Peking under the Ming (1368–1644)', Ph.D. dissertation, Princeton University. (Goldstone, 1991), p.371.

[85] Dennis O. Flynn, 'Fiscal Crisis and the Decline of Spain (Castile)', *Journal of Economic History*, XLII (1982), pp.139–147; Flynn, [1991], 'Comparing the Tokugawa Shogunate with Hapsburg Spain: Two Silver-Based Empires in a Global Setting', in ed. J.Tracy *The Political Economy of Merchant Empires* (Cambridge, 1991), pp.332–59.

[86] Goldstone, (1991), p.367.

Attman, Artur, *American Bullion in the European World Trade, 1600–1800* (Göteborg, 1986).

Atwell, William S., 'Some Observations on the 'Seventeenth-Century Crisis' in China and Japan', *Journal of Asian Studies* XLV, no.2 (1986), pp.223–44.

Bakewell, P.J., *Silver Mining and Society in Colonial Mexico: Zacatecas, 1546–1700* (Cambridge, 1971).

Barrett, Ward, 'World Bullion Flows, 1450–1800', in ed. James D. Tracy, *The Rise of Merchant Empires: Long-Distance Trade in the Early Modern World, 1350–1750* (Cambridge, 1990), pp. 224–54.

Boxer, Charles R., *The Christian Century in Japan, 1549–1650* (Berkeley, 1959).

Boxer, Charles R., *The Great Ship from Amacon* (Lisbon, 1959).

Boxer, Charles R., 'Plata es Sangre: Sidelights on the Drain of Spanish-American Silver in the Far East, 1550–1700', XVIII, no.3 *Philippine Studies* (1970), pp.457–75.

Braudel, Fernand, 'De l'or du Soudan à l'argent d'Amérique', *Annales E.S.C.* no. 1, pp.9–22.

Braudel, F. and Spooner F., 'Prices in Europe from 1450–1750', in eds. E.E. Rich and C.H. Wilson, *The Economy of Expanding Europe in the Sixteenth and Seventeenth Centuries.* Vol. IV of *The Cambridge Economic History of Europe* (Cambridge, 1967), pp.378–486.

Braudel, F., *Civilization and Capitalism, 15th to 18th centuries*, 3 vols. (New York, 1981–1986).

Borah, W., *Early Colonial Trade and Navigation between Mexico and Peru* (Berkeley & Los Angeles, 1954).

Canabrava, A.P, *O comercio português no Rio da Prata, 1500–1640* (São Paulo, 1944).

Cartier, Michel., 'Les importations de métaux monétaires en Chine: Essai sur la conjoncture chinoise', *Annales E.S.C.* no. 36 (1981), pp.454–66.

Chaudhuri, K.N., *The English East India Company* (London, 1965).

Chaudhuri, K.N., *The Trading World of Asia and the English East India Company 1660–1760* (Cambridge, 1978).

Chaudhuri, K.N., 'European Trade with India', in eds. Raychaudhuri, Tapan and Irfan Habib, *The Cambridge Economic History of India, Volume I: c. 1200–c.1750* (Cambridge, 1982), pp.382–406.

Chaudhuri, K.N., *Trade and Civilisation in the Indian Ocean: An Economic History from the Rise of Islam to 1750* (Cambridge, 1985).

Chaunu, Pierre, *Les Philippines et le Pacifique des Ibériques, XVIe, XVIIe, XVIIIe, siecles* (Paris, 1960)

Cross, Harry E, 'South American Bullion Production and Export, 1550–1750', in John F. Richards, *Precious Metals in the Later Medieval and Early Modern Worlds* (Durham, North Carolina, 1983), pp.397–424.

Dale, Stephen Frederic, *Indian Merchants and Eurasian Trade, 1600–1750* (Cambridge, 1994).

Flynn, Dennis O., *World Silver and Monetary History in the 16th and 17th Centuries* (Aldershot, 1996).

Fischer,W., R.W. McInnis, and J. Schneider, eds., *The Emergence of a World Economy, 1500–1914* (Stuttgart, 1986).

Glamann, Kristof, *Dutch-Asiatic Trade, 1620–1740* (The Hague, 1958).

Gaastra, F.S., 'The Exports of Precious Metal from Europe to Asia by the Dutch East India Company, 1602–1795', in ed. Richards, John F., *Precious Metals in the Later Medieval and Early Modern Worlds* (Durham, North Carolina, 1983), pp.447–76.

Geiss, James, 'Peking under the Ming (1368–1644)', Ph.D. dissertation, Princeton University.

Goldstone, Jack, *Revolution and Rebellion in the Early Modern World* (Berkeley, 1991).

Hall, John, 'Notes on the Early Ch'ing Copper Trade with Japan', *Harvard Journal of Asiatic Studies*, pp.446–461.

Habib, Irfan, 'Monetary System and Prices', in eds. Raychaudhuri, Tapan and Irfan Habib, *The Cambridge Economic History of India, Volume I: c. 1200–c.1750* (Cambridge, 1982), pp.360–81.

Habib, Irfan, 'A System of Trimetallism in the Age of the 'Price Revolution': Effects of the Silver Influx on the Mughal Monetary System', in ed. John F. Richards, *The Imperial Monetary System of Mughal India* (Delhi, 1987), pp.137–70.

Hamashita, Takeshi, 'The Tribute System and Modern Asia', *Memoirs of the Research Department of the Toyo Bunko*, no.46 (Tokyo, 1988).

Hamilton, Earl J., A*merican Treasure and the Price Revolution in Spain, 1501–1650* (Cambridge, 1934).

Hayami, Akira, 'A 'Great Transformation': Social and Economic Change in Sixteenth and Seventeenth-Century Japan', in ed. E. Pauer, *Silkworms, Oil, and Chips: Proceedings of the Economics and Economic History Section of the Fourth International Conference on Japanese Studies, Paris, September 1985* (Bonn, 1986).

Hazan, Aziza [1969], 'En Inde au XVIe siècle: trésors américains, monnaie d'argent et prix dans l'empire mogol', *Annales E.S.C.* no.4 (1969), pp.835–59.

Helmer, Maria, 'Comércio e Contrabando entre Bahia e Potosí no Século XVI', *Revista de Historia* IV (1953), pp.195–212.

Herbert, Eugenia W., 'Aspects of the Use of Copper in Pre-Colonial West Africa', *Journal of African History* XIV, no.2 (1973), pp.179–94.

Hogendorn, Jan and Johnson Marion, *The Shell Money of the Slave Trade* (Cambridge, 1986).

Huang, Ray, *Taxation and Government Finance in Sixteenth-century Ming China* (Cambridge, 1970).

Inalcik, Halil and Quataert, Donald, eds. *An Economic and Social History of the Ottoman Empire, 1300–1914* (Cambridge, 1994).

Innes, Robert L., 'The Door Ajar: Japan's Foreign Trade in the 17th Century', Ph.D. diss., University of Michigan.

Kellenbenz, H., ed., *Precious Metals in the Age of Expansion* (Stuttgart, 1981).

Kellenbenz, H., 'Final Remarks: Production and Trade of Gold, Silver, Copper, and Lead from 1450 to 1750', in ed. H. Kellenbenz, *Precious Metals in the Age of Expansion* (Stuttgart, 1981), pp.307–61.

Liang, Fang-chung, *The Single-whip Method of Taxation in China* (Cambridge, Mass., 1970).

Magalhães Godinho, Vitorino, *Os Descobrimentos e a Economia Mundial*, 2 edn., IV (Lisboa, 1984).

Molina, Raul, *Las primeras experiencias comerciales del Plata: El comercio marítimo, 1580–1700* (Buenos Aires, 1966).

Morineau, Michel, *Incroyables gazettes et fabuleux métaux. Les retours des trésors américains d'après les gazettes hollandaises, XVIe–XVIIIe siècles* (London, 1974).

Munro, John H., 'The Central European Silver Mining Boom, Mint Outputs, and Prices in the Low Countries and England. 1450–1550', in ed. Eddy H.G. Van Cauwenberghe, *Money, Coins, and Commerce: Essays in the Monetary History of Asia and Europe* (Leuven, 1991), pp.119–83.

Pamuk, Şevket, 'Appendix: Money in the Ottoman Empire, 1326–1914' in eds. Halil Inalcik and Donald Quataert, *An Economic and Social History of the Ottoman Empire, 1300–1914* (Cambridge, 1994), pp.947–81.

Parker, Geoffrey, 'The Emergence of Modern Finance in Europe 1500–1730', in ed. C.M. Cipolla *The Fontana Economic History of Europe. Vol 2: The Sixteenth and Seventeenth Centuries* (Glasgow, 1976), pp. 527–89.

Perlin, Frank, 'Money-use in Late Pre-Colonial India and the International Trade in Currency Media', in ed. John F. Richards, *The Imperial Monetary System of Mughal India* (Delhi, 1987), pp.232–74.

Prakash, Om, 'Precious Metal flows in Asia and World Economic Integration in the 17th Century', in ed. R. Wolfram Fischer, Marvin McInnis and Jurgen Schneider, *The Emergence of a World Economy, 1500–1914. Papers of the IX International Congress of Economic History* (Stuttgart, 1986).

Raychaudhuri, Tapan and Habib, Irfan eds. *The Cambridge Economic History of India, Volume I: c. 1200–c.1750*, (Cambridge, 1982).

Rich, E.E. and Wilson, C.H. eds. *The Economy of Expanding Europe in the Sixteenth and Seventeenth Centuries vol IV of The Cambridge Economic History of Europe* (Cambridge, 1967).

Richards, John F., ed. *Precious Metals in the Later Medieval and Early Modern Worlds* (Durham, North Carolina, 1983).

Richards, John F., ed. *The Imperial Monetary System of Mughal India* (Delhi, 1987).

Schurz, William L., *The Manila Galleon* (New York, 1959).

Spooner, Frank C., *L'Economie mondiale et les frappes monétaires en France, 1493–1680*, (Paris, 1956).

Steensgaard, Niels, *The Asian Trade Revolution of the Seventeenth Century: The East India Companies and the Decline of the Caravan Trade* (Chicago, 1973).

Tashiro, Kazui, 'Foreign Relations during the Edo Period: Sakoku Reexamined', *Journal of Japanese Studies* VIII, no.2 (1982), pp.283–306.

TePaske, John, 'New World Silver, Castile and the Philippines 1590–1800', in ed. John F. Richards *Precious Metals in the Later Medieval and Early Modern Worlds* (Durham, North Carolina, 1983), pp.425–46.

Tracy, James D., ed. *The Rise of Merchant Empires: Long-Distance Trade in the Early Modern World, 1350–1750.* (Cambridge, 1990).

Tracy, James D., ed. *The Political Economy of Merchant Empires* (Cambridge, 1991).

Van Cauwenberghe, Eddy H.G., ed. *Money, Coins, and Commerce: Essays in the Monetary History of Asia and Europe* (Leuven, 1990)

Van der Wee, Herman, 'World Production and Trade in Gold, Silver, and Copper in the Low Countries, 1450–1920', in ed. H. Kellenbenz *Precious Metals in the Age of Expansion: Papers of the XVIth International Congress of the Historical Sciences*, (Stuttgart, 1981), pp.79–86.

Van der Wee, Herman, *The Low Countries in the Early Modern Period* (Aldershot, 1993).

Vilar, Pierre *A History of Gold and Money, 1450–1920* (London, 1976).

Vogel, Hans Ulrich, 'Central Chinese Monetary Policy, 1644–1800', *Late Imperial China* VIII, no.2 (1987), pp.1–52.

Von Glahn, Richard, *Fountain of Fortune: Money and Monetary Policy in China, 1000–1700* (Berkeley, Calif., 1996a).

Von Glahn, Richard, 'Myth and Reality of China's Seventeenth Century Monetary Crisis', *Journal of Economic History* LVI, no.2 (1996b), pp.429–54.

Yamamura Kozo and Kamiki Tetsuo, 'Silver Mines and Sung Coins: A Monetary History of Medieval and Modern Japan in International Perspective'. in ed. John F. Richards *Precious Metals in the Later Medieval and Early Modern Worlds* (Durham, North Carolina, 1983).

Yang, Lien-sheng, *Money and Credit in China* (Cambridge, Mass., 1952).

1
Silver Production in Central Europe, 1450–1618

John U. Nef

THE history of silver is one of the few topics treated by economic historians which has caught the attention of civil servants, of other administrators, and of students of economics. Any general interest in economic history is welcome, for the development of this subject in its broader aspects, in its relations to other branches of history, can help pave the way for a fresh understanding of the history of civilization. The public interest in the history of silver can be explained to a considerable degree by its relation to the history of money and prices. Its importance for the recent history of civilization is small. But in the Middle Ages and at the beginning of the modern age, when credit was relatively little developed, the supplies of silver played a greater part than they do today in determining the wealth and political power of sovereign states. During the Middle Ages and until the thirties of the sixteenth century most of the supplies for all the European countries had come from Germany and from other parts of the disintegrating Holy Roman Empire to the south and east. With the discovery of rich mines in South and Central America, the chief source of supplies shifted abruptly from central Europe to the New World, at a time when months were usually required to make the hazardous journey from Germany across the Atlantic.

Adolf Soetbeer's enterprising work on the output of precious metal since the discovery of America,[2] has been the accepted source for figures concerning the production of silver in central Europe for some

[1] I was led to make the calculations contained in the following pages in connection with an inquiry into mining and metallurgy in medieval Europe, undertaken as part of the projected *Cambridge Economic History* (Vol. II). I had not thought of treating this subject of silver production in a separate article until my friend, Professor E. F. Gay, suggested to me that, in view of the wide attention paid to Soetbeer's statistical tables, it might be of more general interest to scholars than I had supposed.

[2] *Edelmetall-Produktion . . . . seit der Entdeckung Amerikas bis zur Gegenwart* (Gotha, 1879).

sixty years now. As a result of the public interest in the topic, Soetbeer's statistical estimates have been widely reproduced. Exception has been taken to some of his methods, and scholars who have examined his pages critically know how heavily he leaned on guesses in compiling many of his estimates. Nevertheless, his statistics for the output of silver in Europe in early modern times continue to be printed as if they were subject to no serious question. His most prominent German critic regarded his estimates for Germany and the old Austro-Hungarian empire as very satisfactory.[3]

Since the publication of Soetbeer's tract in 1879, many books and articles have dealt with the history of mining in central Europe. Data concerning silver production, not available when he wrote, have been collected by German scholars. A study of the new material available in print and a re-examination of Soetbeer's work show that his statistical tables misrepresent, in at least three respects, conditions in central Europe during the age of the Renaissance and the century following the Reformation. In the first place they considerably understate the annual output of silver when the prosperity of the mines reached its zenith in the first half of the sixteenth century. Second, they do not fully reveal the rapid increase in production during the last half of the fifteenth century and the first two or three decades of the sixteenth. Third, they do not show at all adequately how sharp was the decline in production during the last five or six decades of the sixteenth century, when the states of Europe found their alternative source of supplies in America.

The period in the economic history of the German peoples and their Slavic neighbors which began about 1450 and ended in 1618 is of special interest today, when the expansive traditions of the German race are once more asserting themselves, and with greater force than ever before in history. The year following the first of the two dates, 1451, marked the Duke of Saxony's grant of the right to use a new invention for separating silver from rich argentiferous copper ores with the help of lead.[4] More than any other invention, this accounts for the great prosperity of mining and metallurgy during the century which followed.

The second date marks, of course, the outbreak of the Thirty Years' War. By that time the industrial leadership assumed by the Germans,

[3] W. Lexis, "Beiträge zur Statistik der Edelmetalle," *Jahrbücher für Nationalökonomie und Statistik*, XXXIV (1879), 364.

[4] Cf. Walter Mück, *Der mansfelder Kupferschieferbergbau* (Eisleben, 1910), I, 57.

the Bohemians, and the Hungarians in mining and metallurgy in the age of the Renaissance, was already on the wane. England was becoming the chief power in heavy industry.[5] The economic influence of the Germans outside central Europe was on the point of eclipse. As the strong industrial position obtained temporarily in the early sixteenth century by the peoples of central Europe was bound up to a considerable degree with the production of silver and copper, the course of silver output from 1450 to 1618 is interesting, not only in itself, but for the light it throws on the whole course of German economic history during this period.

The historical importance of silver production justifies a critical note concerning Soetbeer's estimates. Even though it is not possible to substitute new statistical tables for those contained in his tract, it may be helpful to suggest that his tables give so incomplete a picture of the flood and ebb of mining prosperity in Continental Europe that the historian should always make allowances when he consults them.[6] As guides to the course of German economic history in early modern times, they conceal more than they reveal.

I. PRODUCTION AT ITS ZENITH, 1526–35

Soetbeer's tract contains statistical estimates for the annual world-output of silver in the following periods down to 1620: 1493–1520, 1521–44, 1545–60, 1561–80, 1581–1600, 1601–20. According to his tables, production reached its maximum in the countries of the nineteenth-century Austro-Hungarian empire during the period 1521–44 and in Germany during the period 1545–60. For central Europe as a whole (adding the figures which Soetbeer gives for Germany and Austria-Hungary), the annual output is represented as 47,000 kilograms in the period 1521–44, and 49,500 kilograms in the period 1545–60.

[5] Cf. J. U. Nef, *The Rise of the British Coal Industry* (London, 1932), II, 322; "A Comparison of Industrial Growth in France and England from 1540 to 1640," *Journal of Political Economy*, XLIV (1936), 662, 666; "Prices and Industrial Capitalism in France and England, 1540–1640," *Economic History Review*, VII (1937), 157; "Industrial Europe at the Time of the Reformation (*ca.* 1515–*ca.* 1540)," *Journal of Political Economy*, XLIX (1941), 222–23.

[6] As the present article is the by-product of another piece of research, it is probable that some recently published statistical material bearing on the subject of silver production in sixteenth-century Europe has escaped my notice. Doubtless more is preserved in various German archives. My article is not intended to be a final word on the subject, though I am somewhat dubious about the wisdom of ever trying to draw up new statistical tables to replace the estimates of Soetbeer (see below, p. 591).

The new evidence made available since the appearance of Soetbeer's tables indicates that the mines of central Europe were most fruitful somewhat earlier than he supposed, probably during the decade 1526–35, just on the eve of a very great increase in the imports of American treasure into Spain, which were apparently more than twice as valuable in the five-year period 1536–40 as in the five-year period 1531–35. The increase in imports in the thirties was due mainly to the opening of silver mines in the New World. Hitherto, practically all the treasure from America had come in the form of gold. Henceforth, silver imports overshadowed gold imports, substantially in value and overwhelmingly in weight.[7]

The most prominent of all the mining towns of central Europe at the end of the fifteenth and the beginning of the sixteenth centuries were Joachimstal in Bohemia, Mansfeld in the county of that name, Schneeberg, Annaberg, and Marienberg in Saxony, and Schwaz in the Tyrol. At no other place—with the possible exception of Neusohl in Hungary—was anything like a yearly average output of 40,000 marks of silver ever achieved for a decade in early modern times. Joachimstal was nearly twice as productive of silver in the decade 1526–35 as in any other decade before or after.[8] Mansfeld and Annaberg had their most productive years in the thirties of the sixteenth century.[9] For Schneeberg and Schwaz the period of greatest fruitfulness was considerably earlier, in the decade 1476–85 at Schneeberg and in the five-year period 1485–89 at Schwaz.[10] Beginning in the late twenties of the sixteenth century, the Turkish invasion and the growing dislike of the local population for the German capitalists who had gained partial control of the local enterprises, apparently brought about some decline in the output of silver at Neusohl and most other Hungarian mines.

Let us compare the statistical material now available for an estimate of the average annual output of silver, during the decade 1526–35, in Germany and the countries which comprised nineteenth-century

[7] Earl J. Hamilton, *American Treasure and the Price Revolution in Spain, 1501–1650* (Cambridge, Mass., 1934), pp. 34–35, 40–42.

[8] Kasper Sternberg, *Umrisse einer Geschichte der böhmischen Bergwerke* (Prague, 1836), I, Part I, 415–17.

[9] Mück, *op. cit.*, I, 58–59; Soetbeer, *op. cit.*, p. 17.

[10] Oswald Hoppe, *Der Silberbergbau zu Schneeberg bis zum Jahre 1500* (Freiberg, 1908), p. 112; Stephan Worms, *Schwazer Bergbau im fünfzehnten Jahrhundert* (Vienna, 1904), pp. 85–86, 173; M. R. Wolfstrigl-Wolfskron, *Die tiroler Erzbergbaue, 1301–1665* (Innsbruck, 1903), p. 35.

Austria-Hungary, with the figures which Soetbeer gives for his two most productive periods, 1521–44 and 1545–60. Soetbeer divided Germany into Saxony, the Upper Harz, and the remaining districts.

Saxony.—Soetbeer estimated the average yearly silver output of Saxony at 19,020 pounds during the period 1521–44 and at 26,300 pounds during the period 1545–60.[11] Our knowledge of silver production in Saxony in the sixteenth century has not increased appreciably since Soetbeer wrote. But a consideration of the data that he presents in support of his estimates,[12] leads us to doubt whether he was right in assuming that the yearly output was greater after 1545 than it had been during the preceding twenty years. Soetbeer had complete records of output only for the mines at Freiberg. The course of production there differed greatly from that in central Europe generally,[13] and Soetbeer seems to have leaned too heavily on the Freiberg records. There the output was more than twice as great in the fifties of the sixteenth century as in the thirties, when the other chief mining towns—Annaberg, Marienberg, and Schneeberg—were far more prosperous than in the fifties.

In the decade 1526–35 the yearly output of silver at Freiberg averaged rather less than 9,000 marks. At Schneeberg, which had seen its best days in the late fifteenth century, the yearly output averaged only a shade more than 4,000 marks, though it increased to nearly 14,000 marks during the five-year period from 1535 to 1539. The output at Annaberg in the decade 1526–35 is quite uncertain, but, judging from the dividends paid by the mining companies there in the years of greatest prosperity, 1536 and 1537, it may have been several times as great as at Freiberg. At Marienberg, where the mines reached a zenith of prosperity in the four years 1538–41,[14] the output of silver may well have equaled that at Annaberg during the decade 1526–35. In their best years Annaberg and Marienberg almost certainly each produced upward of 50,000 marks (25,000 pounds)—a figure never approached at Freiberg at any time in the sixteenth century. At both places and also at Schneeberg there was in the forties and fifties a great falling-off in productivity, which probably more than offset the gains at Freiberg. It seems likely that the output of the Saxon mines in the twenties and thirties was substantially greater than in the forties and fifties. It is

[11] *Op. cit.*, p. 21. [12] *Ibid.*, pp. 15–18. [13] See below, p. 587.

[14] Ernst Kroker, "Leipzig und die sächsischen Bergwerke," *Schriften des Vereins für die Geschichte Leipzigs*, IX (1909), 51–52.

possible that the output of silver in Saxony from 1521 to 1544 (including the production at other places less important than Annaberg, Marienberg, Freiberg, and Schneeberg) was nearly twice as great as Soetbeer supposed.

The Upper Harz.—According to Soetbeer, the annual output of silver in this district averaged 7,000 pounds from 1521 to 1544 and 6,800 pounds from 1545 to 1560.[15] The largest single source of silver supplies was the mines of argentiferous lead and copper ores of the Rammelsberg, near Goslar, the oldest of the important mining centers in Germany. Neuburg, the chief authority on the mines of the Rammelsberg, who published his book thirteen years after Soetbeer's tract, estimates the production of silver there at some 8,000 or 9,000 marks in 1526.[16] Soetbeer apparently allowed only 4,000 marks for the Rammelsberg mines.[17] Silver was produced at a number of other places in the Upper Harz (where several new mines were opened between 1521 and 1537),[18] though in some cases it left no trace in the statistical records.[19] So Soetbeer's estimate of 10,000 marks for the remaining mines is probably also an understatement.

The "remainder of Germany."—It is here that Soetbeer's figures prove to be least satisfactory. He took the complete records of German silver production for 1850 as a guide. At that time the production of Saxony and the Harz combined accounted for between four-fifths and five-sixths of the total. Soetbeer made the hazardous assumption that the relative positions of Saxony and the Harz, on the one hand, and the rest of Germany, on the other, in silver production were the same between 1493 and 1850 as in 1850. He therefore assigned the "remainder of Germany" a yearly output of 5,160 pounds from 1521 to 1544 and 6,600 pounds from 1545 to 1560.[20]

The mines of the Leber-Tal, in Alsace, are the only German mines outside Saxony and the Upper Harz for which Soetbeer gave any sixteenth-century figures. These mines were most productive during the three years 1534–36, when the annual output of silver was approximately 9,000 marks on the average.[21] According to another authority,

[15] *Op. cit.*, p. 21.

[16] Clamor Neuburg, *Goslars Bergbau bis 1552* (Hannover, 1892), p. 241.

[17] *Op. cit.*, p. 18.

[18] *Ibid.;* Helen Boyce, *The Mines of the Upper Harz from 1514 to 1589* (Menasha, Wis., 1920), pp. 47–50 *et passim.*

[19] Boyce, *op. cit.*, p. 50.

[20] *Op. cit.*, pp. 20–21.

[21] *Ibid.*, p. 20.

not used by Soetbeer, the yearly yield between 1528 and 1545 from the principal mines in the valley—those in the territory of the counts of Rappoltstein—was not under 6,500 marks.[22] The far more important silver-bearing mines of cupriferous slate at Mansfeld, where Luther's father had worked with pick and shovel, also reached a high point of productivity in the thirties. In good years the output of copper amounted to some 2,000 tons and the output of silver to some 50,000 marks. The annual production of silver at Mansfeld during the decade 1526–35 possibly averaged almost 40,000 marks.[23]

In Silesia, another district neglected by Soetbeer, silver was obtained in a number of places, chiefly from argentiferous lead ores. The annual production for Silesia as a whole is quite uncertain, but it was clearly worthy of notice in the thirties and forties of the sixteenth century. Among the most productive mines were the argentiferous lead-ore group in the east at and near Tarnowitz. There more pits were sunk in 1536 than in any other year. During the three years 1561–63, when the yield had fallen off somewhat, the annual output of silver at this group of mines averaged 3,542 marks.[24] Some additional silver was obtained at a number of other places, among them Waldenburg, Gottesberg, and also Reichenstein, famous chiefly for its gold. At Reichenstein, as at Tarnowitz, the mines were most fruitful in the thirties and forties.[25]

Other regions in Germany produced silver in small quantities in the sixteenth century—for instance, the Spessart (in lower Franconia and Hesse-Nassau), the Black Forest, the Siegerland (in Westphalia), and Thuringia. But it is not possible to ascertain the output in any of these regions. Even if we make no substantial allowance for the silver produced in them, it is clear that Soetbeer's figures for silver production outside Saxony and the Upper Harz fall greatly short of the truth. During the decade 1526–35, the yearly output in "the rest of

[22] Koch, "Geschichtliche Entwickelung des Bergbaues und Salinenbetriebes in Elsass-Lothringen," *Zeitschrift für Bergrecht*, 1874, p. 170.

[23] Mück, *op. cit.*, I, 58–59.

[24] Aemil Steinbeck, *Geschichte des schlesischen Bergbaues* (Breslau, 1857), II, 165–66, 230, *et passim*.

[25] Karl Pflug, "Zur Geschichte des Bergbaues im waldenburger Berglande," *Zeitschrift des Vereins für Geschichte und Altertum Schlesiens*, XLIII (1909), 75–78; Steinbeck, *op. cit.*, II, 51 *et passim;* E. Fink, "Die Bergwerksunternehmungen der Fugger in Schlesien," *Zeitschrift des Vereins für Geschichte und Altertum Schlesiens*, XXVIII (1894), 312–13.

Germany" could hardly have amounted to less than 60,000 marks, or 30,000 pounds. It may well have been considerably larger.

Bohemia.—Soetbeer's selection of periods in connection with his statistical tables prevented him from bringing out fully the importance of silver mining in Bohemia in the late twenties and the early thirties of the sixteenth century. For the years 1521–44 he estimated the annual output of the country at 57,730 marks and for the years 1545–60 at 40,455 marks.[26]

If we use the same sources as Soetbeer and take instead the decade 1526–35, we get a considerably larger total. On his own showing, the figures given by Kaspar Sternberg indicate that at this time Joachimstal was probably the most productive silver-mining center in Europe. A maximum output of 87,500 marks of silver was obtained in the year 1533. The annual average during the decade 1526–35 was more than 54,000 marks.[27] Kuttenberg produced an additional 21,000 marks or so.[28]

Soetbeer's figure for the remaining mines of Bohemia is also too low. The mines at Elischau, Welhartitz and Muttergottesberg produced 3,138 marks of silver per annum on the average from 1528 to 1541. The mines at Bergreichenstein and Příbram produced some 3,000 marks yearly in the thirties; those at Budweis and Rudolfstadt perhaps as much; those at Tabor, Ratiboritz, and Hlasowa nearly 1,000 marks.[29] It is difficult to believe that the annual average of the various Bohemian mines, other than Joachimstal and Kuttenberg, fell below 10,000 marks in the decade 1526–35. So the annual output of silver for Bohemia as a whole during that decade could hardly have been less than 85,000 marks.

The Tyrol and Salzburg.—Soetbeer's figures are 50,000 marks a year on the average from 1521 to 1544 and 54,800 marks from 1545 to 1560. Since he wrote, a good deal of new statistical material on mining in the Tyrol has been published. It relates particularly to Schwaz, east of Innsbruck, where the mines were more productive of silver than at any other place in the Eastern Alps. The new data suggest that in the

[26] *Op. cit.*, p. 27. These are marks, Vienna measure, containing 288.668 gr. Soetbeer's figures for Germany are in marks of Cologne measure, containing 233.8555 gr.

[27] *Ibid.*, p. 26; cf. Sternberg, *op. cit.*, I, Part I, 415.

[28] Soetbeer, *op. cit.*, p. 27.

[29] Sternberg, *op. cit.*, I, Part I, 182, 245–48; Part II, 59–60, 64; Soetbeer, *op. cit.*, p. 25.

decade 1526–35 the yearly output of these mines came to almost 36,000 marks, or some 5,000 marks less than Soetbeer counted.[30] But he did not allow nearly enough for the remaining Tyrolese mines. Agricola, the greatest German mining and engineering authority of the early sixteenth century, regarded the mines just south of the Brenner at Sterzing-Gossensass as almost as important as those at Schwaz. Account should be taken also of silver production in the southern Tyrol at Windisch-Matrei, Primör, Brixen, Persen, and Taufers, and in the northern Tyrol in the Zillertal and especially at Rattenberg and Kitzbühel.[31] Soetbeer noticed only the last two of these mining centers. After Schwaz and Sterzing-Gossensass they were the most important; but Soetbeer was mistaken in assuming that their importance began with the discovery, apparently in 1539, of the Rörerbühel in the lordship of Kitzbühel.[32] At Rattenberg, probably a much more productive center for both silver and copper than Kitzbühel, the most fruitful period was the beginning of the sixteenth century, when the silver output amounted to many thousands of marks per annum. There and also at Kitzbühel mining was carried on at least as early as the middle of the fifteenth century.[33] While no accurate estimate of the total output of the Tyrolese mines in the decade 1526–35 is possible, it is by no means improbable that the yield of silver from all the other mining centers combined exceeded that at Schwaz.

We assume that Soetbeer's figure of 8,000 marks of silver for the annual production at Salzburg is about right.[34] The output of Salzburg and the Tyrol together could hardly have fallen much short of 80,000 marks a year during the decade 1526–35, or more than half again as much as his estimates suggest.

The remainder of Austria-Hungary.—Soetbeer allows for an annual output of only 7,270 marks from 1521 to 1544 in Hungary, Transylvania, and all parts of the nineteenth-century Austro-Hungarian empire other than Bohemia, the Tyrol, and Salzburg.[35] This figure is a pure guess. In so far as Soetbeer had any information at all to guide him it was for 1480 and 1552. The Hungarian mines prospered mainly in the interval between these dates. Soetbeer did not follow the same method here as he adopted for Germany—that of assuming that the

[30] Wolfstrigl-Wolfskron, *op. cit.*, p. 35; Soetbeer, *op. cit.*, p. 29.

[31] Wolfstrigl-Wolfskron, *op. cit.*, *passim.*

[32] *Op. cit.*, p. 30.

[33] Wolfstrigl-Wolfskron, *op. cit.*, pp. 155–59.

[34] *Op. cit.*, p. 30.

[35] *Ibid.*, pp. 27, 31.

distribution of production between the various regions was roughly the same as in the mid-nineteenth century. If he had done so, he would have had a larger output for Hungary than for Austria, instead of a very much smaller one. There is little doubt that his estimate for Hungary in the early sixteenth century falls greatly short of the mark, in view of the evidence discovered since his time that the Fuggers obtained from Hungary most of the supplies of argentiferous copper ore for their great metallurgical factory in Thuringia. The ore came mainly from the rich mines at Neusohl, and silver was an important by-product of the manufacture.[36] Some of the most powerful and ingenious engines for draining mines at the time of the Reformation had been apparently invented in Hungary. As the capital invested in mining, not only at Neusohl, but also at Sohl, Schemnitz, and Kremnitz, rivaled that in some of the principal centers of Germany and Austria, it is probable that as late as the twenties of the sixteenth century the annual output of silver obtained from Hungarian ores amounted to several tens of thousands of marks.[37] Some silver was also produced in Transylvania, Carniola, and Carinthia, where the Fuggers had another great plant near Villach for separating silver from argentiferous copper ore.[38] It is possible that Soetbeer's figure for the "remainder of Austria-Hungary" falls even further short of the truth than his figure for the "remainder of Germany." But in the present state of our knowledge even rough estimates for the production of silver in the countries to the east of Bohemia, the Tyrol, and Salzburg cannot be given without resort to the wildest kind of guesswork.

Our survey of the mines of central Europe in the decade 1526–35 suggests that the annual production of silver in Germany amounted to upward of 150,000 marks, Cologne measure (233.8555 gr.). It suggests that in the nineteenth-century Austro-Hungarian empire the annual production amounted, at the very least, to from 175,000 to 200,000 marks, Vienna measure (280.668 gr.). So the combined yearly total

[36] Ernst Koch, "Das Hütten- und Hammerwerk der Fugger zu Hohenkirchen . . . . , 1495–1549," *Zeitschrift des Vereins für thüringische Geschichte*, XXVI (N.F., 1926), 295, 313, 317–18, 322–23.

[37] Jakob Strieder, *Studien zur Geschichte kapitalistischer Organisationsformen* (2d ed.; Munich, 1925), pp. 8–12, 15, 26–27, 40.

[38] Friedrich Dobel, "Ueber den Bergbau und Handel des Jacob und Anton Fugger in Kärnten und Tirol, 1495–1560," *Zeitschrift des historischen Vereins für Schwaben und Neuburg*, IX (1882), 194–95; Koch, "Das Hütten- und Hammerwerk der Fugger zu Hohenkirchen . . . . ," *op. cit.*, pp. 295–96.

was perhaps in the neighborhood of 3,000,000 ounces. It may well have exceeded that figure considerably. Soetbeer's tables give the erroneous impression that the average annual output did not reach a maximum until the period 1545–60, when it amounted, according to his tables, to a shade more than 1,700,000 ounces.[39]

It was more than three hundred years before an output of 3,000,000 ounces was again obtained from central Europe. Not until the fifties of the nineteenth century, when the German people were beginning once more to play a leading role in European economic and political life, did the silver mines in this part of the world again become as fruitful as at the time of the Reformation.[40]

II. THE GROWTH IN PRODUCTION, *ca.* 1450–*ca.* 1540

It has long been known that the silver mines of central Europe reached their zenith of prosperity in the sixteenth century after a period of rapid growth. But Soetbeer's tables do not help us to estimate the magnitude of this growth, to determine when it began and when it was most rapid. The statistical material of various kinds that has been accumulated by scholars since 1879 is sufficient to make possible rough and tentative conclusions on all three points. The production of silver in Bohemia, Saxony, and most of the mountainous districts peopled by the Germans, Slavs, and Magyars probably reached a low point during the Hussite wars in the twenties and thirties of the fifteenth century. The expansion which followed took two forms—the revival with the help of improved technical methods of drowned or buried mines and the discovery of seams of ore in many old mining centers and in some new places.

It is not possible to determine exactly when the production of silver began to increase. But the growth was certainly not impressive before 1450,[41] although some rich seams were discovered in the first half of the fifteenth century, notably at Schwaz.[42] In the sixties and even more in the seventies and eighties the expansion in the output of silver became rapid. Very rich ore deposits were discovered at Schneeberg in the fifties. There the output rose from, at the most, a few hundred

[39] *Op. cit.*, p. 107.

[40] *Ibid.*, pp. 111–12.

[41] Cf. Albert Jäger, "Beiträge zur tirolisch-salzburgischen Bergwerks-Geschichte," *Archiv für österreichische Geschichte*, LIII (1875), 339.

[42] Worms, *op. cit.*, p. 17.

marks a year around 1450, to about 31,000 marks during the period 1470–76, and to about 44,000 marks from 1476–85.[43] At Schwaz, and probably at Mansfeld,[44] the growth of mining enterprise at this time was hardly less astonishing. In the seventies and eighties the output at Schwaz more than tripled, rising from an annual average of 14,622 marks from 1470 to 1474, to 26,147 marks from 1475 to 1479, to 34,651 marks from 1480 to 1484, and, finally, to 45,336 marks from 1485 to 1489.[45]

By the last decade of the fifteenth century, and perhaps earlier, the silver output in Hungary had begun to grow with similar rapidity. The great mines at Annaberg were first opened up in the nineties. In the second decade of the sixteenth century those at Joachimstal and Marienberg were started. Meanwhile, the Hungarian mines continued to prosper. By the late twenties the ancient mines of the Rammelsberg, near Goslar, and of Kuttenberg, in Bohemia, had recovered most of the prosperity they had once had in the thirteenth and early fourteenth centuries. The output was increasing rapidly at Freiberg.

Around 1450 there was probably not a single mining center yielding 10,000 marks of silver a year. Around 1530 there were eight or more centers producing from 10,000 to upward of 50,000 marks. The output of central Europe as a whole could hardly have increased less than fivefold; it may have increased considerably more. The growth in production was apparently most striking between about 1460 and about 1530. In the Eastern Alps there was a marked slackening in the rate of growth in output at the turn of the fifteenth into the sixteenth century. In the Erz Gebirge, the Sudetic Mountains, and most parts of Germany the rate of growth was almost, if not quite, as rapid during the first three decades of the sixteenth century as during the last four decades of the fifteenth. As Soetbeer's first statistical table is for the period 1493–1520, when production was already on the point of reaching a maximum, it does not bring out the phenomenal nature of the growth in silver output at the end of the Middle Ages.

III. THE DECLINE IN PRODUCTION, *ca.* 1540–*ca.* 1618

According to Gustav Schmoller, the period from 1480 to 1570 was the second brilliant age in the history of silver mining in central Eur-

[43] Hoppe, *op. cit.*, pp. 7, 112, 156–57.

[44] Mück, *op. cit.*, I, 5, 45.

[45] Worms, *op. cit.*, pp. 85–86, 173.

ope, after an earlier brilliant age in the thirteenth century.[46] This may be true in the sense that around 1570 the output of silver was greater than it had been around 1480, though even this much cannot be taken for certain. The statement is not true in the sense that these ninety years were a time of continually increasing productivity. The rapid growth in silver production began rather sooner than Schmoller suggested in his excellent synthesis of the history of mining in early Germany, which was published in 1891. The decline in silver production also began rather sooner.

The erroneous impression conveyed by the work of Soetbeer concerning the decline in silver production, which accompanied the great increase in the inflow of American treasure, is easy to explain. In the first place it arises from the fact that his tables understate the actual output in the twenties and thirties of the sixteenth century. His figures for the last decades of that century and for the early seventeenth century are certainly much less of an understatement for the output of central Europe; indeed, it may be doubted whether they are an understatement at all. The mining centers for which Soetbeer had the most complete data turn out generally to be those where output was best maintained after the Reformation. He had no records for Hungary, where a serious decline in mining prosperity first set in, probably in the late twenties of the sixteenth century. For Germany itself, as we have seen, he leaned heavily on the statistics for Freiberg. These statistics are completely unsatisfactory as a guide, because there was no other important silver-mining center in Europe where output followed the same course during the sixteenth and seventeenth centuries as at Freiberg. More silver was produced there in the decade 1611–20 than in the decade 1531–40. The annual output was greater at the end of the Thirty Years' War, from 1641 to 1650, than from 1524 to 1530.[47]

The contrast is striking between the course of production at Freiberg and that at some of the other leading silver-mining centers of central Europe, for which Soetbeer had only fragmentary statistics or for which he had no statistical evidence at all. At Mansfeld the output in the sixties of the sixteenth century was not much more than half as great as in the thirties. During the first two decades of the seventeenth century the yearly yield of silver had fallen to one-fifth or less

[46] Gustav Schmoller, "Die geschichtliche Entwickelung der Unternehmung," *Jahrbuch für Gesetzgebung, Verwaltung und Volkswirtschaft im Deutschen Reich*, XV (1891), 963 ff.

[47] Soetbeer, *op. cit.*, p. 17.

of what it had been at the zenith of productivity.[48] As we have seen, Schneeberg had largely lost its importance as early as the Reformation. A slight revival in the late thirties was followed by a new decline in the forties of the sixteenth century.[49] Production probably declined pretty steadily at Annaberg and Marienberg from the high points reached between 1536 and 1540. By the seventies both places had lost most of their importance as mining centers.[50] Beginning in the forties Joachimstal underwent a collapse as complete as that at Schneeberg, Annaberg, and Marienberg. The annual output fell from about 54,000 marks in the decade 1526–35 to some 17,000 marks in the decade 1555–64, to between 8,000 and 9,000 marks in the decade 1565–74, and to some 3,000 marks during the twenty years 1601–20.[51]

These figures for Joachimstal were known to Soetbeer, but he had fewer statistics for Schwaz than are now available to us. There the annual output declined from nearly 36,000 marks in the years 1520–44 to about 24,000 marks in the years 1545–69 and to a little less than 11,000 marks in the years 1595–1619.[52] In Silesia the chief silver-producing mines, those in the neighborhood of Tarnowitz, lost almost all their importance during the last half of the sixteenth century. During the three years 1561–63, when the annual output is said to have been considerably less than in the thirties, it amounted to 3,542 marks. During the three years 1578–80 it amounted to 1,444 marks; in 1592 and 1593 it had fallen to 614 marks; in 1601 and 1602 to 520 marks.[53] In the Leber-Tal, in Alsace, the annual output declined from a high point of some 9,000 marks during the three years 1534–36 to some 3,000–4,000 marks during the last half of the sixteenth century and to some 1,500 marks during the early years of the seventeenth.[54]

Silver production at almost all the chief mining centers of central Europe had fallen off by the beginning of the seventeenth century to anywhere from one-third to one-twentieth of its maximum amount.

[48] Mück, *op. cit.*, I, 117–19.

[49] Soetbeer, *op. cit.*, p. 16.

[50] *Ibid.*, pp. 17–18.

[51] Sternberg, *op. cit.*, I, Part I, 415–17; Soetbeer, *op. cit.*, p. 27.

[52] Wolfstrigl-Wolfskron, *op. cit.*, p. 35.

[53] Steinbeck, *op. cit.*, II, 230.

[54] Soetbeer, *op. cit.*, p. 20. At some of the smaller centers the collapse in silver-mining operations was still more precipitate. At Přibram, in Bohemia, the output of silver fell from 2,132 marks in 1553 to 841 marks in 1558, to 117 marks in 1568, and to 20 marks in 1574 (Sternberg, *op. cit.*, I, Part II, 64).

Apart from Freiberg, the only important places at which the silver output was fairly well maintained were two of the other most ancient mining centers of Europe, the Rammelsberg, in the Harz, and Kuttenberg, in Bohemia.

During the forties, fifties, and sixties of the sixteenth century the decline in the prosperity of most of the principal mining districts was offset to some extent by the discovery of seams of argentiferous ores in new spots. But none of the fresh mines approached in fertility the principal ones opened up during the previous eighty years or so. After the sixties few new seams were uncovered and worked. The collapse in prosperity was in full swing during the half-century preceding the Thirty Years' War.

According to Soetbeer's figures, the annual output of silver in the whole of central Europe on the eve of that war—during the first two decades of the seventeenth century—amounted to about 750,000 ounces.[55] These figures, unlike those which he gives for the time of the Reformation, seem about right. Production in the decade preceding the Thirty Years' War was almost certainly less than one-third, probably not more than one-fourth, as great as it had been during the decade 1526–35.

In the interval Europe had been deluged with shipments of American silver. The imports into Spain alone reached 700,000 ounces or so a year as early as the forties of the sixteenth century. In the last decade of the century they reached some 10,000,000 ounces or more a year.[56] These imports helped to send the prices of most commodities soaring, and silver fell from one-half to one-third of its former value. Thus the imports were one of the factors which led to the abandonment of silver mining in all European countries. There was a superabundance of supplies and consequently little incentive to search for or dig more at home at a high cost.

Soetbeer's estimates suggest that in the European countries outside of the Holy Roman Empire the output of silver fell only slightly during the sixty years following the opening of the mines of Potosi, about 1545.[57] But his figures are based on actual statistics of production only in the case of Sweden. There, on Soetbeer's own showing, the annual output declined rapidly and was only about one-fourth as great in the second half of the sixteenth century as in the first.[58] In view of

[55] *Op. cit.*, p. 108.

[56] Hamilton, *op. cit.*, pp. 34, 38, 40–42.

[57] *Op. cit.*, pp. 107–8.

[58] *Ibid.*, p. 34.

the abundant evidence concerning central Europe and the figures for Sweden, such as they are, the burden of proof will certainly rest upon one who contends that the silver output was much better maintained in other European countries for which no respectable evidence at all is available. Soetbeer's tables probably minimize the fall in the production of these countries even more than they minimize the fall in the production of central Europe. In the twenties and thirties of the sixteenth century the silver production of Europe as a whole may well have reached 3,500,000 ounces or more.[59] In the first two decades of the seventeenth it had been reduced in all probability to less than 1,000,000 ounces.[60]

IV. CONCLUSION

The downward curve of prosperity in European silver mining seems to have been an almost perfect converse of the upward curve. Output increased at a very rapid rate in central Europe from about 1460 to about 1500 or 1510. After that output continued to grow, but at a diminishing rate, down to the thirties of the sixteenth century. After the thirties output began to decline, for a time rather slowly. After about 1560 the decline became pronounced; during the last three decades of the sixteenth century it was very rapid. On the eve of the Thirty Years' War, some eighty years after the zenith was reached, the mines were apparently little more productive than they had been in the middle of the fifteenth century, on the eve of the great European silver rush. During the Thirty Years' War the annual output did not perhaps greatly exceed that during the Hussite wars, two centuries before.

While Soetbeer's tabular estimates of European silver production before the Thirty Years' War do not give a satisfactory picture of the growth and decadence of silver mining in central Europe, his figures for the total production of silver in Europe during the sixteenth and seventeenth centuries as a whole probably fall much less short of the truth than his figures for production in central Europe in the twenties and thirties of the sixteenth century. The text of his tract—the result

[59] *Ibid.*, pp. 34–38, and see above, p. 585.

[60] Soetbeer's tables give the annual output in Europe from 1601 to 1620 as a shade more than 1,000,000 ounces (*op. cit.*, p. 108). But his figure of 8,000 kilograms for the countries outside the Holy Roman Empire is, in all probability, too high. There is no evidence that silver supplies of any consequence were being obtained in any of these countries, unless perhaps in European Turkey.

of extensive and solid research—supplies much information that is reliable. It is still very helpful to the historian.

Aristotle remarks that an educated man will not expect greater precision in answering any question than the nature of the subject matter allows.[61] Even today it is doubtful whether, without sacrificing accuracy, the course of silver production in Europe from 1450 to 1618 can be put in the form of statistical tables of the sort Soetbeer tried to construct. In spite of the new data published since he wrote, a great deal still has to be left to guesswork. In all probability the attempt to give precise figures decade by decade, even with the help of all the evidence that is still buried in the archives of central Europe, would leave much to be desired. The educated man will not ask for precise figures. He will content himself with general statements, containing many qualifications, like the statements put forward in tentative and imperfect form in this article.

These statements indicate that the shock to mining and to economic life in central Europe, which accompanied the imports of vast quantities of American treasure, was even more violent than has been supposed. If the silver mines of Germany, Bohemia, Hungary, and the Eastern Alps had proved richer and if the silver mines of South and Central America had proved poorer, the economic eclipse of the Germans in the late sixteenth and seventeenth centuries might have been somewhat less complete than it was. During that age the Germans lost their leadership in heavy industry. The economic, and particularly the industrial, balance of power shifted from the Continent (from central Europe, Italy, Spain, and the Spanish Netherlands) to Holland, Sweden, and, above all, to Great Britain. That shift is to be understood only in terms of the general history of European civilization, for it was the result of many factors. In considering the factors which contributed to the decline in the economic importance of central Europe, it would be a mistake to neglect altogether the distribution of the sources of precious metal made by Nature, or by Divine Providence, as the Greeks might have said.

[61] *Nicomachean Ethics*, Book i, chap 1.

2
Una forma de oposición: el contrabando

Zacarías Moutoukias

I

El contrabando significó, obviamente, la presencia de los tradicionales adversarios o enemigos del Imperio — Francia, Holanda e Inglaterra — en un ámbito y un espacio sobre los cuales les leyes españolas proclamaban su exclusiva soberanía, las Indias occidentales y su comercio colonial. Este solo hecho, tan evidente que puede parecer innecesario recordarlo, muestra la importancia de examinar el contrabando como parte de las oposiciones al Imperio. Mas la presencia de las potencias no españolas en América y en su sistema comercial, era solo el aspecto más visible y mejor documentado de una compleja red de fenomenos económicos, sociales, y politicos que afectaban tanto a las relaciones de España con dichas potencias rivales, como a los vínculos de las élites americanas con la corona y, por ende, influía sobre las formas del poder imperial.

A pesar de las enormes carencias bibliográficas sobre el tema y los límites del presente texto, es necesario hacer un esfuerzo por abordar — al menos eso también formará parte de mi demostración — el exámen global de esa trama que era el contrabando, y procurar mostrar su lugar en el conjunto del imperio.

> ... la contrebande européo-exotique fut universelle aux XVIIème et XVIIIème siècles. Ainsi s'explique la masse des textes inédits qui s'y rapportent. [...] ils [los ingleses] sont les posseseurs des Indes au point que le roi d'Espagne n'en plus que le seigneur ....[1]

---

1 L. VIGNOLS, *L'ancien concept de monopole et la contrebande universelle*, «Revue d'Historie économique et sociale», XIII, n. 3, 1925, p. 240; H. SÉE y L. VIGNOLS, *La fin du commerce interlope dans l'Amérique Espagnole*, *idem*, p. 303.

Estas citas sintetizan, al mismo tiempo, una realidad histórica solidamente establecida, y el tono de una literatura que se ocupó de estudiarla desde la primera década de este siglo.[2] Aunque presentaba muchas limitaciones, dicha literatura comenenzó a desbrozar un camino que luego fue abandonado.

Excede mis propositos tratar este interesante aspecto de la historiografía económica, y más aún desarrollar un estado de la cuestión sobre el comercio atlántico durante los siglos XVI a XVIII. Sin embargo, creo necesario recordar dos cuestiones a fin de establecer el contexto de mi argumentación. La primera: son bien conocidas las obras y las conclusiones generales de los autores que trazaron las grandes lineas del comercio legal entre España y América durante esos siglos aplicando métodos cuantitativos — Hamilton, Chaunu, Garcia-Baquero y Garcia Fuentes —. Según ellos, entre 1510 y 1778, el volumen de dicho tráfico, medido en toneladas, atravesó una etapa de expansión hasta 1620, continuada por otra de sucesivas contracciones entre 1620 y 1705-'10; a partir de este nadir el movimiento inició una paulatina recuperación hasta 1740-'50, seguida por otra fase de acelerado crecimiento hasta 1778. Si observamos sus cifras sobre el

[2] No es mi intención presentar una bibliografía completa; solo me limito a recordar los títulos, a mi juicio, más representativos. Ademas de los autores ya citados en la nota anterior, ver también: H. Sée, *Bosquejo de las relaciones comerciales de Holanda con España y Portugal a fines del siglo XVII*, «Anuario de Historia del Derecho Español», t. II, Madrid, 1926; y *L'activité commerciale de la Holande a la fin du XVIIe siècle*, en «Revue d'Histoire économique et sociale», XIV, 1926, n. 2; L. Vignols, *El asiento francés (1701-1713) e inglés (1713-1750) y el comercio franco-español desde 1700 hasta 1730, con dos memorias francesas de 1728 sobre esos asuntos*, «Anuario de Historia del Derecho Español», t. V, Madrid, 1928 (la lista completa de trabajos de estos autores, relativos al contrabando, es bastante más larga); A. Aiton, *The asiento treaty as reflected in the papers of Lord Shelburne*, «Hispanic American Historical Review» (en adelante «H.A.H.R.»), vol. VIII, 2, 1928; G. de Artiñano y Galdácano, *Historia del comercio con las Indias durante la época de los Austrias*, Barcelona, 1917; V. Lee Brown, *The South Sea Company and the contraband trade*, «American Historical Review», n. 31, 1925-26; Idem, *Contraband trade: a factor in the decline of Spain's Empire in America*, «H.A.H.R.», vol. VIII, 1928, 2; A. Bermejo de la Rica, *La Colonia del Sacramento*, Madrid, 1920; E.W. Dahlgren, *Les relations commerciales et maritimes entre la France et les côtes de l'Ocean Pacifique*, Paris, 1909, t. I; C. Haring, *The Buccaneers in the West Indies in the seventeenth century*, New York, 1910; Idem, *Comercio y Navegación entre España y las Indias en la época de los Habsburgos*, México, 1979 (primera edición en inglés, Mass. 1918). Aunque algo más tardios. los siguientes y muy conocidos trabajos continuan la problemática de los anteriores: A. Girard. *Le commerce français a Seville et Cadix au temps des Habsbourg. Contribution a l'étude du commerce étranger en Espagne aux XVIe et XVIIe siècles*, Paris, 1932; A. Christelow, *Contraband trade between Jamaica and the spanish main, and the free porte act of 1766*, «H.A.H.R.», vol. XXII. 1942. 2; Idem, *Great Britain and the trades from Cadiz and Lisbon to Spanish America and Brasil. 1759-1783*, «H.A.H.R.», vol. XVII. 1947, 1; el clásico libro de A.P. Canabrava apareció en otro contexto, pero, por sus característiсas, está más cerca de esta historiografía que de cualquier otra: *O comercio Portugues no Rio da Prata (1548-1640)*, Saô Paulo, 1944.

valor del comercio — cuyo único indicador para todo el periodo es la importación de metales preciosos en España —, éstas muestran una evolución bastante próxima a las del volumen.[3]

Ahora bien, dichos autores tratan al comercio ilícito de manera colateral; sea para fundamentar el uso de una fuente que no lo registra, sea para demostrar que la existencia del contrabando no afecta sus conclusiones generales. En consecuencia, aun cuando generaron una verdadera ruptura y un salto en los métodos y en las perspectivas con las que se estudia el comercio atlántico en su conjunto, repecto del contrabando no reemplazaron lo esencial de las hipótesis y explicaciones del primer grupo de trabajos citado. Solo recientemente se ha intentado profundizar el estudio de la actividad mercantil no autorizada incorporando esos fundamentales avances historiográficos.[4]

Un pequeno párrafo aparte merece el libro de Morineau. Sus resultados modificaron los terminos en los que se venía debatiendo sobre la evolución del comercio entre España y América y sus efectos a ambos lados del océano. El historiador francés demuestra brillantemente que no había correlación entre el valor y el volumen del tráfico, como lo prueba el hecho de que las llegadas de metales preciosos, lejos de disminuir, durante la segunda mitad del siglo XVII alcalzaron sus niveles más elevados.[5] Las correcciones que introduce a las series anteriormente existentes, corresponden a fraudes, subregistros, omisiones o simplemente a movimientos que las otras fuente hasta entonces utilizadas no consignaban; sus datos se refieren fundamentalmente al interior del sistema español, aún si a partir de 1690 distinguen entre la plata americana que llegaba a España de la que lo hacía directamente a otros puertos de Europa. Es decir que comparte con los autores más arriba mencionados la óptica, la mirada del tráfico desde los puertos del monopolio, en donde la presencia extranjera se apoyaba en la dependencia de España respecto de

[3] E. Hamilton, *El tesoro americano y la Revolución de los precios en España*, Barcelona, 1975, p. 47; P. Chaunu, *Séville et l'Atlantique*, Paris, 1955-'59, 13 vols., t. VI, pp. 64 y 65 y t. VII, p. 59; L. Garcia Fuentes, *El comercio español con América*, Sevilla, 1980, p. 209 a 236 y pp. 288-9 (en realidad el autor señala una contradicción en la evolución del valor — creciente — con la del volumen — decreciente — de las exportaciones españolas hacia América entre 1662 y 1678, pero no incorpora esto a sus conclusiones; ver pp. 236 y 381); A. Garcia-Baquero, *Cádiz y el Atlántico*, Sevilla, 1976, 2 vol., t. I, p. 535 a 556, t. II, gráficos 3, 7 y 13; y *Andalucia y la carrera de Indias (1492-1824)*, Sevilla, 1986, p. 55 y p. 87 a 100.

[4] C. Malamud, *Cádiz y Saint Maló en el comercio colonial Peruano (1698-1725)*, Cádiz, 1986; Z. Moutoukias, *Contrabando y control colonial, Buenos Aires y el espacio peruano en el siglo XVII*, Buenos Aires, 1989.

[5] M. Morineau, *Incroyables gazettes et fabuleux métaux*, Paris, 1986, pp. 321 y 563. No puedo extenderme aqui sobre las ideas de este autor que modificó los paradigmas alrededor de los cuales se discutía sobre los aspectos centrales del sector externo de la economia colonial. Para sus críticas sobre los cálculos de tonelaje, ver: *Jauges et méthodes de jauges anciennes et nouvelles*, Paris, 1966.

sus manufacturas. Se trata de una perspectiva legítima, deliberadamente adoptada para resolver problemas de la historia económica europea. Mientras que si nos ubicamos en cualquiera de los puertos americanos, los problemas son diferentes. Alli poco importaba que un navio del sistema oficial hubiese sido, antes de su partida de España, cargado legalmente o gracias a los servicios de algún 'pasador' clandestino. En cambio la presencia masiva de navios no españoles — frequentemente mayoritaria en distintas época y lugares — afectaba las relaciones de las élites locales con una corona que podia verse reducida a la condición de «Señor» de posesiones que no poseia. Pero me estoy adelantando.

El segundo aspecto que deseo subrayar a los efectos de mi argumentación es el de la presencia del contrabando en cantidad de libros y artículos que lo abordan en el cuadro de una historia de las instituciones económicas, o como parte de las 'causas' o los 'efectos' en una historia política o de las relaciones internacionales.[6] Debido a lo expuesto arriba, dicha historiografía, con frecuencia de excelente calidad, se apoya en los aportes de aquel mismo grupo de publicaciones, algunas de las cuales — como vimos — tienen más de sesenta anos.

Entre estas últimas, el trabajo que, en mi opinion, mejor presenta y sintetiza el conjunto de ideas comunmente aceptadas sobre el tema del contrabando, es el libro de Haring. Según su análisis, el sistema del monopolio resultaba de una alianza entre la Corona, por un lado, y por el otro, poderosos comerciantes sevillanos, mexicanos y limeños, agrupados en sus respectivos Consulados. Estos últimos interesados en mantener un control fiscal de la actividad económica mediante un intervencionismo que acababa por desalentarla. El régimen impuesto marginaba — siempre según Haring — a mercaderes medios y a regiones excéntricas al sistema de Flotas y Galeones. Esto creaba una «demanda» de contrabando en los sectores y áreas desplazados. El comercio intérlope, tanto en España como en América, significó el fracaso del monopolio andalúz. Las causas se encontrarían en la bancarrota industrial de la peninsula, que facilitaba la presencia extranjera, mientras la Corona insistía en una política económica inadecuada. La persistencia de las irregularidades se habría debido a la profunda corrupción de los funcionarios, consecuencia de los bájos salarios y la venta de cargos públicos. Se encuen-

[6] Algunos ejemplos: M.E. Rodriguez Vicente, *El Tribunal del Consulado de Lima en la primera mitad del siglo XVII*, Madrid, 1960; G. Walker, *Política española y comercio colonial, 1700-1789*, Barcelona, 1979; J. Israel, *The Dutch in the Hispanic world. 1606-1661*, London, 1982; Idem, *A conflict of Empires: Spain and the Neetherlands*, «Past and Present», 76, 1977; C.C. Goslinga, *The Dutch in the Caribean and the Wild Coast. 1580-1680*, Assen, 1971.

tran, entonces, contenidos lo que en no pocos trabajos aun hoy siguen siendo los principales tópicos sobre el tema del comercio ilegal: fragilidad económica de España frente a la expansión de las otras potencias europeas; la actitud favorable de algunos comerciantes españoles y un sector deplazado de las élites americanas; la corrupción de los funcionarios; y la irracionalidad del sistema.[7]

Es en relación a este cuerpo más o menos establecido de ideas que trataré de examinar la trama de acontecimientos que constituían el contrabando. Se trataba de un fenómeno estructuralmente vinculado a la vida económica del mundo indiano, que afectó no solamente a regiones marginales y sectores subalternos, sino también — o sobre todo — a las élites de los núcleos capitales de las posesiones españolas. La importancia del mismo tornan insuficientes las cifras del comercio entre España y América como indicador del comportamiento del sector externo hispanoamericano. Asociado a la crisis y debacle del régimen original de la «Carrera» — no a su fracaso, idea que indicaría irracionalidad en los principios —, se sostuvo gracias a la capacidad no satisfecha de la oligarquias indianas, así como a la corrupción: un sistema de relaciones y comportamientos que articulaba prácticas económicas y el anudamiento de dichas oligarquias con el aparto local de la Corona, dandole continuidad al fenómeno. En particular procuraré mostrar que, en este contexto político, aquel comportamiento independiente del sector externo hacia posible una relación exclusiva con España. De forma tal que el contrabando fué un importante elemento de oposición al Imperio, por el cual las élites coloniales fortalecían su posición relativa y autoconstruían su sitio en el conjunto.

II

Conforme se iba organizando el régimen que regulaba las relaciones comerciales entre España y América, instaurando el monopolio andaluz, los limites fijados por la legislación eran rápidamente desbordados de distintas maneras. Para mediados del siglo XVI, está documentada la introducción ilegal de manufacturas importadas utilizando autorizaciones para navegar entre las Islas Canarias y puertos de América; en éstos últimos navios españoles solían fondear sin autorización, y vendían sus car-

[7] Haring, *Comercio y navegación...*, *cit.*, cap. VI; Artiñano y Galdácano, *op. cit.*, p. 122; ver también Rodríguez Vincente, *op. cit.*, pp. 254-57.

gas amparados en el derecho a pedir refugio en caso de «... tiempos contrarios, necesidad de bastimentos y otras causas ...» — el mismo mecanismo, la «arribada maliciosa», que más tarde practicarían de ordinario navios no españoles —. Hacia fines del siglo XVI, la presencia de extranjeros era regular, no solo como corsarios, mas también como activos comerciantes en los centros neurálgicos del tráfico.[8]

De modo general, entonces, el contrabando fue tan temprano como el monopolio mismo; si bien con este termino se comprenden, en realidad, tres fenomenos diferentes que conviene distinguir, debido — entre otras importantes razones expuestas más abajo — a que cada uno tuvo su propia dinámica y peculiar cronología, a pesar de haber estado fuertemente vinculados entre ellos. Estos eran: a) el contrabando en el interior de la «Carrera» oficial; b) lo que Vignols y Malamud han llamado «comercio o contrabando directo» de las potencias no españolas en América; c) la navegación intercolonial e interprovencial.

El primero de dichos fenómenos abarcaba un abanico de operaciones que iban desde el fraude en los registros de mercancias y tesoros, hasta la sistemática transgresión a las normas que reglamentaban la introducción de productos importados y la participación de mercaderes extranjeros en el tráfico. A este último aspecto me referiré abajo con mayor detalle. En cuanto a la generalización del sub-registro, fue la respuesta a una creciente imposición fiscal; y al contraerse la base imponible se generaba el caracteristico círculo vicioso de aumento de la presión y mayor evasión. La relación entre fiscalidad, guerra, expansion militar o defensa del Imperio es demasiado evidente como para detenernos sobre el tema. Sobre todo, tratandose del tesoro americano, cuya importancia estratégica para el estado y ejercito modernos ya fue abundantemente destacada.[9]

El estudio de Céspedes del Castillo sobre la avería ilustra muy bien la cadena de fenómenos que iba desde la oposición externa al Imperio hasta el fraude en los registros. Este impuesto *ad valorem* estaba destinado a financiar los gastos de defensa de las flotas y era pagado por el comercio — su administración corría por cuenta del Consulado —. En conse-

[8] *Recopilación de Leyes de Indias*, reproducción facsimilar de la edición de JULÍAN PAREDES de 1681, Madrid, 1974; libro IX, título XXXVIII, ley V, y título 41, ley 15; *Colección de Documentos inéditos relativos al descubrimiento, conquista y organización de las antiquas posesiones españolas de ultramar*, Madrid, 1885-1900, 2da serie, 13 vols., t. X, p. 561; HARING, *op. cit.*, pp. 22 y 23, p. 143 y ss.

[9] F. BRAUDEL, *Méditerranée et le monde méditerranéen à l'époque de Philippe II*, Paris, 1968, 2 vol., t. II, p. 13 y ss.; ver tambien, A. DOMINGUEZ ORTIZ, *Los caudales de Indias y la politica exterior de Felipe IV*, «Anuario de Estudios Americanos», t. XIII, Sevilla, 1963.

cuencia, el monto dependia de dos variables, el costo de la «armada de guardia», y el valor declarado de las cargas. Con el inicio de la ofensiva holandesa en 1621, aumentaron las escoltas, mientras disminuían los registros legales de mercancias y metales preciosos. La situación se agravó debido a las confiscaciones de tesoros a cambio de juros, a las que recurrieron Felipe III y Felipe IV, que procuraban a los comerciantes razones adicionales para ocultar el metálico. Hacia 1630 la avería comenzó a recaudarse a una tasa del 35 por ciento. A partir de 1641 el Consulado se negó a renovar el contrato, y el Estado se vio obligado a administrarla directamente. Desde 1660 la Corona reemplazó el cobro del impuesto por una contribución fija de los Consulados de Sevilla, México y Lima y derogó la obligación de registrar los tesoros, lo que puso fin al sistema original. En estos mismos años se generalizó el pago de indultos con los que se compraba la remisión de las contravenciones. Negociados entre las autoridades y los particualres implicados, el indulto se convirtió, a ambas orillas del Atlántico, en un verdadero impuesto informal al contrabando y a toda clase de infracciones o delitos de orden económico.[10]

Según dos estimaciones, entre 1630-1640 la cantidad de metales preciosos que escaparon al control de las autoridades habría duplicado al embarcado legalmente; y de 1650 a 1659, el valor de lo ocultado habría oscilado entre el 40% y el 90% del total del tesoro transportado. La información disponible para la navegación legal al Río de la Plata — los Navios de Registro — revela también que los capitanes solo declaraban un 10% a un 20% de los metales preciosos con los que retornaban a España. Y también aqui se apelaba regularmente a los indultos.[11] En todo caso, a partir de 1630, la generalización del fraude dentro de la «Carrera», internalizó la presión militar externa, asociandola a la crisis interna, lo que provocó el progresivo descalabro del sistema original. Este fué reemplazado por otro en el cual la legislación se combinaba con mecanismos informales por todos admitidos, como el del indulto, y con la contribución directa de los Consulados de Sevilla y América a los gastos de la «Carrera». Dentro de este cuadro, la corrupción y la ansiedad fiscal

---

10 G. Céspedes del Castillo, *La avería en el Comercio de Indias*, Sevilla, 1945; J. Lynch, *España bajo los Austria*, Barcelona, 1975 (1ra ed. en inglés, Oxford, 1969), 2 vol., t. II, p. 228 a 243; Dominguez Ortíz, *op. cit.*, pp. 314, 342-352, y 362; Haring, *El comercio…, cit.*, pp. 65-68, 83-104 y 143; Garcia Fuentes, *op. cit.*, pp. 124-158; Z. Moutoukias, *Power, corruption, and commerce: the making of the local administrative structure in the 17th century Buenos Aires*, en prensa «H.A.H.R.», 1988; J. de Veitia Linage, *Norte de la contratación en las Indias Occidentales*, Buenos Aires, 1945 (1ra ed., Sevilla, 1672), pp. 710-711.

11 Chaunu, *op. cit.*; Morineau, *Fabuleux, op. cit.*, p. 242 y ss.; Moutoukias, *Contrabando…. cit.*, cap. III.

se articularon en un sistema que facilitó a las oligarquias americanas el acceso a un espacio más autónomo.

El comercio directo de las potencias no españolas en las costas americanas — el segundo de los fenómenos ennumerados que abarcaba el término contrabando — contribuyó abundantemente a este proceso que afectó a las relaciones entre las élites coloniales y la metrópoli.

Comparando las cifras de Garcia-Baquero y Morineau sobre las llegadas de metales preciosos a España durante la segunda mitad del siglo XVIII, podría pensarse que entonces las autoridades habían logrado controlar el fraude reduciendolo a hechos marginales.[12] Sin embargo, para la época, la presencia de navíos extranjeros era masiva en todos los centros comerciales de las Indias Occidentales. A finales de la década de 1750, desde sus bases en las Antillas, los ingleses dominaban el tráfico semi-clandestino en el Golfo de México así como en la linea costera que iba de Campeche hasta Venezuela; también eran activos franceses y holandeses, asentados — entre otras poseciones — en Haiti y Curaçao respectivamente. La Colonia Sacramento, en el Rio de la Plata, era una puerta abierta para el comercio de Portugal, Brasil; y — sobre todo — Inglaterra, cuyos navios no dejaban de estar directamente presentes en la región. Al Pácifico entraban continuamente embarcaciones francesas e inglesas, y mercancias que seguían la ruta que nacía en el Caribe y atravesaba el Istmo de Panamá o los Andes colombianos hasta el reino de Quito. Si creemos una de las tantas memorias francesas sobre el particular, redactada en 1762, el comercio directo de los extranjeros en América superaba en valor y volumen al que realizaba la metrópoli.[13]

Aunque se inició a mediados del siglo XVI, la presencia de extranjeros comenzaría a hacerse persistente y ordinaria a comienzos de la década de 1630, mientras avanzaba la crisis de la «Carrera» oficial. La historia de los acontecimientos que marcaron el proceso está escrita y se conoce; en cambio su historia económica y social es aún muy fragmentaria. Aqui me limitare a recordar aquellos aspectos que permiten efectuar algunas comparaciones. Como se sabe, hay cuatro fechas que simbolizan la imposición de un albergue de adversarios o enemigos en el nudo americano del sistema comercial: en 1634 los holandeses ocuparon Curaçao; en 1640 los franceses tomaron la minúscula isla de la Tortuga, iniciando un alojamiento en territorio español que culminaría con su instalación en la mi-

[12] Garcia-Baquero, *Cádiz...*, *cit.*, t. II, p. 264 y ss.; Morineau, *Fabuleux...*, *cit.*, p. 519.

[13] *Memoire sur le Commerce des Indes 1762*, Bib. Nat. de Paris, Mss., Fonds français, Papiers de Beliardi (en adelante B.N.P.-FF.FF.-B.), 106769, fol. 3 a 56.

tad de La Española hacia 1659; cuatro años antes, en 1655, los ingleses se habian adueñado de Jamaica; en 1680, los portugueses fundaron Colonia, frente a Buenos Aires, ocupando un paraje comunmente usado por los contrabandistas.

Además del comercio que practicaban con las costas venezolanas, la Compañia holandesa de Indias Occidentales intentó crear una colonia de envergadura en el norte de Brasil — por entonces parte del Imperio —. Para ello desalojó a los portugueses de sus bases negreras en Africa occidental, buscando completar una estrategia en la que articulaba la colonización agricola para la exportación, el tráfico de esclavos, asi como el comercio ilícito, fuente de plata usada en los intercambios con Oriente. Hacia 1640, la colonia holandesa estructurada alrededor de Pernambuco producía el 60% de todo el azúcar del Brasil. Esta ofensiva formaba parte de la guerra por la supervivencia económica que, hasta 1648, enfrentó a España con las Provincias Unidas. Combinado con las operaciones militares, la Corona impuso un bloqueo comercial, con variado éxito, que empujó a Holanda hacia sus posesiones en América. El resultado final — como es sabido — fue favorable a la primera. En consecuencia, a pesar del fracaso de la empresa brasileña, la República surgió de la Paz de Münster como la principal potencia en la práctica del comercio semi-clandestino en América española, principalmente en Cartagena y Buenos Aires.[14]

En este último puerto, la presencia de navios holandeses fue predominante entre 1653 y 1680. En otro trabajo he podido estimar que, desde 1648 hasta 1702, unas 200 embarcaciones sin autorización entraron en el Rio de la Plata provenientes de ultramar; de las cuales han sido efectivamente verificadas 129 — más de dos tercios del total del tráfico atlántico de Buenos Aires, con y sin autorización —. De estas, el 50% eran holandesas; el resto portuguesas, inglesas y francesas, en ese orden de importancia. La presencia de las primeras, en realidad fue mas importante entre 1653 y 1682. Durante esos años, sobre no menos de 128 embarcaciones provenientes de ultramar (de unas 200 a 400 toneladas) que comerciaron con Buenos Aires, solo 20 tenian autorización para hacerlo — los Navios de Registro —; el resto eran «arribadas maliciosas», de las cuales habia por lo menos 59 holandesas, es decir, la mitad del total.[15]

[14] ISRAEL, *A conflict of empires..., cit.*, p. 34 y 74; C.R. BOXER, *Os holandeses no Brasil. 1624-1654*, San Pablo, 1964, p. 7 a 156 y p. 223 a 343; IDEM, *Salvador de Sá e la luta pelo Brasil e Angola*, San Pablo, 1973, p. 55 a 82 y p. 236 a 305.

[15] MOUTOUKIAS, *Contrabando..., cit.*, cap. V, Anexo y Apéndice.

Cuando no se refugiaban en las Islas San Gabriel, frente a Buenos Aires — futuro asentamiento de Sacramento — desde donde podían comerciar clandestinamente con los innumerables establecimientos rurales ubicados en la costa, las naves que llegaban sin autorización alegaban una «arribada forzosa». Ya tengo mencionado este derecho a pedir refugio en circunstancias adversas, cuyo respeto estaba explicitamente contemplado por las paces firmadas en 1648. Al terminar la guerra, la Corona española debió forjar un nuevo tipo de relación politica con Holanda con un doble objetivo; por un lado buscaba contrapesar el creciente poderio de Inglaterra y Francia y, por el otro, moderar el propio impulso comercial de la primera. Este doble proposito se reflejaba en la correspondencia entre Madrid y el Gobernador de Buenos Aires, a proposito de la actitud a seguir ante las «arribadas forzosas», que en realidad siempre eran «maliciosas» y su admisión abría, inevitablemente, el puerto al tráfico no autorizado. El Gobernador recibia, con frecuencia, instrucciones en el sentido de evitar los motivos de reclamo diplomático respentando lo pactado sobre el derecho a pedir refugio, lo que hacia imposible la represión del comercio directo suponiendo que hacerlo estuviera en las intenciones del algún gobernador.[16]

Esto nos lleva a dos conclusiones. En primer lugar, se trataba, entonces, de un comercio semi-clandestino, en el que comerciantes locales, funcionarios-mercaderes, y capitanes extranjeros lograban aprovechar usos aceptados y el equívoco espacio que les ofrecia la legislación. En segundo término, la relación de fuerzas favorable a las Provincias Unidas permitió, en el Río de la Plata y en el resto de América española, una extensión semi-tolerada, o dificil de reprimir, de las arribadas; que luego beneficiaría también a ingleses y franceses. Aunque las cambiantes alternativas de las relaciones diplomáticas no permiten aventurar una generalización acerca de la tolerancia, el fenomeno de las arribadas se instaló en todo el ambito americano para durar.[17]

---

[16] Israel, *A conflict..., cit.*, p. 74; Moutoukias, *ibid.*, cap. IV; R.R.C.C., Madrid 25-VII-1648 (contiene instrucciones y capítulos de las paces); Madrid 18-III-1652; R. Trelles, *Registro Estadistíco de la Provincia de Buenos Aires. 1867*, Buenos Aires, 1868, t. II, pp. 19-20 y 23; ver también: la R.C., Madrid 18-VII-1661, en la que se ordena: «... disponiendo que en las arribadas legítimas y lícitas que hicieren los navios holandeses a los puertos del distrito de ese Gobierno, *se les haga buen pasaje y tratamiento para que no puedan tener ocación de justa queja o sentimiento de que no se les guarda lo capitulado en los tratados de las dichas paces*, pero esto habeis de practicar ajustandos a la observancia de lo dispuesto y declarado ...», Idem, p. 42; y *Carta del Gobernador 15-XI-1677*, A.G.I., Charcas 279.

[17] R.R.C.C., Madrid 18-III-1660 y Madrid 25-IX-1660, al comunicar el fin de la guerra con Inglaterra y Francia, la Corona española reitera las recomendaciones hechas para los holandeses, aunque las alternativas de las relaciones diplomáticas hacen que más tarde se retire expresamente a la primera los beneficios de la paz; R.C. Madrid, 20-III-1669; los tres documentos en *Trelles, op. cit.*, t. II, pp. 35, 37 y pp. 47-48.

Si bien carezco para el Caribe de cifras agrupables como las de Buenos Aires, aquel fue el ámbito en el que Holanda desplegó sus esforzos más importantes, combinando igualmente operaciones ilegales con arribadas. Dominante durante la segunda mitad del siglo XVII, a lo largo de la primera del XVIII conservó un destacado segundo puesto detrás de Inglaterra. Hacia 1702, un viajero español, señalaba su constante presencia en las costas de los dominios españoles a partir de su base en Curaçao: Honduras; Portobelo y otros puertos menores del Istmo que enlazaban con el sistema del Pacífico centrado en Lima; Cartagena; Rio de Hacha, donde comerciantes extranjeros tenian — al igual que en Buenos Aires — tiendas públicas; Caracas ...[18] La documentación encontrada en distintos navios comisados por autoridades espanolas en 1724, 1736, y 1751, revelan la existencia de una tupida red de vinculos comerciales entre Curaçao y Amsterdam, y Curaçao y New York.[19]

La actividad desplegada por los holandeses ilustra como se fue imponiendo la presencia del contrabando directo, gracias a una combinación de agresividad naval y presiones diplomáticas; en un contexto militar desfavorable a España y de crisis del sistema comercial estructurado a mediados del siglo XVI. Visto desde América, el fenomeno se completaba con la actitud propicia de las élites coloniales.

En el caso de los ingleses, el Tratado de Madrid, de 1670, sancionó su dominio sobre las islas y territorios que venían ocupando. Para entonces ya se habían convertido en activas bases del comercio directo. En particular, Jamaica era el principal centro articulador de los tráficos que unían puertos británicos y de Nueva Inglaterra, con las costas de las poseciones españolas en todo el Caribe. En 1690, Robles, el mencionado viajero espãnol, visitó la isla. Más que alguna información concreta, de su relato impresiona el clima general que pintó: una activa comunidad de mercaderes, con frecuencia tambien propietarios rurales, que mantenía un activo comercio con las costas de Nueva España, Tierra Firme y Cuba.[20] Algunos años más tarde, un oficial español hecho prisionero,

---

18 G. de Robles, *América a fines del siglo XVII. Noticias de los lugares de contrabando*, Introducción de Victor Tau Azoategui, Valladolid, 1980, pp. 81 a 98; ver también: P.D. Huet, *Comercio de Holanda, o el Gran Thesoro Historial, y Político de Floreciente comercio que los Holandeses tienen en todos los Estados... etc.*, traducido del francés por Francisco Xavier de Goyeneche, Prólogo de Gerónimo de Uztáriz, Madrid, 1717; R. Arauz Monfante, *El contrabando holandés en el Caribe. Siglo XVIII*, Caracas, 1986; E. Còrdova Bello, *Compañias holandesas de Navegación. Agentes de la Colonización neerlandesa*, Sevilla, 1964.

19 Arauz Monfante, *op. cit.*, p. 49 a 67.

20 Robles, *doc. cit.*, p. 32 a 34.

pudo observar el movimiento naval de Kingstown entre diciembre de 1718 y septiembre de 1719. Durante esos meses habrían zarpado un total de 393 navios; 201 se dirigieron hacia poseciones españolas, 13 a Curaçao, y 58 hacia el océano. De las que navegaron a hispanoamérica, 143 se encaminaron a Cuba, 14 a Portobelo, 15 al Golfo de Honduras, y 24 hacia Santa Marta y Cumaná.[21] Hacia la época se estimaba que las mercancias reexportadas desde Jamaica producían anualmente unos 3 millones de pesos.[22]

En bien sabido que tras la Paz de Utrech, la Compania del Mar del Sur desempeño un papel esencial en el contrabando inglés, desde sus bases operativas en Jamaica, Barbados y Buenos Aires. El derecho al sumunistro exclusivo de esclavos, del que gozó entre 1713 y 1739, encubría un intenso tráfico de mercancias. Según un cálculo, hecho a partir de fuentes inglesas, entre 1730 y 1739 la Comapañia traficó por un valor de 6 millones de libras, mientras que los particulares lo hicieron por 1.500.000 libras.[23] Es decir un total de 32.100.000 pesos y un promedio anual de — obviamente — 3.200.000 pesos. La guerra de 1739 de ninguna manera significó un disminución del comercio directo inglés. Un informe elaborado en 1759, a poco de la coronación de Carlos III, afirmaba que, considerando el conjunto de sus operaciones en América, los ingleses ganaban alrededor de 6.000.000 de pesos anuales.[24]

Igualmente extendido estaba el comercio francés, cuya presencia en el Pacífico fue excelentemente estudiado por Malamud. Obviamente, las alianzas de la Guerra de Sucesión facilitaron, durante algunos años, su participación en la defensa del languideciente sistema comercial español; además de la concesión del Asiento de esclavos a la Compañia de Guinea.[25] Pero el contrabando directo representó lo esencial de sus actividades mercantiles. Operaron principalmente en Veracruz y las costas peruanas. Al inicio del siglo XVIII la presencia francesa fue mayor en el Atlántico, cuyos centros redistribuidores eran La Martinique y Saint Domingue. Entre 1701 y 1707 entraron en Veracruz no menos de 50 navios zarpados de puertos galos, distribuidos del siguiente modo: 1701, 1 navio; 1702, 3 navios; 1704, 2 navios; 1705, 10 navios; 1706, 13 navios; 1707, 21 navios.[26]

---

[21] BROMN, *Contraband Trade...*, *cit.*, pp. 181-182.

[22] M.D. GARCIA MOLLEDA, *El contrabando inglés en América. Una correspondencia inédita de la factoria de Buenos Aires*, «Hispania», t. X, n. 39, p. 354.

[23] G. NELSON, *Contraband Trade under the Asiento 1730-1739*, «American Historical Review», 1946, n. 51, pp. 57 y p. 62-64; ver tambien MALAMUD, *op. cit.*, p. 38 a 41.

[24] CHRISTELOW, *Contraband Trade...*, *cit.*, p. 313.

[25] WALKER, *op. cit.*, Primera Parte.

[26] MALAMUD, *op. cit.*, p. 48.

Durante el último quinquenio del siglo XVII comenzó a hacerse regular el tráfico hacia las costas peruanas gracias al dominio que los franceses lograron de la ruta del Cabo de Hornos. Entre 1698 y 1724 comeciaron con puertos chilenos y, sobre todo, peruanos, 148 navios de ese origen, con la siguiente distribución quinquenal: 1700/4, 12 naves; 1705/9, 41 naves; 1710/14, 62 naves; 1715/19, 11 naves; 1720/24, 12 naves (dos embarcaciones llegaron en 1698, y para 11 se desconoce el año). Estas embarcaciones importaron a Francia, desde 1701 hasta 1725, 54.602.791 pesos en metales preciosos, o sea un promedio anual de unos 2.200.000 pesos.[27]

Está claro, entonces, qtue el comercio directo de las potencias no españolas en América, representaba una parte tan esencial del comercio atlántico, que no es posible extraer conclusiones generales sin tomarlo en consideración. Según las cifras de Morineau, entre 1701 y 1725, se importaron en Europa unos 291 millones de pesos en metales preciosos provenientes de hispanoamérica; de los cuales 196 millones — alrededor de dos tercios — llegaron a España, y el otro tercio — 95,2 millones — entraron por puertos ingleses, holandeses y franceses. El propio autor admite que sus evaluaciones subestiman el contrabando directo tanto de holandeses como de ingleses, en particular el tráfico intérlope de Jamaica.[28] Además, estos datos no toman en cuenta los 54 millones en metálico importados a Francia desde Perú; ni los que pudieron retornar con los 50 navios provenientes de Veracruz, cuya estimación en 25 millones parece razonable. O sea que, durante aquellos 25 años, Hispanoamérica exportó hacia Europa no menos de 370 millones de pesos, de los cuales, por lo menos un 46% a traves del comercio directo. Podría pensarse que semejante proporción estaba relacionada con la Guerra de Sucesión, con las acciones de los enemigos y la defensa coactiva de los amigos. Sin embargo para los años 1714-1725, cifras de Morineau y Malamud reunidas, tenemos que de los 195 millones de metales preciosos que exportaron las colonias españolas, solo 2/3 fueron hacia su metrópoli. En 1740, la guerra reaparece con un juego de alianzas similar y las proporciones vuelven a cambiar, aún si solo nos guiamos por las cifras de Morineau: sobre un total de 94 millones en metálico extraidos de América española hasta 1748, menos de un 60% entró a España.[29] Durante la segunda mitad del siglo XVIII, la guerra fué más frecuente que la paz y los sistemas de alianzas no se alteraron, al menos no en la parte que a nosotros interesa; sin embargo, la misma serie de datos estaría indicando que la

[27] *Ibid.*, pp. 62 y 67.
[28] MORINEAU, *Incroyables..., cit.*, p. 314 a 321, pp. 371-372 y pp. 474-475.
[29] *Ibid.*, p. 372 y ss. y p. 474.

Corona española habría logrado reducir a un mínimo las transferencias de «su» tesoro americano hacia reinos adversarios.[30] Como quiera que haya sido, dos cosas son indudables. La primera es que, al igual que los fraudes en el interior de la «Carrera», el comercio directo, tuvo una intensidad variable, o sea que tuvo su historia. No obstante — y he aqui la segunda aseveración —, la confrontación entre la infomacion de Morineau y Malamud, en el brevisimo periodo en que es posible, así como las estimaciones sobre el comercio inglés que no incluí en la comparación, indican que la citada memoria de Beliardi de 1762 hacia un diagnostico basicamente correcto: el comercio directo de las potencias no españolas con hispanoamérica era por lo menos tan importante como el de éstas con su metrópoli. Lo que nos coloca ante una conclusión un tanto obvia, semejante actividad comercial no puede ser considerada como un fenómeno coyuntural, como una anomalia de larga duración. Estaba estructuralmente vinculada a la vida económica de las Indias Occidentales; y otorgaba una enorme autonomia al comercio externo de las élites coloniales, cuya capacidad para importar y exportar superaba a las de su Metrópoli.

Cualquiera fuera su estatuto legal, el comercio ultramarino era impensable sin una multitud de circuitos que entrelazaban puertos y ciudades mediterraneas del continente americano. Esta red de intercambios se sostenía sobre una diversidad de economias locales en cuyo seno se producían los medios de transporte y los hombres que hacían circular las mercancias. Navegación de cabotaje, recuas de mulas, flotas de carretas, los tráficos locales, casi siempre más importantes en valor y volumen que el comercio ultramarino de una determinada región, fueron los primeros en vincular entre si a los asentamientos españoles, articulando mercados, creando las condiciones para que las mercancias del comercio a gran distancia, manufacturas suntuarias, metales preciosos, esclavos, azúcar, ...pudieran darse cita en algún puerto.[31] Pero estos vaivenes interesan no solo porque sobre ellos debía insertarse el comercio ultramarino, legal o clandestino, para alcanzar la totalidad de sus mercados. En si mismos, los tráficos navales interprovinciales e intercoloniale fueron objeto de restricciones legales cuyo desborde era, al mismo tiempo, un problema específico y uno de los tres fenómenos designados por el término contrabando, arriba ennumerados.

---

[30] *Ibid.*, p. 475.

[31] Para un exámen general del comercio interprovincial, C.S. Assadourian, *Sobre un elemento de la economia colonial: producción y circulación de mercancias en el interior de un conjunto regional*, en *El sistema de la economia colonial*, Lima, 1982; D. Ramos, *Minería y comercio interprovincial en Hispanoamérica. Siglos XVI, XVII, y XVIII*, Valladolid, 1970.

Las princiapales regiones que abarcaron dichos tráficos fueron ... todas; el Caribe, el Pacífico, las costas del Brasil hasta el Rio de la Plata. En los dos últimos casos es donde más claramente la legislación se proponia limitar simultaneamente el cabotaje y el contrabando. Desde Acapulco hasta Valparaiso, la navegación interprovincial en el Pacífico se apoyaba en una verdadera constelación de puertos. Sus centros eran el propio Acapulco, Panamá, Guayaquil y el Callao, cuyos tráficos recibieron atención de historiadores y observadores contemporáneos.[32] Como se sabe esta linea se articulaba con otras dos navegaciones, interocéanicas esta vez: la que comunicaba con las Filipinas y la del Atlántico a traves del Itsmo de Panamá.

Durante la década de 1570 se hicieron regulares los viajes entre Manila y Acapulco, por un lado y, por el otro, las remesas de mercurio peruano, lo que acabó de completar un sistema de intercambios iniciado con productos locales. Muy esquematicamente podemos resumirlo del siguiente modo: Perú exportaba vino, mercurio y plata hacia México y las Filipinas, a cambio de manufacturas suntuarias, especias, hierro y cobre de Oriente, asi como manufacturas suntuarias y tejidos tambien de México. Este, además de sus importaciones desde Filipinas — pagadas con metálico — recibia cacao de Guayaquil y otros productos locales.[33] El rápido crecimiento de este complejo movimiento, y el consecuente drenaje de metales preciosos hacia las Filipinas, impulsó a la Corona a dictar, entre 1587 y 1591, una serie de restricciones destinadas a limitar, al mismo tiempo, las remesas de plata hacia Oriente, y el envio de vinos peruanos a fin de defender les exportaciones españolas. Tan poco efecto tuvieron, que debieron reiterarse hasta la definitiva prohibición del comercio marítimo entre ambos virreynatos en 1634.

Los efectos de esa legislación o, dicho en otros términos, el grado de autonomía que podía alcanzar un determinado espacio de circulación mercantil, ha sido interpretada de diferente manera. Hamilton señaló que una de las causas de la crisis del comercio atlántico durante la segunda mitad del siglo XVII habría sido la cantidad de plata distraida hacia el intercambio con las Filipinas a traves del Galeon de Manila.[34] O sea que

[32] M. Moreyra y Paz-Soldàn, *Estudios sobre el tráfico marítimo en la época colonial*, Lima, 1944; W. Borah, *Early colonial trade and navigation between México and Perú*, Berkeley/L.A., 1956; P. Chaunu, *Les Philippines et le Pacifique des Iberiques (XVIe, XVIIe, XVIIIe siecles)*, Paris, 1960; M. Carmagnani, *Les mécanismes de la vie économique dans une société colonial: le Chili (1680-1830)*, Paris, 1973; J. Juan y A. de Ulloa, *Noticias Secrétas de América*, ed. anotada de L. Ramos Gòmez, Madrid, 1985, 2 vol.

[33] Borah, *op. cit.*, pp. 80 a 95.

[34] Hamilton, *op. cit.*.

podría observarse una coyuntura propia al Pacifico, independiente, que se opondría, compensando y explicando la coyuntura atlántica. En cambio, Chaunu ha buscado demostrar la inexistencia en el Pacífico de un espacio economicamente autónomo. Puesto que sub-colonia de Nueva España, las Filipinas o, más precisamente, su comercio con el Occidente oriental fue, en realidad

> ...le simple prolongement, dans l'ordre des échanges, de l'Atlantique hispanoamericain de Séville...[35]

En consecuencia, copió su evolución: expansión hasta 1620, seguida de una meseta en suave pendiente que, hacia 1630-1640, se fue transformando en la caida que llevó a la fuerte depresión de los años 1650-1680; la recuperación posterior a esa fecha se prolongaría en una nueva fase de expansión hasta fines del siglo XVIII. En otras palabras, los indices del Pacífico aparecerían conformes a los esquemas de la económia atlántica.[36] Sin embargo vimos que, a partir del libro de Morineau, esos esquemas están sujetos a reexámen. Mas, a los fines de mi argumentación, de momento cuenta subrayar que, a pesar de las restricciones, los mismos datos de Chaunu, es decir las cifras oficiales, indican que la plata siguió llegando en fuertes cantidades a Manila. La proveniente de México sumaba: 5,42 millones entre 1621-1630; 3,5 millones en 1631-1640.[37] El fraude parece haber llenado los huecos dejados por el silencio de la documentación. En 1638 el Cabildo de Lima declaraba que todos los años seguían entrando mercancias asiaticas por unos 600.000 pesos al año; y que el giro anual del comercio con las Filipinas se estimaba en unos 2.000.000 de pesos. Hacia 1704 se seguían reiterando las prohibiciones para comerciar con Nueva España; mientras, los propios apoderados de la ciudad de Manila, al intentar explicarlo, admitían implicitamente que los intercambios con aquella excedían lo autorizado.[38] En suma, el tráfico entre Nueva España y Perú, que comprendía productos orientales, estuvo muy lejos de interrumpirse.

Pero este es solo un aspecto en relación a estimar el grado de autonomia de un ámbito de navegación como el Pacífico. Sabemos que mantenía dos articulaciones con el Atlántico. La que comunicaba con el universo del contrabando directo en el Caribe cruzando el Itsmo o por tierra hasta Guayaquil. En ambos casos, la navegación de cabotaje sostenida en una abundante cadena de puertos, aseguraba la distribución final de

[35] CHAUNU, *Les Philipines...*, *cit.*, p. 17.

[36] *Ibid.*, pp. 249 a 266.

[37] *Ibid.*, pp. 136-143.

[38] K.J. ANDRIEN, *Crisis and decline. The viceroyalty of Perù in the seventeenth century*, Albuquerque, 1985, p. 36 a 39; Biblioteca Nacional de Madrid, Mss. de América, 1915, fls 618 y ss.

las mercancias.[39] El otro entrelazamiento con el Atlántico era el correspondiente al sistema de Galeones y la feria de Portobelo, el tráfico entre el Callao y Panamá comenzó a organizarse en un sistema de convoyes escoltádos por la Armada del Mar del Sur.[40] Este sistema también estaba recorrido por el fraude. En 1647 la Audiencia de Panamá ennumeraba las medidas que habia tomado para evitar la llegada a Portobelo de plata sin registrar, no obstante las cuales, cuatro años más tarde estos representarían no menos del 25% del total de caudales transportados.[41] Pero dichos tráficos representaban una mínima parte de la actividad de la flota mercante peruana. Hacia fines del siglo XVI sumaba unos 50 navios, pero solo unos 4 o 5 eran anualmente utilizados para el transporte de caudales; y alrededor de 1680, la flota contaba con más de 70 embarcaciones. Esto nos remite nuevamente a la navegación de cabotaje;[42] sobre cuya vitalidad se sostuvieron los incontenibles desbordes a los limites que la legislación le imponía a la navegación interprovincial en el Pacifíco.

Los signos del ajetreo abundan. El valor del comercio exterior de Santiago — cuyos vínculos con Lima son muy bien conocidos — pasó de un índice 100 para el decenio 1680-'89, a 317 para el de 1730-'39.[43] El tráfico de cabotaje con centro en él Callao fue de 111 unidades en 1698-'99 y 237 en 1701-'4.[44] Asi adquieren significado las citadas descripciones de Juan y Ulloa que subrayaban la multitud de pequeños puertos que enlazaban a los principales.

De modo que, al irrumpir en el Pacifíco, los franceses entraron en un complejo espacio de circulación naval, fuertemente articulado, que iba de Valpariso a Manila, pasando por El Callao, Guayaquil y Acapulco. Sin esta red de base, el comercio directo simplemente no hubiera tenido donde desembarcar. Y los franceses supieron muy bien aprovechar su estructura. Malamud los ha encontrado entregados no solamente al comercio de cabotaje, sino también a un fructuoso intercambio triangular con Oriente, cuyos productos vendian en las costas peruanas.[45]

---

39 Juan y Ulloa, *op. cit.*, pp. 119 a 145 y pp. 509 a 563.

40 P.E. Peres Mallaina Bueno y B. Torres Ramirez, *La Armada del Mar del Sur*, Sevilla, 1987, p. 1 a 6; para un cuadro global de su evolución ver el cp. VI.

41 *Colleción de Documentos inéditos relativos al descubrimiento, conquista y colonización de la colonias españolas en América y Oceanía*, 1ra serie, 42 vols., Madrid, 1864-1884, vol. 21, p. 12; Rodriguez Vicente, *op. cit.*, p. 259.

42 L.A. Clayton, *Trade and navigation in the seventeenth-century Viceroyalty of Perú*, «Journal of Latin American Studies», pp. 1 y 18, vol. VII, n. 1.

43 M. Carmagnani, *Les mécanismes..., cit.*, p. 32.

44 Malamud, *op. cit.*, p. 136.

45 *Ibid.*, p. 134 y p. 138 a 152.

En el Rio de la Plata, el comercio intercolonial con el Brasil comenzó muy poco tiempo despues del definitivo asentamiento español en sus costas. Gracias al cual, sus vecinos, en gran medida de origen portugués, realizaron una lucrativa intermediación intérlope, que ayudó a consolidar un núcleo poblado en Buenos Aires, considerado por la Corona de alto valor estratégico para la defensa del traspatio altoperuano. No fué ésta la menor de las pradojas de este comercio semi-legal, que ponía en movimiento manufacturas europeas y azúcar desde Brasil, esclavos desde las costas africanas y Brasil, a cambio de plata, y algunos productos locales, cueros, sebos, harinas... Entre 1602 y 1622, las autoridades metropolitanas buscaron controlar el tráfico autorizando una parte. Durante esos años una serie de «autorizaciones» — de mercedes — prohibian la extracción de plata, asi como la importación de escalvos — aunque consentian la navegación hacia Guinea —, y limitaban el valor y volumen del comercio. A la sombra de lo que las mercedes permitían, se desbordaron completamente sus límites: los metales preciosos representaron no menos del 80% del valor de las exportaciones, y entre 1595 y 1640 se introdujeron unos 30.000 esclavos...[46]

El comercio ilícito — como cualquier otro comercio — no solo necesitaba de autoridades complacientes en los puertos para comunicar los mercados con los centros de penetración. El sistema de intercambio no era una comunicación directa con los asentamientos mineros, entre región productora y el puerto. En realidad lo metales preciosos seguian un doble movimiento de dispersión hacia las regiones proveedoras de la industria minera, y de concentración en los puertos para su exportación. En otras palabras, la ruta Potosí-Buenos Aires se apoyaba sobre un conjunto de economias regionales, cuyo funcionamiento articulaba una red de mercados locales. Un comerciante porteño viajaba con «ropa» hacia Santa Fe; alli trocaba una parte contra ganado vacuno en pie y yerva del Paragay que conducia a Salta, donde al conjunto era vendido para su reventa en Potosí. Un traficante luso-portugues llegaba a Córdoba con herramientas, tejidos, y azúcar con los que obtenia mulas para conducir al Alto Perú; al producto de la operación lo podía invertir en asociarse con algún terrateniente de la ciudad para importar esclavos desde Buenos Aires o directamente desde Guinea. En fin, ademas de carretas, de Buenos Aires partian hacia Potosí caravanas que conducían yerva mate, esclavos, manufacturas europeas y mulas. O sea que los excedente locales — parte de los cuales eran los mismos medios de transporte — se integraban en

[46] MOUTOUKIAS, *Contrabando...*, *cit.*, cap. 2.

los intercambios a gran distancia. Lo que aseguraba la circulación así como el consumo suntuario de las oligarquias provinciales, esencial para asegurar su condición de élite colonial.[47]

Como tengo dicho, en el Caribe la circulación interprovincial e intercolonial fué particularmente intensa y diversificada. El caso del comercio de cacao entre México y Caracas es el ejemplo mejor estudiado de estos tráficos. Durante el siglo XVII las exportaciones a México adquirieron tanta importancia que éste se convirtió en la principal metrópoli económica de Caracas, si exceptuamos — como veremos — a los holandeses. En el periodo 1620-1650 Venezuela exportó a México 35.512 fanegas de cacao y solo 289 a España. En la última década del siglo XVII, mientras las compras españolas de todos los productos alcanzaban un valor de unos 2.000.000 de reales, las ventas de cacao a Nueva España sumaban unos 17.000.000 de reales.[48]

Aqui también los circuitos se entremezclaban, los ruanes y cuchillos se embrollaban con cueros o harinas. Al igual que en el Rio de la Plata, esta imbricación generaba niveles diferenciados de circulación «monetaria». Nuestro mencionado viajero español observaba con sorpresa como en Jamaica separaban la moneda gruesa y de buena ley para destinarla a la exportación y guardaban las piezas menudas y las deficientes para los intercambios locales.[49] Al igual que en el Pacífico, los mercaderes y marinos extranjeros que fueron entrando en el Caribe, llegaron a un área de navegación cruzado por enredadas lineas de tráficos; a los que rapidamente usufructuaron como via de penetración y como una oportunidad más de ampliar los negocios. Observando la bahía de La Matanza, el mismo testigo encontró que alli

> ...entran nuestras embarcaciones del trato sin dificultad a cargar el tabaco y frutos de la tierra. Como también lo suelen hacer las de otras naciones, y especialmente ingleses y holandeses, que son las más ciudadosas de estos tráficos...;

o, más tarde en Puerto de Plata, en la Española

[47] *Ibid.*, cap. 7; ver también, C. Garzòn Maceda, *Economia del Tucuman. Economia natural y economia monetaria. Siglos XVI-XVII-XVIII*, Córdoba, 1968; C.S. Assadourian, *El Tráfico de esclavos en Córdoba. De Angola a Potosí. Siglos XVI-XVII*, Córdoba, 1967; Idem, *La producción de la mercancía dinero en la formación del mercado interno colonial*, en Florescano (ed.), *Ensayos sobre el desarrollo económico de México y América Latina (1550-1975)*, México, 1979, p. 233 a 235.

[48] E. Arcila Farias, *Comercio entre Venezuela y México en los siglos XVII y XVIII*, México, 1950, pp. 56, 106 y pp. 134-135.

[49] Robles, doc. cit., p. 33; para el Río de la Plata ver: E. Tandeter, *El papel de la moneda macuquina en la circulación monetaria rioplatense*, «Cuadernos de numismática», IV, marzo 1975, n. 14, Buenos Aires.

> ...dos balandras holandesas, y aquella noche llegó otra inglesa...

que comerciaban con cueros, generos, sebo y

> ...fructos de los que se cultivan en la Isla....[50]

La participación de los holandeses en la exportacion de cacao muestra bien el interes de las potencias no españolas en los tráficos locales. Entre principios del siglo XVIII y la fundación de la Compañia Guipuzcoana, comercializaron 2/3 del volumen del cacao exportado desde Caracas.[51] Nuevamente el comercio directo se sostenía sobre una trama de intercambios interprovinciales, solo que aqui es mucho más dificil distinguir los tráficos preexistentes o surgidos espontaneamente, de aquellos que generó el propio comercio directo.

Para establecer un balance de las corrientes comerciales a las que me vengo refiriendo existe una perspectiva europea y otra americana. La primera implica trazar las lineas de la evolución del comercio colonial, calcular su volumen y valor, el ritmo de llegada de metales preciosos a Europa, estudiar sus posibles efectos, etc., e intentar estimar que parte en todo esto le cupo al comercio oficial, y cual al contrabando. El punto de vista americano significa colocarse en un puerto y observar su actividad con un ojo mirando hacia el océano y el otro hacia el interior del espacio. Para seguir su movimiento es necesario integrar en un sola figura todos los tráficos: intercoloniales o interprovinciales, y ultramarinos, legales o no. Los resultados son distintos en cada uno de los casos.

Respecto al comercio colonial en su conjunto, las raras cifras relativas al contrabando directo permiten interesantes comparaciones parciales entre lo oficial y lo «clandestino». Según Garcia Fuentes, de 1650 a 1700, zarparon 313 navios desde España hacia Portobelo, es decir hacia el comercio del Pacífico.[52] Durante esos mismos años entraron al Rio de la Plata unos 200 navios extranjeros, a lo que se agregó, despues de 1680, la actividad de Colonia Sacramento. O sea que, en volumen o, mejor dicho, en orden de magnitud, el tráfico oficial hacia el Pacífico y su competidor en cuanto a los mercados, el comercio directo a traves de Buenos Aires, eran comparables. Son conocidas las dificultades de España para organizar los convoyes de Galeones a Tierra Firme. Entre 1690 y 1723 solo partieron siete: 1690, 1695, 1706, 1713, 1715, 1721 y 1723. Ya vimos que durante ese mismo periodo entraron al Mar del Sur 148 navios franceses. Agreguemos: el contrabando que desde el Caribe llegaba al Perú atravesando los Andes o el Itsmo; nuevamente la actividad de Colonia; y el contrabando de los asientos franceses e ingleses en el Rio de

[50] ROBLES, *doc. cit.*, pp. 30 y 35.

[51] CÒRDOVA-BELLO, *op. cit.*, pp. 55-57.

[52] GARCIA FUENTES, *op. cit.*, p. 213.

la Plata. En otras palabras, si tomamos entre 1650 y 1730 al eje Lima (y El Callao) — Buenos Aires — un sistema cuyo pivote era la mineria del Alto Perú, y que abrazaba al Caribe y al Brasil —, el conjunto mantuvo un comercio colonial sobre el cual no podemos extraer conclusiones si nos limitamos al segmento legal, pues había un claro predominio del contrabando directo. Para Nueva España, no es posible cotejar resultados con los datos que conozco. Solo puedo recordar que mientras desde Jamaica se dirigian a las poseciones españolas en América aquellos 201 navios, lo mismo hacian desde Cádiz unas ...19 unidades.[53] Todo lo cual nuevamente recuerda y amplia las conclusiones del exámen de la memoria de Beliardi.

Hasta aqui la perspectiva europea. Veamos los resultados de agupar los diferentes tráficos de la actividad de Buenos Aires como puerto de las economías del interior, en particular de la economía minera, durante el siglo XVII. Entre 1590 y 1640, el comercio intercolonial con el Brasil y las colonias portuguesas de Africa occidental representó la mayor parte de aquella actividad portuaria. Pero también llegaban grandes navíos directamente desde Europa: españoles, portugueses y holandeses. De este modo se importaban productos suntuarios europeos, azúcar, esclavos, y — como vimos — se exportaban metales preciosos y algunos pocos productos locales — harina hasta 1630, cueros, sebos y cecinas — cuyo valor no representaba más del 15%-20% del total. Durante la primera mitad del siglo XVII, se habrían introducido unos 30.000 esclavos. El comercio atlántico de Buenos Aires de este periodo alcanzó su momento culminante entre 1600 y 1625, para entrar luego en una decadencia que se aceleró a partir de 1640. La actividad portuaria se recuperó rápidamente gracias a la llegada masiva de navíos holandeses. En los 30 años que van desde 1650 hasta 1680, la entrada de embarcaciones europeas de gran calado — en particular las arribadas holandesas — dió la coloración dominante, sin desplazar completamente al tráfico intercolonial. El tipo de mercancias casi no diferían de las del periodo precedente, y durante esos 30 años se introdujeron no menos de 10.000 esclavos. A partir de 1675-'80 el ritmo global de llegada de navios fue disminuyendo, aunque aumentó el de embarcaciones legalmente autorizadas; pues estas últimas solo representaban, entre 1648 y 1702, el 30% del movimiento transatlántico del Rio de la Plata. Pero en 1682 la fundación portuguesa de Colonia do Sacramento creó un nuevo eje de comercio clandestino. A pesar de ese importante estímulo, las evidencias indican que los años

[53] García-Baquero, *Cádiz...*, *cit.*, t. II, p. 123.

de 1657-1665 fueron los de mayor actividad portuaria durante todo el siglo XVII.[54]

Observando de este modo los intercambios de un puerto, es evidente que la sola consideración del comercio legal, o la extrapolación sobre cada región de la cronologia de crecimiento y contracción del comercio entre España y América, trazaría un cuadro con la perspectiva en falsa escuadra. En el ejemplo del Rio de la Plata, que en esto no difiere del Pacífico y del Caribe, la superposición de los distintos tráficos crearon un perfil de actividad portuaria propio, totalmente independiente de los datos de la «coyuntura atlántica», es decir de la crisis española de la segunda mitad del siglo XVII y primera del XVIII.

III

Desde el punto de vista europeo, el contrabando directo fue un episodio de la creciente colonización del comercio atlántico por parte de las potencias no españolas. En este proceso se sucedieron o se combinaron la presencia de colonias extranjeras en Sevilla, Cádiz, o puertos cantábricos, la ocupacion territorial y el contrabando de aquellas potencias en los puertos americanos. Durante el siglo XVII, como vimos, las alternativas y el resultado del conflicto entre la Corona española y las Provincias Unidas constituyeron una faceta particularmente importante del proceso.

En cuanto al primero de los aspectos ennumerados, las colonias de extranjeros datan de, cuanto menos, mediados del siglo XII. En 1665 habría en Sevilla unos 7.000 extranjeros, de los cuales 2.000 eran de origen portugués. A medida que crecia la importancia de Cádiz en la «Carrera», también lo hacía la presencia de comerciantes no españoles. Tanto, que a fines del siglo XVII la aristocracia mercantil vinculada al tráfico colonial estaba mayoritariamente integrada por extranjeros, que operaban a traves de intermediarios españoles. Los franceses eran la comunidad más numerosa.[55]

Esta influencia alcanzada por los extranjeros se sustentaba sobre uno de los rasgos estructurales de la vida económica peninsular: la fuerte dependencia de España respecto de las mercancias importadas para abastecer el comercio con América. Algunas estimaciones contemporaneas cor-

[54] MOUTOUKIAS, *Contraband...*, cps. 2, 5 y 7.

[55] HARING, *Comercio y Navegacion...*, *cit.*, p. 137 a 144; GARCIA FUENTES, *op. cit.*, p. 30 a 54; GIRARD, *op. cit.*, pp. 87-88, 323, 415, 445 y 455; LYNCH, *op. cit.*, t. II, p. 239 a 241 y 270-71.

roboran esta afirmación. Se trata de las conocidas y frequentemente citadas memorias francesas. Según una, en 1670 la flota proveniente de México trajo consigo 8,5 millones de pesos en metálico, de los cuales por lo menos la mitad pasó a manos de mercaderes extranjeros: 1,5 a los genoveses, 1,3 a los franceses, 700.000 a los holandeses, etc. Más tarde, en 1682 otro documento reiteraba la descripción del mismo fenómeno. De los 22.809.000 millones que transportaban los galeones llegados a Cádiz, más de 13 habrían transferido al extranjero; 4,5 a Génova, 3,5 a Holanda, 2,5 a Inglaterra y 2,5 a Francia. En 1691 se describía un panorama similar.[56] Otros cálculos revelan que una parte importante de dichas transferencias pagaban las mercancias cuya reexportación mantenía el comercio entre España y América. En 1686, hacia ese destino se habrían embarcado mercancias por un valor de 43,24 millones de libras tornesas, entre las cuales solo habia 2,4 millones en productos españoles, o sea en 5,5%. Por el resto, las importaciones desde Francia contribuyeron con el 39,41%, desde Génova con el 16,88%, de Inglaterra el 14,53% y desde Holanda con el 11,95%.[57] Un siglo despues, series más completas de datos muestran una situación más atenuada, que por lo mismo revela una imposibilidad en el largo plazo. Hacia 1780, y por el resto del siglo, las mercancias importadas representaban algo más que la mitad del valor total de las exportaciones de España hacia las Indias Occidentales.[58]

Una explicación a este fenómeno es que España no pudo responder a la modificación en la estructura de la demanda americana, que sobrevino como consecuencia de la combinación de varias circunstancias. El comercio atlántico concernía fundamentalmente a los que consumian como europeos, y podían pagar ese consumo. Del lado español, las principales mercancias exportadas al comienzo de su dominación eran productos agrícolas. Ahora bien, las élites hispanoamericanas fueron satisfaciendo sus

[56] Dahlgren, *op. cit.*, p. 77, nota 3; Haring, *Comercio y navegación..., cit.*, p. 140 a 145; Malamud, *op. cit.*, pp. 116 y 122; H. Sée, *Documents sur le commerce de Cadix*, «Revue d'Histoire des colonies françaises», Paris, 1926, p. 509.

[57] Morineau, *Incroyables..., cit.*, p. 265 a 269.

[58] J. Ortiz de la Tabla, *Comercio exterior de Veracruz, 1778-1821. Crisis de dependencia*, Sevilla, 1978 p. 232 y ss.; para Buenos Aires mis propios resultados para el periodo 1778-1796 en base a documentación de: A.G.I., Audiencia de Buenos Aires, legajos 384, 385 y 392 a 401; datos globales en J. Fisher, *Commercial relations between Spain and Spanish America in the era of free trade, 1778-1796*, University of Liverpool, 1985, p. 46. Obviamente me estoy refiriendo al comercio oficial. La presencia directa de Inglaterra en América, por caso, fue — como es bien sabido — más fuerte que nunca en el periodo que comenzó con la Paz de Paris y culminó con la ocupación francesa.

necesidades de dichos productos recuriendo crecientemente a su propia producción local o a los tráfico interprovinciales e intercoloniales. Paralelamente, la expansión de la mineria y de las economias regionales, a pesar de sus vaivenes coyunturales, fortalecía a las capas dominantes, cuya población, además, creció hasta por lo menos mediados del siglo XVII. Los considerados españoles quizás sumaran unos 220.000 habitantes hacia 1570 y alrededor de medio millón por 1630.[59] Como se sabe, a pesar de la caída de la población aborigen, fue aumentando el número de mestizos e indígenas integrados a los circuitos de la economia hispanocriollas.[60] De modo que, los tres elementos señalados — la producción local de una agricultura europea, el crecimiento económico, y la evolución demográfica — se conjugaron, entonces, para modificar la demanda americana en el sentido de un mayor poder de compra, de una creciente capacidad de absorver mercancias, volcada cada vez más sobre manufacturas que España no producía, al menos no en cantidad suficiente.

Los productos agrícolas españoles dominaron las importaciones de Nueva España hasta 1580-1600. Pero hacia esa misma época, los vinos del norte de Lima alcanzaron mercados mexicanos a través del Pacífico, Zacatecas se abastecía de viñedos situados al norte del distrito (así como Potosí lo hacia con los vinos de Arequipa), para no tomar sino un ejemplo.[61] A mediados del siglo XVII el predominio de las manufacturas, en particular los tejidos, en las importaciones americanas ya estaba totalmente consolidado y el proceso se hizo irreversible. Esto a pesar de que Garcia-Baquero señala el peso de los productos agrícolas en las exportaciones españolas entre 1720 y 1751, medidas en toneladas de aforamiento.[62] Pero se trata de volúmen y no de valor y durante esos años el comercio directo era masivo.

Otras argumentaciones intentan explicar la dependencia española respecto de las importaciones nord-europeas y francesas, poniendo el acento no en la demanda americana, sino en la "oferta" peninsular, es decir en la decadencia de su industria, supuestamente resultado de la competencia extranjera y de una inadecuada política de la Corona.[63] El tema es

[59] N. Sanchez Albornoz, *La población de América Latina*, Madrid, 1977, pp. 89-90 y 122.

[60] J. Israel, *Razas, clases sociales y vida política en el México colonial, 1610-1670*, México, 1980 (1ra ed. en inglés, Londres, 1975), p. 37.

[61] P.J. Bakewell, *Mineria y sociedad en el México colonial. Zacatecas (1546-1700)*, México, 1976, p. 107; P. Chaunu, *Séville et l'Amerique*, Paris, 1977, pp. 132 y 199; K. Davies, *La tenencia de la tierra y la sociedad rural arequipeña*, «Histórica», vol. 1, n. 2, Lima, 1977, pp. 185-188 y 190-191; Lynch, *op. cit.*, t. II, p. 258 a 272.

[62] Garcia Fuentes, *op. cit.*, p. 234 a 326; Garcia-Baquero, *Cádiz..., cit.*, t. II, gráf. Nos. 14 y 15.

[63] Garcia Fuentes, *op. cit.*, p. 230 a 236; Garcia-Baquero, *op. cit.*, p. 67 a 86.

bien complejo y excede los límites del presente texto. Pero conviene puntualizar que los citados análisis sobre la ruina de la industria peninsular se refieren a centros urbanos; mientras que la mayor parte de las exportaciones de los Paises Bajos, Inglaterra y Francia estaban vinculadas al desarrollo de una industria rural especializada en la producción para mercados interregionales y ultramarinos, el fenómeno de la llamada protoindustrialización, de débil o inexistente localización en España.[64]

En realidad cabría preguntarse si alguna vez estuvo España equipada para escoltar con sus suministros una colonización ultramarina cuyo rumbo y magnitud era imposible siquiera de imaginar en los inicios. Más aún, cabría interrogarse si, antes de la segunda mitad del siglo XVIII, alguna región ó Estado de Europa tenía los recursos suficientes para erigirse en metrópoli economicamente monopólica de colonias como las hispanoamericanas. Algunas cifras indicarían que no. En 1699-1701, solo el 16% de las exportaciones inglesas estaban destinadas a mercados no europeos, en América, Asia y Africa; para 1772-1774 la proporción ascendería a un 55%.[65] Durante la década de 1680, la media anual de mercancias necesarias para alimentar a la «Carrera» sumarían unos 21 millones de libras tornesas, equivalentes a 2/5 del valor total de las importaciones y poco menos de 2/3 del de las exportaciones de Amsterdam en 1667-1668. La explotación de América a traves del sistema oficial aparecia como un enorme negocio en el que participaba la mitad de Europa, desde Hamburgo hasta Génova, con ramificaciones hacia el corazón de Silesia.[66]

Los comerciantes, la Corona y los responsables de la administración y funcionarios de todo nivel se ajustaron, como mejor pudieron, a la circunstancia de una metrópoli que debía reexportar mercancias extranjeras hacia sus poseciones, compensando el desequilibrio en la balanza comercial así generado con los metales preciosos que aquellas producian. A lo sumo se intentó fortalecer la función de intermediaria que España ejercia. En principio, las potencias no españolas aceptaban el reparto, pues de ese modo podían participar en el tráfico con las Indias Occidentales sin correr con el costo de sostenimiento. En ese sentido Malamud ha subrayado la existencia de una correlación inversa entre la marcha del comercio francés en Cádiz y el contrabando directo de estos en ultramar.[67]

---

64 P. Kriedte, H. Medick, J. Schlumbohm, *Industrialization before industrialization*, Cambridge, 1981 (primera ed. en aleman, 1977), p. 1 a 37.

65 *Ibid.*, p. 36.

66 Morineau, *Incroyables...*, *cit.*, p. 263.

67 Garcia-Baquero, *Cádiz...*, *cit.*, pp. 82 y 83; Garcia Fuentes, *op. cit.*, p. 54; Malamud, *op. cit.*, p. 281; en la citada documentación de Beliardi también aparece con frecuencia la idea de que el interes de Francia consistia en un fortalecimiento de España respecto de su adversaria común, Inglaterra, que consolidase el comercio entre ambos reinos borbónicos.

El caso fué que la intermediación se resquebrajaba progresivamente, lo que indujo a la presencia regular de los extranjeros en América.

Como es bien sabido, alrededor de mediados del siglo XVII un intricado tejido de factores malograban la capacidad de respuesta del Imperio: crisis económica y demográfica, crisis política, desenlace adverso de los conflictos armados, tensiones fiscales, etc. En este contexto, la crisis de intermediación tuvo, entre otras, una manifestación concreta que es conveniente subrayar, el progresivo deterioro de las comunicaciones entre España y América que se daba en el cuadro de una irreversible decadencia de la industria naval. Si durante el siglo XVI, entre el 80% y el 90% de los navios de la «Carrera» eran de construcción española, el porcentaje bajó al 31% para la segunda mitad del siglo XVII; el resto, un 22% de naves americanas y un 14% de extranjera. Durante el siglo XVIII el predominio de las embarcaciones de construcción extranjera era total: 73,59%, mientras que las españolas representaban un 22,15% y las americanas apenas un 4,26%.[68] Evidentemente con esta fragilidad los recursos, las comunicaciones estaban sometidas a cualquier contingencia. Esto se tradujo en una creciente dificultad de España para organizar el tráfico con sus dominios americanos, patente de manera dramática durante la última década del siglo XVII y primera del XVIII.

Sin embargo, el uso de una u otra de las alternativas — comerciar con las Indias a traves de España o directamente en sus costas — no fué una simple sucesión de dos modalidades, sino una tensión constantemente renovada entre dos opciones que coexistían. Si por un lado las conveniencias de la primera posibilidad están demostradas, por el otro, los enfrentamientos de Holanda, Inglaterra y Francia con el Imperio español debilitaba la capacidad de intermediación de este. A medida que se acentuaba la crisis del sistema oficial, se generalizaba el comercio directo, mientras los extranjeros controlaban una proporción creciente de un decreciente tráfico oficial. La diferente intensidad con que se recurría a una u otra alternativa cambiaba según las distintas épocas o potencias. Estamos, entonces, ante un vasto comercio de Europa con América en relación al cual España desarrolló una intermediación cuya eficacia fué declinando. Conforme avanzaba el siglo XVII el movimiento global del comercio atlántico era el resultado del entrecruzamiento de tráficos que formaban una red cada vez más compleja. Sus centros fueron Amsterdam, Saint Malo, Londres, Génova, Sevilla, Cádiz, Las Canarias, las costas africanas y brasileñas, Veracruz, Portobello, Cartagena, Jamaica, Curaçao, el Rio de la Plata, el Pacífico...

[68] GARCIA FUENTES, *op. cit.*, p. 205 a 207; GARCIA-BAQUERO, *op. cit.*, p. 230 a 239.

Al recoger los argumentos con los que vecinos de Cuba explicaban su participación en el contrabando, el viajero Robles repite las razones con los cuales los vecinos de Buenos Aires justificaban un pedido, hecho en 1668, para que se autorizase nuevamente los intercambios con el Brasil: carestía y escasez de los mercancias importados por el sistema oficial y desvalorización de la producción local en relacion a lo practicado en los intercambios en cualquiera de las dos alternativas semi-clandestinas. Cuando examinamos el caso de Buenos Aires vimos que es difícil distinguir entre contrabando y comercio legal. Por sus circuitos, sus mecanismos y los hombres que lo realizaban, los diferentes tráficos que vimos constituían una única red de actividades portuarias; más amplia y diversificada que la que surgía del sistema oficial, y a la que éste se integraba a pesar de las diferencias señaladas. La superposición de tráficos creaba un moviento mercantil con perfil propio, tanto en el Río de la Plata como en los demás puertos. En consecuencia, aquellos argumentos pueden interpretarse como que, para las élites hispano-criollas, el comercio directo significó una extención del espacio de oportunidades económicas, sumado al que estas mismas generaban con los tráficos intercoloniales e interprovinciales, y no una opción excluyente del comercio legal. Dicha ampliación, al ofrecer una alternativa a la crisis de la metrópoli fortalecia la posición relativa de las élites coloniales en el conjunto imperial.

Ahora bien, si la presencia de navios extranjeros se explica por la crisis de la intermediación de España — y el conjunto de factores arriba analizados —, el comercio directo se sostenía en los recursos de las oligarquias hispanoamericanas, que les permitían abosorver mercancias más allá de la oferta metropolitana. En otras palabras, la ampliación de las oportunidades económicas era una construcción de las élites coloniales, una activa oposición al lugar que el Imperio les asignaba, y una resistencia al destino que les cabía dentro de la crisis general de la Peninsula ibérica.

Pero el comercio directo no modificó — salvo en circunstancias y momentos bien preciosos — la estructura de intercambios ultramarinos del sistema colonial de Antiguo Régimen, basado, por un lado, en la exportación de metales preciosos y algunos productos agrícolas, y por el otro, en el consumo suntuario, los mercados compulsivos, y la dependencia de insumos importados como el hierro y los esclavos. En este sentido, estaba estructuralmente incorporado a la vida económica de Hispanoamérica. Esto resulta obvio, pero es menester recordarlo. Esta ambigua oposición no se trasformó en enfrentamiento, en alternativa al Imperio, pues el sistema colonial era un fenómeno interno; por lo cual la dominación española podia adquirir la forma de una contrabandista extranjero.

Más aún, al fortalecer su situación en el conjunto Imperial, la construcción de la elites coloniales ayudó a conservar los vinculos imperiales, aunque reformulados.

Retomando un comentario citado en la introducción, la Corona siguió siendo Señor de las élites coloniales, pero no las poseía, al menos no de manera exclusiva, y esta autonomía construida prolongó la vida del matrimonio. Para comprender esto debemos examinar el fenomeno de la corrupción.

IV

Visto desde Hispanoamérica, el contrabando fué un aspecto de la corrupción — un sistema de relaciones y comportamientos que actuaban sobre los vínculos entre la Corona y las élites coloniales. Si el comercio directo estaba estructuralmente asociado a la economía colonial, como acabamos de ver, también estaba entelazado a un complejo político que explica su continuidad. Dicho complejo cristalizó la ambigüa oposición de las élites bajo la forma de un espacio de mayor autonomía, conquistado gracias a la influencia local, sobre las estructuras imperiales que generaba la currupción, y financiado por el fortalecimiento económico relativo de esas oligarquías.

El tema de las relación entre oligarquías locales y la administración colonial ocupa un sitio antiguo en la historiografía americanista. Contra una idea tradicionalmente aceptada, la bibliografía de los últimos quince años ha subrayado la existencia de fuertes lazos entre ambos, así como la influencia que ejercían las élites coloniales sobre las magistraturas. Esto aparece como el resultado de varios factores: a) la venta de cargos y/o nombramientos facilitó a criollos poderosos, no sólo el control de corporaciones municipales sino también el acceso a las más altas magistraturas — como la Audiencia y a veces alguna gobernación; b) los altos funcionarios eran rápidamente incorporados, por alianza o cooptación, a una estructura informal de relaciones personales, cuya coexistencia con la estructura formal del poder imperial creaba las condiciones de una doble lealtad: hacia las autoridades metropolitanas y hacia los grupos de intereses locales. Dicha red de relaciones personales es presentada como la base del fenómeno de corrupción, el cual — a su vez — era fuente de influencia política y flexibilizaba las estructuras administrativas.[69]

[69] Solo cito algunos de los trabajos más recientes: J. Barbier, *Elites and cadres in Bourbon Chile*, «H.A.H.R.», vol. 52, n. 3, agosto 1972; D. Brading, *Miners and merchants in Bourbon Mèxico*, Cambridge, 1967; Idem, *Government and Elite in Late Colonial México*, «H.A.H.R.», vol.

Existe un acuerdo general sobre la cronologia del fenómeno de influencia política local sobre las estructuras administrativas del Imperio español en América: fuerte aumento durante la segunda mitad del siglo XVII, continuación a lo largo de la primera mitad del siglo XVIII. Muro Romero llega a hablar de emancipación informal entre fines del siglo XVII y principios del XVIII.[70] Según dicho autor, el cuadro general en que se desarrollaría este proceso sería el de una reformulación, perceptible hacia las décadas centrales del siglo XVII, de lo que se da en llamar el pacto colonial. La monarquía no habría decaido hasta el punto de no poder adaptarse a su propia crisis. Empezaría a desarrollar, de manera más o menos consciente, una readaptación forzada en base a las crecientes necesidades de la debilidad imperial, que se tornaría menos exigente con el orden americano colaborando con la instauración de un pacto renovado, que afectó como es lógico a las instituciones del gobierno.[71]

Los autores más arriba comentados, a pesar de incluir al fenómeno de la corrupción en un contexto más amplio, la consideran como un abuso respecto de un ideal. Abuso de mas o menos larga duración, nacido de las combinación de dos factores: bajos salarios de los funcionarios y un comportamiento tradicional en relación al aparato administrativo. En cambio en dos articulos, Vincente Vives y Pietschman consideran a la corrupción como un sistema, retomando en este sentido los análisis y conclusiones de Van Klaveren. Para el primero de ellos

> ...la corrupción se produce cuando la administración intenta dominar un hecho económico importante con una legislación cerrada y arcaica...

En consecuencia, si arraigó la corrupción fue porque

> ...la administración tenía que hacer funcionar el mecanismo del comercio americano a pesar de las leyes....[72]

Pietschman retoma esta idea junto a las arriba expuestas. Subraya la definición de Van Klaveren del magistrado corrupto como aquel que considera al oficio público como un negocio cuyo rédito e ingreso procura maximizar, transformando asi al cargo en una «unidad maximi-

53, agosto 1973, n. 33; M. Bourkholder y D.S. Chandler, *From Impotence to Authority: The Spanish Crown and the American Audiencias. 1687-1808*, Mo & London, 1977; L. Campbell, *A colonial establishment: creole domination of the Audiencia of Lima during the late eighteenth century*, «H.A.H.R.», vol. 52, feb. 1972, n. 1; D. Ladd, *The Mexican Nobility at Independence. 1780-1826*, Austin, 1976; F.J. Moreno, *The Spanish colonial system: a functional approach*, «Western Political Quaterly», junio 1967, n. 20; J. Phelan, *Authority and Flexibility in the Spanish Imperial Burocracy*, «Administrative Sciences Quaterly», n. 5, junio 1960; Idem, *The Kingdom of Quito in the seventeenth Century*, Madison, 1967.

70 F. Muro Romero, *Administración y sociedad en América Española hasta 1750*, [...], p. 448; ver también Lynch, *op. cit.*, t. II, pp. 273-274.

71 Muro, *ibid.*, p. 450.

72 J. Vincens Vives, *Estructura administrativa y estatal en los siglos XVI y XVII*, *Coyuntura económica y reformismo burgués*, Barcelona, 1969 (1ra ed. del artículo, Estocolmo, 1960), pp. 137-138.

zante». En consecuencia la corrupción en hispanoamérica debe ser considerada como un sistema, que se explica en términos de una tensión permanente por la distribución del poder y la riqueza entre el estado español, la burocracia colonial y la sociedad colonial.[73]

Así considerada, desde el punto de vista de los funcionarios, la corrupción consistió en la infracción regular de un repertorio de normas que impedían el desempeño como empresarios, es decir, que limitaban su integración en las oligarquías locales. No se trataba solamente de funcionarios que, envueltos en una telaraña de vínculos primarios, flaqueaban en su lealtad hacia la Corona. El movimiento que acoplaba a los representantes de la Monarquía y los sectores dominantes de las sociedades hispanoamericanas tenia doble dirección; aquellos se autotranformaron entrelazandose con éstos, al punto de integrar una sola red de notables. Con frecuencia, Oficiales Reales, Oficiales de las guarniciones, Corregidores y Alcaldes Mayores,incluso miembros de las Audiencias y Gobernadores se dedicaban al comercio, invertian en tierras, etc.; se asociaban a los miembros de la élite al reclutarlos, por alianza o por cooptación, a una estructura informal de relaciones personales...[74] Ya vimos que la Corona se adaptó a esta situación en un cuadro de debilidad cuyas causas, obviamente, eran externas al mundo hispanoamericano pero estaban relacionadas con el comercio directo, la alternativa colonial a la crisis económica de España. En consecuencia, la corrupción fué un sistema de relaciones que, en la cuspide de la sociedad, articulaban un complejo de prácticas económicas y de poderes políticos, acrecentando la autonomía de las élites coloniales, pero conservando el cuadro general del Imperio.

Algunos de los rasgos del contrabando en el Río de la Plata — estudiado en los trabajos ya citados — permiten ilustrar y ampliar estas ideas. La introducción y blanqueo de las mercancias requería importantes recursos y ponía en marcha complicados mecanismos: propiedades rurales en las orillas del Rio, depósitos y tiendas para las entradas completamente clandestinas; también depósitos, remates fraudulentos u otras formas de venta pública de las mercancías «comisadas» en las arribadas; en ambos casos medios de transporte, carretas, o mulas, para comunicar con los mercados del interior. En esto el Río de la Plata no difería de cualquier otra región de América. Más que de «pasadas» clandestinas y complicidades esporádicas, se trataba de una red de contactos y asociaciones

[73] H. Pietchman, *Burocracia y corrupción en Hispanoamérica. Una aproximación tentativa*, «Nova Americana», n. 5, Turin, 1982, pp. 30-31.

[74] Sobre la idea de la autotransformación de las élites, J. Gelman, *Economie et administration local à Buenos Aires au XVIIe siecle*, tesis de 3er ciclo, E.H.E.S.S., Paris, 1983, pp. 237 y 303 a 311; Idem, *Cabildo y Elite local. El caso de Buenos Aires en el siglo XVII*, *HISLA*, n. 6, 2do semestre, 1985.

regulares, de un verdadero aparato de distribución. Su funcionamiento implicaba el entrelazamiento del comercio y la estructura administrativa y militar, que adquirió la forma — repito — de una imbricación entre la élite local y las autoridades.

Frecuentemente, oficiales del fuerte de toda graduación, así como algunos soldados, aparecían implicados en diferentes operaciones comerciales. Llegaron a constituir un grupo importante de «personas dedicadas regularmente al comercio», y las fuentes las señalan practicando las más diversas actividades: dueños de tiendas y pulperías, proveedores de los navíos, socios de capitanes extranjeros y españoles, etc. Una descripción parecida puede hacerse de los Oficiales Reales, miembros de la Audiencia y Gobernadores. Durante el siglo XVII a la casi totalidad de los Oficiales Reales se les demonstró su participación en el contrabando. En realidad estos funcionarios llegaron a desempeñar todo tipo de actividades económicas. Al igual que sus congéneres militares, formaban entre sí, y junto con otros funcionarios o comerciantes, agrupamientos más o menos estables cuyas operaciones iban desde el contrabando hasta las remesas de ganado en pie hacia el Alto Perú. Si seguimos el camino en sentido inverso, llegamos al mismo ámbito. Dueños de tierra y señores de esclavos, los comerciantes contrabandistas culminaban sus carreras y acceso a la dignidad con algún cargo en el Cabildo, escribanía, nombramiento como oficial en al guarnición o en las milicias, o como teniente de gobernador.

Confluyendo sobre un mismo espacio de actividad empresarial, los actores señalados integraban una trama de vínculos primarios, de parentezco, alianzas y lealtad personal, liderada por patriarcas y dignitarios. Al estimar la actividad del puerto de Buenos Aires, vimos que esta era una combinación de tráficos que formaban parte de un mismo fenómeno, el funcionamiento de la ruta Potosí-Buenos Aires, dentro del cual resultaba artificial distinguir entre comercio legal y clandestino. Simetricamente, dicha actividad fué la obra de aquel único núcleo, la *clique* de comerciantes, terratenientes y dignatarios. De modo que los representantes de la Corona que debían reprimir al contrabando no formaban un segmento socialmente diferenciado del de los protagonistas del comercio directo. Esto no significaba que aquellos jamás recurriesen a la represión, lejos de ello; nunca faltaban contrabandistas en algún calabozo. No obstante, las autoridades civiles, militares y eclesiásticas actuaban más bien integradas a facciones en pugna antes que enfrentando, como «aparato estatal», a supuestos delincuentes. Sea como, fuere, llama la atención el hecho de que el discurso de la crítica al contrabando y la defensa de la legislación fuese utilizada también por quienes más mezclados estaban en el comercio ilícito. Dicha conducta no fue exclusiva de los mercaderes porteños: los miembros del Consulado de Lima vinculados al contrabando de los franceses — por ejemplo — actuaban de la misma manera.

En la formulación hecha por Muro, la creciente autonomía conquistada por las élites coloniales aparece como resultado de la readaptación forzada de la Monarquía, de lo cual aquellas fueron beneficiarias pasivas. En cambio lo expuesto muestra un proceso activo de oposición de las oligarquías, que afectaba permanentemente su relación con una Corona en situación también cambiante. Por un lado tenemos una guarnición militar y un grupo de funcionarios que, ambos, constituían los medios por los cuales la Corona podía alcanzar sus objetivos. Por otro lado, que dichos funcionarios y militares integraban una élite local cuya principal actividad económica era el contrabando, el cual fortalecía su situación económica relativa. La adaptación de la Corona consistió en que financió su aparato administrativo y militar local gracias a las actividades económicas de dicha élite, en particular la más rentable — el contrabando —. En otros términos, la Corona logró montar un dispositivo militar en Buenos Aires, enviar tropas a Chile, y mantener comunicaciones regulares utilizando los recursos que le ofrecía el comercio atlántico del Río de la Plata, tanto legal como clandestino. Además se servía del crédito local para paliar el retraso en los situados, asociando a los comerciantes con la guarnición.

Las prácticas económicas de la élite colonial fortalecían al conjunto, es decir a las élites mismas, a la Corona y a sus representantes. Y simetricamente, le acción de la Corona consolidaba a los grupos dominantes locales. Esto equivale a ver el surgimiento de la administración colonial no como el resultado de la subordinación de los grupos dominantes locales, sino como aspecto central del pacto entre éstos y la Corona, el cual también incluía a la «corrupción». Un pacto colonial reformulado, para retomar los términos de Muro, en un sentido más favorable a la oligarquía local. Esto, gracias a la capacidad de resistencia y autonomía que le brindaron los recursos del contrabando.

En el estado actual de la investigación es algo arriesgado generalizar las observaciones y conclusiones que acabo de exponer, sin embargo se pueden intentar algunas comparaciones. El Caribe y el Río de la Plata compartían la circunstancia de ser una región de frontera y dificil asentamiento, atravesada por una ruta de intercambios a gran distancia. Claro que las caracteristicas y dimenciones eran completamente diferentes, como lo fueron las respuestas de la Corona en uno y otro caso. En el segundo, como acabamos de ver, logró montar un dispositivo militar con un mínimo de recursos, el cual enfrentó con bastante éxito a la expansión portuguesa, cada vez que la diplomacia de Madrid así lo necesitaba. En el Caribe, en cambio, la Armada de Barlovento — construcción imperial por excelencia — drenó caudales de seis Cajas Reales para mantener una maquinaria que nunca funcionó adecuadamente.[75]

En cuanto al desempeño de los funcionarios como empresarios, el fenómeno es bien conocido, estaba difundido por toda América española y está tan ampliamente documentado que lo dificil es seleccionar ejemplos: en 1650 el Alcalde Mayor de Portobelo, socio de mercaderes peruanos, descaminó plata que le fue decomisada; hacia fines del siglo XVII se denunciaba la constante participación de los soldados de la guarnición de La Habana en granjerias y descaminos; en 1726 se denunciaba la activa y regular intervención en el contrabando, de los gobernadores de Trinidad, Margarita y Cumaná, y los testimonios prolongaban los hechos hasta 1740.[76] En fin, la venta de los nombramientos en la Real Hacienda de Lima a partir de la década de 1630, aceleró la formación de un grupo de Oficiales fuertemente unido a los intereses locales y frecuentes socios en toda suerte empresa ilegal.[77]

Al igual que los holandeses en el Rio de la Plata, los franceses se relacionaban con los principales comerciantes peruanos, los únicos que podían ofrecerles una infraestructura adecuada para la introducción de sus mercancias. Incluso llegaban a formar con ellos companias para comerciar con Oriente.[78] También similar fue la actitud de funcionarios de todo rango, como los Corregidores de Concepción y Arica a principios de la decada de 1710. El primero de ellos tenía tienda púbica, autorizó el desembarco de navios franceses, cuyas mercancias utilizaba para adquirir trigo destinado a los soldados, que remitia hacia Lima; el segundo era cabeza de un denso tejido de negocios, que comprendían tráficos entre dicho puerto y Potosí, con productos de la tierra y efectos provenientes del comercio directo.[79]

Ejemplar fue el caso de Juan Andrés de Ustáriz, uno de los tantos citados por Malamud. Natural de Vizcaya, vecino y del comercio de Sevil-

[75] B. Torres Ramirez, *La Armada de Barlovento*, Sevilla, 1980, caps. VIII y IX.

[76] *Colección de Documentos inéditos..., op. cit.*, 1ra serie, t. XI, p. 16; R. Konetzke, *Coleción de documentos para la historia de la formación social hispanoamericana*, Madrid; Arauz Monfante, *op. cit.*, p. 112.

[77] Andrien, *op. cit.*, p. 103 a 119; ver tambien, Idem, *The sale of Fiscal Offices and the decline of Royal Authority in the Viceroyalty of Perù, 1633-1700*, «H.A.H.R.», vol. 62, 1982, n. 1.

[78] Malamud, *op. cit.*, pp. 181 y 193.

[79] *Ibid.*, p. 254 y p. 265 a 270.

la, decidió recuperarse de la bancarrota que había sufrido comprando el nombramiento de Gobernador de Chile. En 1709 llegó a Valparaiso con un séquito de 15 parientes y allegados, entre ellos tres hijos — dos de los cuales fueron corregidores y uno capitán del ejercito — y dos sobrinos. Adecuadamente colocados, estos establecieron solidos lazos locales y permanecieron en América, asi como otros membros de la parentela. Los negocios del clan llegaban a Mendoza, Potosí y otros puntos del Alto Perú y sus relaciones con el comercio directo de los franceses comprendia toda suerte de operaciones.[80]

La Audiencia de Lima y el Consulado se acusaban mutuamente de participar en el contrabando y con toda seguridad, ambos tenían razón. Respecto al primero, no hay dudas de que la mayoría de sus miembros lo hicieron, aunque como cuerpo la corporación mercantil reiterase las denuncias sobre los efectos negativos del comercio directo.[81] En cuanto al tribunal, no pocos de sus integrantes estaban emparentados con los principales mercaderes limeños y ellos mismos eran miembros acaudalados de la élite. Sabemos que en este organismo se cruzaban importantes tramas de poder local: los Oidores casados con peninsulares fueron minoria hasta 1780, y el predominio de los peruanos era absoluto entre 1705 y 1712, así como despues de 1740.[82]

Todo lo cual confirma el conocido cuadro trazado por Juan y Ulloa a mediados del siglo XVIII. La cadena de la corrupción integraba agentes económicos y funcionarios, y entre estos los eslabones iban desde el virrey hasta los corregidores, pasando por la Audiencia y los Oficiales de la Real Hacienda.[83] En suma, el caso del Rio de la Plata era uno más dentro de las Indias, aunque el peso de la guarnición en relación a los recursos locales acentuase algunos rasgos peculiares, en particular la función del contrabando en el mantenimiento del dispositivo militar. Según afirma Brading en su articulo citado, los colonos criollos fracasaron en sus intentos de constituir un poder político autonomo que establizase su encumbramiento.[84] Esto parecería menos evidente si pensamos en las élites coloniales como una conformación unitaria, en la que la oposición criollo-peninsular era uno entre otros rasgos de heterogeneidad. Lo expuesto sobre la articulación entre el sistema de la corrupción y la practi-

---

[80] *Ibid.*, p. 259 a 265.

[81] *Ibid.*, p. 195 y p. 192 a 211; PARQUER, *op. cit.*, p. 176.

[82] CAMPBELL, *op. cit.*, p. 10 a 15; G. LOHMAN VILLENA, *Los ministros de la Audiencia de Lima (1700-1821)*, Sevilla, 1974, p. LXIX a LXXII y XCVII a CVI.

[83] JUAN y ULLOA, *op. cit.*, p. 367 a 391.

[84] BRADING, *Elites...*, *cit.*, p. 402-403.

cas económicas, muestra que en realidad aquellas sí habian logrado construir un eje político autonomo que defendía la marcha cotidiana de los intereses de las oligarquias indianas.

Las respuestas a los estímulos del contrabando creaba al propio estímulo reproduciendo el fenómeno y sus condiciones. Ya lo vimos al mostrar la oposicion al Imperio en el plano económico, que atraía al comercio directo al integrarlo en un espacio alternativo de actividad mercantil. En el plano político, los protagonistas locales del contrabando formaban parte del cuadro de poder que aseguraba su continuidad. Por un lado, se incorporaban a su represión disolviendola en los conflictos internos de la élite. Por el otro, la corrupción reproducía dicha estructura política.

El mecanismo funcionaba porque era uno de los rasgos más importantes de los sectores dominantes en las Indias Occidentales. Estos constituían una conformación social elitista de asentamiento urbano, estructurada internamente como comunidad, es decir en base a un sistema de vinculos primarios, de parentezco, alianza y lealtad personal, que encuadraban el acceso a la riqueza y al poder. El entrelazamiento con los aparatos formales de la Corona o del gobierno local era fuente de estratificación interna, de modo que en la cuspide se encontraba una red de patriarcas dignatarios. Así articuladas, las élites coloniales efectuaban una apropriación rentista de los excedentes, gracias al control simultaneo de un abanico diversificado de actividades económicas y los mecanismos internos del comercio a gran distancia.[85] El contrabando les permitió realizar la renta colonial mucho más alla de los limites, legales o no, de una relación univoca con España. Fué el instrumento, entonces, de una ambigua oposición, de una autonomia dentro del imperio en constante tensión.

V

Al comenzar decía que el contrabando era una red de fenómenos económicos, sociales y políticos, y en mi exposición he intentado recorrer el conjunto. El sistema comercial español instaurado a mediados del siglo XVI — llave atlántica del Imperio — se fué desmoronando a partir

[85] Ver: S. Blank, *Patrons, brokers and clients in the families of the elite in colonial Caracas. 1595-1627*, «The Americas», vol. XXXVI, July 1979, n. 1; Ladd, *op. cit.*, cps. II a IV; R. Lindley, *Haciendas and economic development. Guadalajara, México, at Indipendence*, Austin, 1983; F. Tonnies, *Comunidad y asociación*, Barcelona, 1979 (8va ed. alemana corregida por el autor, 1935), p. 33 a 66; R. Syme, *Colonial élites. Rome, Spain and the Americas*, Londres, 1958, cps. I y II.

de la década de 1630, bajo el doble efecto de, por un lado, la oposición externa y, por el otro, la reacción a la presión fiscal provocada por esa oposición — que al mismo tiempo la internalizaba —, combinada con la crisis económica general de España. Se lo sustituyó, también paulatinamente, por un régimen en el cual los indultos ocupaban el lugar de una fiscalidad informal y los Consulados americanos financiaban con aportes directos una parte de los gastos de la «Carrera».

Esta tranferencia de funciones Imperiales hacia América formaba parte de un contexto peculiar. La crisis de intermediación de España provocó la presencia masiva y regular de navios extranjeros, estimulada por la capacidad no satisfecha de absorver mercancias importadas que tenian las oligarquias indianas. Estas incorporaron al comercio directo en la construcción de un espacio de actividad mercantil, junto a los tráficos interprovinciales e intercoloniales y el mismo comercio legal. De tal suerte que la evolución del sector externo de una región dada era independiente del de la «Carrera», sobre todo independiente de su crisis. Al aumentar así su capacidad de realizar la renta colonial más alla de las posibilidades que ofrecía la metrópoli, las élites coloniales fortalecieron su situación relativa; lo cual alimentaba, con estimulos y recursos, el anudamiento entre estas y el aparato local de la Corona, que la corrupción articulaba. El poder político así conquistado otorgaba autonomía informal y reproducía el sistema. Ambos, autonomía y reproducción, transcurrían en una tensión con el Imperio constantemente renovada.

Tambien decia — y he repetido luego — que, así visto, el contrabando fue un elemento importante de oposición al Imperio. Para las élites coloniales fue un instrumento clave en su constitución como sector dominante y, por lo mismo, de la autoconstrucción de un lugar en el Imperio, según sus propios rasgos. Tan propios fueron esos rasgos que duraron hasta mediados del siglo XIX, cuando el Imperio ya se habia desmoronado.

3

The Production and Uses of Gold and Silver in Sixteenth and Seventeenth-Century Japan

A. Kobata

I. *The Progress of the Mining Industry*

From about the middle of the sixteenth century to the seventeenth century, the history of mining was marked by the sudden opening of gold and silver mines throughout the country and a great increase in the production of gold and silver. After about the middle of the seventeenth century the output of gold and silver declined and in their place the mining of copper showed a sudden rise.

The development of precious metal mines was undertaken as a result of the zealous plans of the warring *daimyos* for increasing their own resources. Gold and silver, as military funds and as rewards for warriors, gradually demonstrated their use as a measure of high monetary value. As the system of warring feudal estates gave way to a system of national political unity, this development in the use of precious metals became the basis of great developments in the exchange of goods. Compared to the copper coins which had been previously the sole form of metal currency, gold and silver had a much greater value as monetary exchange, and came gradually into common use.

Though the *daimyos* of the *Sengoku* period (sixteenth century) strove to utilize the value of gold and silver, their collection of these precious metals cannot be compared to the accumulation and utilization of gold and silver by Hideyoshi, and later, by Ieyasu. If we look at Hideyoshi's extensive military activities and construction works, this is easily understandable. In the establishment of the Tokugawa *Bakufu*'s hegemony, the main gold and silver mines, as part of the domain of the *Shogun*, became an important financial foundation. Thanks to the gold and silver currency system of the Tokugawa family, minting was substantially unified by the latter half of the seventeenth century and became almost an exclusive prerogative of the *Bakufu*.

The sudden increase in the production of gold and silver, particularly of silver, after the sixteenth century, was closely connected with new developments in foreign trade. Anticipating the end of the Muromachi *Bakufu*'s system of trade in licensed ships (*Kango sen*), the *daimyos* of south-west Japan sought freer trade, but all their efforts were thwarted by the policy of the Ming Empire in China. However, about that time the arrival of the merchant ships of Ming became frequent. Their purpose was to carry away the silver produced in abruptly increasing amounts in Japan. The development of trade with Japan by Portuguese ships and the consequent increase of profits were in fact a result of the intermediary trade consisting of the exchange of Japanese silver

The article has been translated from the Japanese by W. D. Burton.

for Chinese raw silk and other commodities. From the end of the sixteenth century to the seventeenth century, Holland and England made their appearance in the Far East, fought Portugal and Spain and struggled with each other attempting to preserve 'absolute' supremacy in trans-oceanic expansion under a mercantilist system which emphasized the establishment of colonies to secure precious metals and a supply of products for the mother country. The goods distributed to the colonies were industrial and manufactured goods of the mother country, but in answering the demand in Far Eastern countries, these European products were not always suitable. One pivotal point of Far Eastern trade was China, but in China and the other countries of the area, the greatest demand was for silver. Part of the silver flowing into China was that of the new continent, brought by way of Manila, but from the middle of the sixteenth century into the first half of the seventeenth century Japanese silver played a more important part. This was the reason why Japan seemed so important in Far Eastern trade, even to the Western Europeans. It is true that an increase occurred in Japanese consumption of the products of southern countries, such as sugar, spices, medicines, and clothing materials, but the largest import was of raw and woven silk from China and Indo-China. For this reason the success or failure in obtaining Chinese goods was closely related to the success or failure of English and Dutch trade with Japan. It was the main purpose of the trade licensed by Hideyoshi to secure Chinese commodities in a third country because of the Ming policy of forbidding landings of foreign ships, especially of Japanese ships, on the mainland. For about a century after the middle of the sixteenth century, Japanese foreign trade enjoyed a brilliant period of development, and it was in this period that the production of precious metals in Japan reached its most flourishing peak.

It is said that in world history the middle of the sixteenth century witnessed sudden changes in the history of the production of pecious metals. At the beginning of that century, in Europe, Germany and Austria opened new gold and silver mines, but especially important was the exploitation of mines in America, particularly the beginning of silver mining at Potosi, Mexico, and the use of the amalgam method of refining, the result being a sudden increase in the output of silver. The production of gold throughout the sixteenth century was no more than about average, but production of silver after the middle of the century showed a conspicuous rise. Without considering the historical effect of the production of precious metals in the New World, we cannot understand the development of capitalism in Europe.

The almost instantaneous increase in Japan's output of precious metals paralleled the concurrent increase in world output. This was not because the geographically isolated Far Eastern islands of Japan had remained long removed from economic relations involving gold and silver. Japanese silver had been exported to China in particular, and to all the countries of South-East Asia. On the other hand gold from China, the Philippines, Sumatra, and Indo-China had been imported to Japan. Although Japanese silver was not shipped directly to Europe, Japanese trade occupied an important position in the Far Eastern trade of all the Western European countries, sustained by the export of Japanese silver, which matched the import of gold from China and other countries, a major aspect of all trade in the Far East. The trend of Japan's production of precious metals is therefore closely related with the history of colonization and the Far Eastern trade of Europeans. It is probable that the bringing of huge amounts of silver from Mexico for the China trade, the similar import into China of silver from India, and the carrying of gold from South-East Asia to Europe were also basic factors in the development of relationships between the values of precious metals in the sphere of Far Eastern trade. Considering the large production of precious metals and the extensive trade of Japan, we can see that their ramifications must have been great. Principally in Japan and China, and also in the other countries of South-East Asia, the values of gold and silver were at first almost unrelated, and subject to great disparities, but from the latter half of the sixteenth century to the beginning of the following century the gradual establishment of a unified balance can be discerned.

According to the research of European scholars[1] the average annual amount of silver produced in the world (excluding the Far East) in the sixteenth and seventeenth centuries is as follows:

1561–1580 ...	... 299,500 kg.
1581–1600 ...	... 418,900
1601–1620 ...	... 422,900
1621–1640 ...	... 393,600

The relative amount produced by the Potosi silver mine can be seen by looking at average annual output between 1581 and 1600:

Potosi	254,000 kg.
Mexico (other than Potosi)	74,300
Peru	46,000
Austria	17,000
Germany	14,300

The production of gold per annum averaged as follows:

1521–1544 ...	... 7,160 kg.
1545–1560 ...	... 8,510
1561–1580 ...	... 6,840
1581–1600 ...	... 7,380
1601–1620 ...	... 8,520
1621–1640 ...	... 8,300

Unfortunately it is now impossible to calculate the production of gold and silver in Japan at that time. We can only record such instances as the fact that an influential mine operator from Sado Island contracted with the Nanbu (Iwate Prefecture) gold mine for the transport of one year's output of gold,

[1] A. Soetbeer, *Edelmetall-Producktion und Wertverhältniss zwischen Gold und Silber seit der Entdeckling Amerikas bis zur Gegenwart*, 1879. This study does not include any estimates of Japan's gold and silver production.

6,500 pieces, i.e. 975 kg.[1] that at the end of the sixteenth century the amount of silver transported from the Ikuno silver mine (Hyogo Prefecture) as tribute to Hideyoshi came to 10,000 kg.;[2] the silver taken as tribute to Ieyasu near the beginning of the seventeenth century from only one shaft of the Iwami silver mine (Shimane Prefecture) amounted to 12,000 kg.; [3] and finally, about the same time, the production of silver of the Sado mine can be presumed to have reached between 60,000 and 90,000 kg. per year.[4] This large Japanese production of precious metals, through export and import, must have had an important direct connexion with world history. If we make a bold conjecture as to the export of Japanese silver at the beginning of the seventeenth century, it would very likely have amounted to 200,000 kg. annually. Lasting for about a century, this level of production of precious metals in Japan and its economic ramifications are certainly not to be overlooked in the history of the world's precious metals.

II. *The Development of Foreign Trade in Gold and Silver*

The export of gold and gold dust to China is discernible since the time of trade with the Tang and Sung dynasties, became considerable from about the twelfth century, and continued in trade with the Ming dynasty. Though gold was exported, Japanese records mentioning certain types of high-quality Chinese silver (*Nantei* and *Nanryô*) show that from the end of the Heian period through the Kamakura and Muromachi periods Chinese silver was imported to Japan. Pieces of silver valued at 50 *taels* were very common. This *tael* is one weighing 10 *monme*. One instance may be found in the correspondence of the Ming envoy who crossed to Japan in 1434 with the priest Mansai of the Sambôin (Daigo Temple near Kyoto), as follows: 'One piece of *Nantei* silver weighs 48 *taels*.' [5]

[1] A. Kobata, *Hohonoki Kinzan, Tohoku Chihō Kinzan Keiei no Ichi Keitai* (Research concerning Gold Mining in the North-East Region of Japan in the Tokugawa Era), in *Uozumi Sensei Koki Kinen Ronso* (Studies of Japanese History in honour of S. Uozumi), 1959.

[2] A. Kobata, *Ikuno Ginzan no Kenkyū* (A Study of the Ikuno Silver Mine), in Memoirs of the Dept. of Literature, Kyoto University, no. 3, 1954.

[3] Y. Ōga, *Iwami no Kuni Ginzan Kyūki* (Old Records of the Iwami Silver Mine), 1816. See also, T. Yamane, *Iwami Ginzan ni Kan suru Kenkyū* (Study concerning the Iwami Silver Mine), 1932, and A. Kobata, *Iwami Ginzan no Kenkyū* (Study of the Iwami Silver Mine [in the 16th century]), in Shirin (Journal of History), Kyoto University, vol. 18, no. 3, 1933.

[4] Memorial to the Governor of Sado Island from Miners in 1655. In Ichizaemon Shizume's term as Governor of Sado from 1618 to 1627, this memorial records that the average annual tribute was 30 tons of silver. The tribute was levied on each shaft of the mine according to conditions in each, and amounted to as much as one-half of the total production of a shaft, but generally was one-third or less. The total production of silver per annum must have been more than 60 tons and was possibly over 90 tons. Concerning the Sado Silver Mine, the *Sado Nendai Ki* (Annual Records of Sado) and *Sado Fūdoki* (Descriptive History of Sado) and other records compiled in Sado Island at the end of the Edo Period, and, more recent, S. Fumoto's *Sado Kinginzan Shiwa* (Chronicle of the Gold and Silver Mine of Sado), 1956, may be consulted. There are records giving estimates of the amounts of gold and silver produced on Sado after 1614, but as they are based on misinterpretations of original documents, they cannot be accepted.

[5] A. Kobata, *Nihon no Kingin Gaikoku Bōeki ni Kan suru Kenkyū* (Study concerning Japanese Gold and Silver in Foreign Trade), Part 1, in Shigaku-Zasshi (Historical Journal of Japan), Tokyo University, vol. 44, no. 10, 1933. Reference is made to the trade of gold and silver between Japan and China before the sixteenth century.

The policy of the *Lee* dynasty of Korea was to prohibit all foreign trade in gold and silver other than that carried out under the auspices of the government. In trade with Japan the export of Korean silver was strictly forbidden, and the import of the so-called '*Wa-Kin*' (Japanese gold) was at first prohibited in principle. Gold and silver were therefore exchanged in secret in trade done by private persons. Because proscribed items continued to be handled in private trade, there was even a time about the middle of the fifteenth century when private trade was completely prohibited and all trade was done by officials. But because of the excesses of the officials, the arguments for allowing or preventing private trade soon reappeared. The crux of the debate, as noted in a proclamation of Seisô, concerned the exchange by '*Wa-jin*' (Japanese) of gunpowder and gold for silver. Since gold continued to be imported from Japan and exchanged in secret trade for silver, there are many examples of the detection of violators of the laws. Restriction was concerned primarily with the exchange of Korean silver for Japanese gold, and the import of a certain amount of '*Wa-kin*' was probably not so severely limited. In fact, in the early half of the fifteenth century when Korea was obliged to send a tribute of gold to the Ming Court, an emissary was sent to Japan to try to arrange the purchase of gold. Later, after the latter half of the fifteenth century, the import of Japanese gold became remarkable. In the year 1480 Sô Sadakuni of Tsushima sent an envoy with gold amounting to 45 *tei*, each weighing 42 *monme* and including two round pieces, seeking to exchange it and copper for cloth and textiles. Every year after this for eight years gold was exported, totalling 23,894 *monme*. At 42 *monme* per *tei*, this would amount to 572 *tei*. The ships of the *Bakufu*, the Ôuchi family, and others, also took part in the export of gold. When the ruler Enzan-Kun replaced Seisô the inflow of Japanese gold had become large. All the gold deposited in the Korean treasury was gold imported from Japan.[1]

Japanese gold was also exported to China by way of the Ryukyus. In 1436 the chief minister of the Ryukyus, Kaiki, sent tributary ships to the Taoist headquarters seeking a licence for trade (*gofu*) for the king and himself, and among the valuable gifts were 20 packets of gold sand weighing 80 *taels*, from the king, and 10 packets weighing 40 *taels*, from the minister. (One packet weighed 4 *taels*, or 40 *monme*.) When the Portuguese took possession of Malacca at the beginning of the sixteenth century, they heard reports of certain people of '*Gore*' who had formerly brought great wealth in the form of gold and gold sand to trade in Malacca. It can hardly be doubted that the ships of these people of *Gore* were ships of the Ryukyus. And it could hardly be wrong to assume that this gold was transported from Japan. João de Barros, in the third edition of his work, *Da Asia*, records that the ships of Fernand Peres, on arriving at *Tamau* island, probably near Canton or Macao, in 1517, met ships from the Ryukyus which had brought, among other valuable goods, a large amount of gold. Seeing this, Peres was impressed with the possibility of dealing with

1 A. Kobata, *Chūsei Kōhanki ni okeru Nissen Kingin Bōeki no Kenkyū* (Study of the Trade in Gold and Silver between Japan and Korea in the latter half of the Muromachi Period), in Shigaku-Zasshi, vol. 43, nos. 6 and 7, 1932. This article concerns the trade between 1393 and 1566.

the Ryukyus more amicably than with China and sent Jorge Mascarenhas to investigate the country. In current records of the Portuguese, it is frequently written that a great amount of gold was produced in the Ryukyus, but this was deduced from the fact that the people of the Ryukyus transported gold in their ships, and is not a verifiable fact. The fact that the gold exported from Japan had weights of 42 or 40 *monme* shows that it was *Inaka-me* gold, i.e. gold valued in districts outside Kyoto at 10 *taels* per unit. In case of refined gold at 10 *taels* per *tei*, it must have been produced in some special form, with round pieces included. According to the Portuguese, the gold carried by the '*Gores*' contained some flat, elongated pieces with the imprint of the king's seal. Though they concluded that it was the king's seal by analogy with their own coins in use in Europe, it is more likely that the mark was one signifying gold weighing 10 *taels*. This was the so-called '*han-kin*' (stamped gold) of Japan, and it must be noticed that this *han-kin* appeared early as a form of gold used in trade.[1]

It is necessary to pay some attention to the relationships of actual and relative values of gold and silver in foreign trade before the first half of the sixteenth century. Of course, the following estimations are very rough. Probably the value of gold in Japan from about the thirteenth century to the beginning of the sixteenth century was 10 *taels* (45 *monme* by *Kyo-me* measure, i.e. as weighed in Kyoto, the capital) or around 30 *kan* (copper coins in current use). Similarly, silver (weighing 43 *monme*) can be safely estimated at a value of 5 or 6 *kan*. Consequently, the values of gold and silver were in the ratio of 1 to 5 or 6. According to the researches of Dr Shigeru Kato, the value of gold in China at the end of the eleventh century was 10 *kan* per *tael* (Taishô gold) and later, with a gradual rise, at the middle of the twelfth century was 30 to 40 *kan*, while silver about the middle of the twelfth century was around 2 *kan* per *tael*, and rising in value, in the first half of the thirteenth century, became more than 3 *kan*. That is, from the twelfth to thirteenth centuries the values of gold and silver in China were in the ratio of 1 to 13, more or less. In the Kamakura period (thirteenth and fourteenth centuries), because silver had a somewhat higher market value, measured in copper currency, in China than in Japan, and because gold had a value four or five times as high, the export of gold from Japan and the import of copper coins was profitable. The words '*Wa-gin*' can be found in Chinese records of the thirteenth century, and because of the condition of trading and small changes in market prices, Japanese silver was probably exported. With the advent of the Ming dynasty, from the end of the fourteenth century to the beginning of the sixteenth century the relative values of gold and silver in China were in the ratio of 1 to 5 or 6 and silver was gaining value in relation to gold. In relative value, Japanese gold and silver were not much different. In this period it cannot be supposed that Chinese silver was imported to Japan in quantity, but gold was an important export

[1] A. Kobata, *Chūsei Nantō Tsūkō Bōeki Shi no Kenkyū* (Study of the History of Trade via Southern Islands in the Middle Ages [from the 14th to the 17th century]), 1939, pp. 264–372. This section refers to trade between Ming China and the Ryukyu Islands. The export of Japanese gold by the *Gores* is dealt with in section 3 of chapter 3 concerning commercial relations between the Ryukyu Islands and Malacca.

for Japan, and was exchanged for raw silk, silk materials, etc., in China.[1]

In Korea, the relative values of gold and silver in 1432 were in the ratio of 1 to 11·1–11·7, and in 1435, 1 to 11·4–12·5. This was because Ming China waived the tribute in gold and the Korean government lowered the purchase price of gold. But in 1436 once again, responding to the market price, a drop occurred and the ratio became 1 to 6·7–7·5. If stated in terms of copper currency in the market, the value of gold had dropped from one-third to almost one-fifth or one-sixth, and calculated in terms of Japanese weights and measures, pure gold (rated at 10 on a scale of purity) was valued at 40,500 *mon* (a common coin) per 10 *taels*, pure silver (purity 10) at 5,160 *mon* per 10 *taels*, purity 7 gold at 13,500 *mon*, and purity 7 silver at 1,935 *mon*. We may assume that gold exported from Japan was generally close to pure gold (purity 10). In the latter part of the fifteenth century, the export of Japanese gold greatly increased and 30 *hiki* (1 *hiki* being about enough for one kimono) of cotton were given in exchange for 1 *tael* of gold. This being the rate fixed according to a decision in 1436, it also shows that the gold was pure gold rated at 10 on the scale. But, because of the increase in export of gold from Japan the exchange rate dropped to 25 *hiki* of cloth. In the time of Enzan-Kun, there was an increase in the production of silver because of the opening of the Tansen (Korea) silver mine and others, and though strengthening the tendency towards export of silver to China, it caused the price of silver to rise. About the year 1538 the inflow of Japanese silver to the Korean peninsula began and the value of silver at that time was set at the rate fixed in 1438, i.e. at 4 *hiki* of cotton per *tael* of silver. In the early part of the Lee period, it would not be far wrong to say that the relative values of gold and silver were generally in the ratio of 1 to 10, more or less. In 1469, by means of a translator, a Korean merchant sought to trade 40 *taels* of silver to a Japanese person for 8½ *taels* of gold. Since at this time the rate in Japan was 4 or 5 to 1, we can understand why the merchants of Korea even evaded strict laws and 'brought silver to trade for gold'.[2]

From the middle of the sixteenth century the production of gold and silver in Japan increased suddenly, and from about the beginning of the seventeenth century it leapt ahead even faster. Let us look briefly at the quantities and values involved. As previously shown, *Kyo-me* measure was 4·5 *monme* per *tael* of gold, but after the latter half of the sixteenth century the *binryo* method, in which 4 *monme* 4 *bu* were equal to 1 *tael*, spread from the Kinai area (around Kyoto) and was adopted throughout the country. The reason is not clear, but possibly it was because the *tael*, *bu*, *shu* system (40 *shu* = 4 *bu* = 1 *tael*) was used at the same time as the *monme*, *bu*, *rin* system (100 *rin* = 10 *bu* = 1 *monme*) and a convenient correspondence was necessary. Since gold came to be widely used

[1] S. Kato, *Nihon no Kingin Kakau oyobi sono Bōeki ni tsuite* (Concerning the Trade and Prices of Japanese Gold and Silver), in Shakaikeizaishigaku (Socio-Economic History Society), vol. 3, no. 3, 1933. This article compares the prices of gold and silver in Japan and in Sung China during the twelfth and thirteenth centuries on the basis of the trade in gold and silver, but gives a mistaken evaluation of the prices of Japanese gold and silver. See A. Kobata, *Chūsei no Kingin no Kakaku oyobi sono Nihon Bōeki* (Japanese Trade and the Prices of Gold and Silver in the Kamakura Period), in Shakaikeizaishigaku, vol. 3, no. 6 for a criticism of Dr Kato's evaluation.

[2] *Chūsei Kōhanki ni okeru Nissen Kingin Bōeki no Kenkyū.*

as a weighed form of currency, the change in the method of weighing was probably a response to the needs of the society. As far as silver was concerned, the rate of 4 *monme* 3 *bu* per *tael* did not change but, in this period the *kan*, *monme* system came into common use. In the early part of the *Tenmon* period (1532–1555), continuing the earlier rate, the price of gold was 30 *kan* per 10 *taels*, more or less, and gradually declining after about 1541, from the *Tenmon* through the *Ganki* and *Tensho* periods, i.e. during the latter part of the sixteenth century, it was 12–16 *kan* pieces per 10 *taels*. The relative value of gold and silver, which had been in the ratio of 1 to 5 or 6, underwent remarkable changes along with changes in market price in the latter half of the *Tenmon* period, and the drop in the value of silver was particularly large. This was the result of the sudden, and at the same time comparatively large, increase in the production of silver, till then a relatively meagre quantity. The market price of silver consequently falling in relation to gold, in the latter half of the sixteenth century the ratio was generally about 1 of gold to 10 of silver. These changes brought about great changes in trade based on gold and silver.[1]

The export of large amounts of Japanese silver to Korea began about 1538. In the eighth month of that year, in a memorial of a high Korean official, it is recorded that the '*Wa-jin*' (Japanese) just arrived had only silver and had brought no other trading goods. In the tenth month of the same year, in another official memorial, it is apparent that *Wa-jin* brought 350 *kattes* of silver, the trading value of which was more than 480 *do* (a *do* being equal to 50 *hiki*), and only one-third of which was permitted to be traded under the auspices of officials, the remainder being considered to be under the general prohibition of conveyance of silver to Korea. However, the secret import of *Wa-gin* continuing unabated, so that the markets were said to be full of Japanese silver in circulation, and even some being transported to China, the officials were forced to undertake rigid control. In the fourth month of 1542, the priest Anshin Tôdô, said to have been the *Bakufu*'s envoy, stated on his arrival in Korea, 'In the Hokuriku area of our country there is a mine, the name of which is Kaneyama, producing much pure silver in recent years, one of the marvels of the world'. He had taken 80,000 *taels* (800 *kan*) with him. As a result of the controversy concerning trade in Korea, 20,000 *taels* were to be bought, but because at that time the value of silver was dropping, the rate of 3 *taels* of silver for 2 *hiki* of cotton was proposed. But Tôdô, obstinately seeking the old price set in 1533, was willing to trade only 10,000 *taels* at the current market price and wanted 700 *do* of cotton for the rest, basing his demand on the 1533 price. Holding out for 20,000 *taels* worth of trade, Tôdô seems to have finally settled on the procurement of 1,200 *do* for 15,000 *taels*, which is about the rate set in 1533. After this, in 1553 Anshin Tôdô once more went with a large amount of silver, and in 1554 Tenpu Tôdô, also probably an envoy of the *Bakafu*, took 30,000 *taels* of silver with him. Altogether it seems that, at

[1] A. Kobata, *Nihon Kahei Ryūtsū Shi* (History of the Circulation of Japanese Currency), 2nd ed. 1943, pp. 380–410. These pages contain an analysis of the price fluctuations and weighing methods of gold and silver in the sixteenth century.

the old rate of 1 *tael* of silver per 4 *hiki* of cloth, about 3,000 *taels* were exchanged under official sanction.

After 1540 first merchant ships from Chuan-chow and Chang-chou in Fukien province, then other Chinese ships came to Japan to start trade. It was their main purpose to carry away the silver produced in Japan in such abundance. After Portuguese sailors had been thrown ashore on Tanegashima in 1543 and had come to know Japan for the first time, the Portuguese determined to profit from trade with Japan. Presently their ships were coming to many of the ports in Kyushu almost every year, and after 1550 they entered Hirado harbour every year until 1561. After this they transferred their port of call to Yokoseura, Fukudaura, and others, and in 1570 Nagasaki was opened. The main commodities of the Portuguese ships' trade with Japan were raw silk, silk goods, and other Chinese products as import and Japanese silver as export. Thus they made intermediary trade between Japan and China their main support. In the record of the voyage in 1563 of the Venetian merchant Cesare Frederici, it is recorded: 'Every year one ship loaded with silk goes from China to Japan to trade it for silver bullion.' In the log of Jan Huygen van Linschoten on a voyage to the East Indies, the following is written: 'The commodity taken from Macao to Japan is silk, while only silver is brought from Japan, the profit on silver being considerable.' The London merchant Ralph Fitch, reporting affairs of the 1580's, said that the Portuguese carried nothing besides silver from Japan, but that every year the silver amounted to more than 600,000 *cruzados*, and this, along with more than 20,000 *cruzados* from India, was used in trade in China. Padre Sebastião Conçalves states that the Portuguese ships every year took on 1500 *picos* (1 *pico* = 100 *kattes*, 60 kg.) of raw silk, velvet, silk damask, etc., and traded them for 500,000 *cruzados*, and in the writings of Alessandro Valignano, we can read that the Portuguese carried raw silk and other goods from China to Japan and almost every year took away silver worth 500,000 *ducats*. Since *ducats* were also called *cruzados*, and the *cruzado* corresponded to 10 *liras*, and since this Spanish *lira* was calculated at 1 *masu* (1 *monme*), it can be seen that 1 *cruzado* was equal to 1 *tael* (a *tael* weighing 10 *monme*). We can probably assume therefore that around 1580 Portuguese ships exported something like five or six thousand *kan* of silver every year from Japan.[1]

Unlike silver, which replaced it as an export from Japan, gold now came to be an import. In the 1580's, as reported by Ralph Fitch, in a consignment of goods sent from Macao to Japan Chinese gold was next in importance to raw silk and silk textiles. In 1590 annals concerning the Chinese Empire and other political affairs were published in Macao in Latin dialogue form. Among a wealth of instructive articles, one states that in that year 2,000 pieces of gold bullion, valued at about 100 *ducats* per piece, were exported from China to

[1] *Nihon no Kingin Gaikoku Bōeki ni Kan suru Kenkyū*, Pt. 2, vol. 44, no. 11. This article concerns the trade of gold and silver from 1543 to 1640. The 1547 expedition of ships, trading with Ming under the Kangō system, ended the Muromachi *Bakufu*'s official contact with Ming China. About this time the trading ships of Ming first came to Japan. The results of the ensuing trade between Ming and Japanese merchants are noted in the author's *Chūsei Nisshi Tsūkō Bōeki Shi no Kenkyū* (Study of the History of Trade and Communications between Japan and China), 1941, pp. 450–98.

Japan. An incident is recorded in the annual report of Frois concerning the breaking of the Administrator's Law by a Portuguese ship which entered Nagasaki harbour in the seventh month of 1591 carrying Chinese gold in the form of 'ship money' (*insu kin*) for exchange, more than 30,000 *ducats* being destined for purchase by the merchants of the capital (Kyoto) and Sakai (near Osaka). If this gold bullion weighed about 12 ounces or 100 *monme* per piece, in all it would amount to 300 pieces. According to a résumé of prices of goods exported from Canton to Japan, which is presumed to date from the turn of the century, about three to four thousand *taels* of gold was exported to Japan. If this may be said to have been *insu kin* at 100 *monme* per piece, there must have been three or four hundred pieces. In 1605 Ieyasu ordered *insu kin* to the amount of 10,000 pieces at 100 or 105 *monme* per piece from Portuguese ships.

In China from the latter half of the sixteenth century to the seventeenth century the market prices of gold and silver were in the ratio of 1 to 7 or 8. According to the above-mentioned résumé of prices for Japanese goods in the trade of the Portuguese at Canton, 1 *tael* of pure gold and 7 *taels* of pure silver had the same price. According to Dutch data, in China (Fukien) the calculated relative values of pure gold and silver were in the ratio of 1 to 8 in the 1620's, 1 to 10 in 1635, and 1 to 13 from about 1637 until the 1640's. In the daily records of Ku Yen Wu it is stated: 'During the period from 1629 to 1644 gold was worth 10 times the amount of silver, and 13 times the amount south of the Yangtze.' And probably the advancing price of gold south of the Yangtze river was due to the export of gold to Japan and other countries by the Dutch and Chinese. In Japan during the latter part of the sixteenth century the relative values of gold and silver were generally in the ratio of 1 to 10. In his history of Japan, Frois asserts that: 'One piece of silver has a weight of 4 *taels* 3 *masu*, and in our currency has a price of 4 *cruzados* 6 *vintem*, while one piece of gold has a price of 43 *cruzados*.' And we can probably say that from the end of the *Keicho* period into the *Kanei* period (before the middle of the seventeenth century) the relative values of gold and silver of comparable purity were in the ratio of about 1 to 12 or 13. The price of gold imported into Japan in the 1610's, and again in the 1630's, for example, was about 13 of silver for 1 of gold. It was possible to realize a profit of 60 per cent on the import of gold from China in the 1610's but about ten years later the profit did not exceed 30 per cent, and after about 1640 the advantageous relationship of value in the exchange of gold for silver had already been lost.

In the latter part of the sixteenth century the value of gold in relation to silver was less in the Philippines than in China. For this reason, the Chinese very early transported gold from the islands, but it is not true that they brought silver with which to buy it. After the establishment of a base at Manila by the Spanish the import of Mexican silver coins increased, and soon the Chinese were making silver from the New World an important export from the Philippines to China. The fact that Japanese came to Luzon and Mindoro, taking out gold and honey, can be found in the correspondence of Miguel Lopes de Lagaspi as early as 1567. In a letter of 1575 from Joan Pacheco Mardnâdo to Philip II of Spain, it was written: 'This Japan from which is brought ever-increasing amounts of silver is about three hundred (*sic*) leagues from Luzon island. Every year Japanese ships come loaded with trading goods, but most important is the exchange of silver for gold, at the rate of 1 *mark* of gold for between 11 and 11·5 *marks* of silver.' At the end of the same century, the Chinese brought raw silk goods, etc., which became the main commodities of trade with the Japanese in Luzon. Gold was also keenly sought, and as exchange for these commodities silver was the most important cargo.[1]

At the beginning of the seventeenth century the gold of Indo-China, Sumatra, Siam, and other such countries was imported to Japan by the Dutch and English. According to the records of William Keeling in Siam, in 1608 gold was exchanged for silver at the rate of 1 to 3. In a letter sent from Ayutthaya (Siam) in September of 1617, Cornelis van Neijenroode reported to the Amsterdam office of the Dutch East India Company that one Hautman, a factor at Ayutthaya, had bought a small amount of the gold sent to Japan and had realized a profit of 35 or 40 per cent. In Indo-China until about 1633 the ratio was certainly 1 of gold to less than 10 of silver, and this became the market price after the value of gold had risen sharply as a result of export to foreign countries, including Japan, and in that year for the same reason gold rose once more by 20 per cent.

About 1610 in northern Sumatra 1 *masu* of gold was worth 6 *reals*, which was computed as equal to 6 *monme* of '*chôgin*', silver currency issued by the *Bakufu*. At the end of the sixteenth century, in the Philippines, Sumatra, Siam, and Indo-China, the price of gold relative to silver was lower than in China, and this gold was imported to Japan, but as the price gradually rose as it had done in China, by about 1640 we can presume that the relative values of gold and silver had almost reached equilibrium throughout all the countries of East Asia.

The Japanese trade of Chinese and Portuguese ships, with import of Chinese products and export of Japanese silver as the main cargoes, made steady progress. At the end of the sixteenth century and beginning of the seventeenth century, Holland and England strove to open Far Eastern trade, and with the State in the background, both formed East India companies and sent commercial fleets to the Far East repeatedly, working to secure their hold on trade centred in Java and other East Indies islands. Not only did they compete with Portugal and Spain, but with each other. But the best manufactured articles of the European mother countries were not always the things in demand in the Far East. China was one focal point of Far Eastern trade, and in China particularly, as well as in other countries of the Far East, the product most in demand was silver. Silver from the New World was brought into China by way of Manila, but it is often overlooked that silver from Japan was even more important as far as trade from the middle part of the sixteenth century into the first part of the seventeenth century was concerned. Though the con-

[1] Concerning trade between the Philippine Islands and Japan, as well as the material noted for p. 496, see the author's *Philippine no Kin Gin* (Gold and Silver in the Philippines), in Nanpō Dozoku (Journal of the former Taihoku Imperial University), vol. 2, no. 2, 1933.

sumption in Japan of sugar, spices, medicines, clothing materials, and other products of the south greatly increased in this period, the largest imports were Chinese raw silk, silk goods, and for a short time, gold, and next in importance came raw silk, silk goods, etc., from Indo-China. Thus, because the successful acquisition of Chinese commodities had such a close relationship with the success of trade with Japan, this fact cannot be ignored in considering the development of Dutch trade in Japan. As the approach of Japanese ships to the mainland was absolutely forbidden by Chinese policy, inevitably the goods of China were sought in third countries and registered Japanese ships (*Shûinsen*) took advantage of such a system.[1] Until the closure of the country to foreign trade, what amount of silver was exported? This is an interesting and important problem, but its solution is singularly difficult. It seems uncontestable that in the total value of all exports, silver formed the greatest amount. Although in the 1630's Portugal's Japan trade was already passing its peak of prosperity, in 1635 three ships exported 1,500 cases of silver, in 1636 four ships exported 2,350 cases, in 1637 six ships exported 2,600 cases, and in 1638, 1,250 cases. One case contained 10 *kan*, i.e. weighed 1,000 *taels*.

It is significant that about 1640, with the establishment of a balance in the relative values of gold and silver in the Far East and the loss of the relative advantage of gold over silver in Japan, a period began when the export of Japanese gold was possible. However, with the closure of the country, only the trade of China and Holland flourished. And they continued the heavy export of Japanese silver. In 1668 the export of silver was prohibited, and for sometime only China was favoured, being excluded from this prohibition. Finally, in 1685, owing to restrictions on the amount of trade and a concurrent remarkable increase in the export of copper, silver lost its place to copper at the head of the list of exports.[2]

III. *The Progress of Gold and Silver as Currency*

The use of gold sand as a tribute to the emperor and retired emperor at the investiture of a *Shogun*, or as a stipend given to the officials in charge at such a ceremony, remained customary from the Kamakura period into the Edo period. After the sixteenth century, at the investiture of a *daimyo*, or as acknowledgement of the bestowal of name and rank by the *Shogun*, there are examples of gold being given to Emperor and *Shogun*. The exchange of gold and silver as presents also took place among nobles, shrines and temples, and military officers.

[1] S. Iwao, *Shuinsen Bōeki Shi no Kenkyū* (Study of the History of Trade by Ships under the *Shuin* System), 1958, pp. 320–30. This book also deals with Japan's foreign trade at the beginning of the seventennth century and with the export of silver.

[2] For an outline of the trade in gold and silver after the closure of the country in 1641 see the author's *Nihon no Kahei* (Japanese Currency), 1958, pp. 134–6. For Holland's trade in Japanese gold, silver, and copper in the seventeenth century, see also Oskar Nachod, *Die Beziehungen der Niederländischen Ostindischen Kompagnie zu Japan im siebzehnten Jahrhundert*, Leipzig, 1897. Though this is an old account, it is still valuable. For a more complete analysis of the export of copper from Japan, see the author's *Kinsei Zenki no Dō Bōeki to Sumitomo* (Sumitomo and the Trade in Copper in the Early Tokugawa Period), in Senoku Sōko (Sumitomo Group Journal), no. 9, 1957.

The opening of gold and silver mines came about as a means of financing the *daimyos* of the *Sengoku* (sixteenth century) period, and gold and silver first showed their efficacy as mediums of high value in their hands. As military resources with great buying power and easily transportable or storable, gold and silver soon demonstrated their convenience. As rewards for meritorious service they were suitable in place of land, supply of which was limited.

When Uesugi Kenshin died, there were as many as 2,500 *mai* of gold deposited in his treasure house at Kasuga-san castle, Niigata. Anayama Baisetsu of the House of Takeda controlled the manor of Kôchi in Kai (Yamanashi prefecture) which produced gold, and with the downfall of his lord in the spring of 1582 he took 2,000 *mai* of gold and capitulated to Nobunaga. In the subjugation of the House of Takeda, Nobunaga gave one of his leading retainers 50 *mai* of gold to be converted into more than 8,000 *hyo* (straw sacks) of rice for the army. It was Hideyoshi who accumulated huge amounts of gold and silver and used it for extensive military operations. In the spring of 1587 when he departed for Kyushu, 'twelve horses loaded with gold and silver' were part of the company. In the year before the subjugation of Kanto, making Natsuka Masaie the chief provisioner of the army, Hideyoshi contracted for 200,000 *koku* of rice from his own domains and in the following spring (1590) sent it quickly by sea to Ejiri and Shimizu in Suruga (Shizuoka prefecture). At the same time he ordered 10,000 *mai* of gold to be converted into rice provisions from the provinces of Ise, Owari, Mikawa, Totomi, and Suruga (all along the Pacific coast) which were to be delivered to wharves in the neighbourhood of Odawara (on Sagami Bay.) Ten thousand *mai* of gold would have been equivalent to 500,000 *koku* of rice. In the spring of 1592, as preparation for the sending of an army to take part in the Korean expedition, the Chôsôgabe family of Shikoku ordered the collection of gold from each of the *bushi*, the value of which was compensated with land. During the years of Bunroku (1592–96), silver from Iwami (Shimane prefecture) was cast into coins for military use, at Hakata in Kyushu, and it is said that after the Korean expedition the circulation of silver currency in Korea began.[1]

Gold and silver gradually came into use in the payment of tribute and taxes. From the middle of the sixteenth century great quantities were sent from the areas where gold and silver were produced. According to the records of the temple Honganji (Osaka), almost every year after 1536, gold was sent from Kaga (Ishikawa prefecture) to the temple; in the spring of 1547, 1 *mai* (10 *taels*) of *Inaka-me* gold was sent from a believer in Shinano (Nagano); in the fourth month of 1553, 3 *mai* of *Kyo-me* gold from the Noto peninsula as felicitation on the recovery from an illness of the chief priest Shônyô; and in 1542 a contribution of gold came from Hitachi (Ibaraki prefecture). The annual tribute of the Todai Temple's domains of Niirei and Kokuga in Suhô province (Yamaguchi) was paid in the form of rice or *zeni* (copper coins) during the years from 1532 to 1555. (After 1570 silver was used.) In the same province, the Tokuji domain of the Tofuku Temple (Kyoto) paid its tribute between

[1] *Nihon Kahei Ryūtsū Shi*, pp. 437–44. These pages describe the use of gold and silver as funds for military campaigns.

1555 and 1558 in *zeni*, but from the Temple accounts of monies collected after the New Year of 1565, which record the sale of Tokuji silver coins at the Tofuku Temple, it can be seen that payment in silver had begun. In the Kanto area, the Hojo's, after 1558 or so, prescribed that tribute must be paid in good copper *zeni*, or, when such good coins did not suffice, should be supplemented by equivalent payments of rice and grain, or gold.

In Hideyoshi's time, the domains near Kyoto in which Hideyoshi's storehouses were situated sometimes paid tribute in gold. According to the records of the Kannon Temple in Ômi (Shiga prefecture), the tax in rice of Hideyoshi's domains in Ômi was fixed at a value of 30 *mai* of gold, and in the seventh month of 1585 at a rate of 30 *koku* per *mai* the tribute was to be 900 *koku*, but in the following month at a rate of 32 *koku*, 960 *koku* of rice were paid. In the spring of 1583, when Maeda Toshiie took the field in support of Shibata against Hideyoshi, he sent a letter to his caretaker in the domain stipulating that the tribute should be collected at the rate of 100 *hyo* (straw sacks) of rice per *mai* of gold, levied in the form of gold, but that changes in the market price should be taken into account. In Noto (Ishikawa prefecture) in the districts of Fugeshi and Suzu, the rates of 100 sacks in 1585, 120 sacks in 1586, and 140 sacks in 1587 were used in the payment of tribute in gold.

In such cities as Kyoto, gold and silver were also collected as tax. According to the accounts of the southern part of Kyoto in the sixth month of 1573, 1 *cho* (a small section of the city) was required to pay 13 *mai* of gold, the 54 *cho* in the 6 *kumi* (larger sections of the city) being assessed 30,186 *monme* (a *mai* weighing 43 *monme*), of which 979 *monme* was remitted as exemption, leaving a total tax of 29,206 *monme* 4 *bu*. Seventeen temples in this southern part of Kyoto paid a total tax of 1,045 *monme* 6 *bu* of silver. In the fourth month of 1588, inviting the emperor Go-Yozei to his palace, the Jûrakutei, Hideyoshi gave him, among other tributes, taxes from the city of Kyoto amounting to more than 5,530 *monme*. In the daybook of the Tamon-in, under the tenth month of 1589, it is noted that: 'Recently the taxes on Kyoto establishments have been collected in gold and silver, and now it is ordered that they be collected in this way throughout the capital, the scales being heavily weighted against us.' The licence fee for the guilds of Hyogo went up by 5 *monme* of silver in the latter half of 1583, and from the following year 10 *monme* was collected from them each year. In the early years of Keicho (after 1596), the city of Sakai paid 250 *mai* of silver as licence fees for its guilds and 1000 *mai* as land tax to Hideyoshi's government in Fushimi.[1]

With the spread of the circulation of gold and silver, the landlords, shrines, and temples made regulations concerning their exchange values. The Kasuga Shrine Records, which are thought to have been written at the end of the Tenmon period (1532–1555), states that rates were set at 20 *kan* of *zeni* per 10 *taels* of gold and 2 *kan* per 10 *taels* of silver, and all taxes and tribute were to be paid at these rates. In the sixth month of 1568 the Hojos' order fixing the rates of tax gives, along with the prices in *zeni* of rice and wheat, the rate of

1 *Nihon Kahei Ryūtsū Shi*, pp. 450–7, concerning payment of taxes and levies in gold and silver.

1,500 pieces of *zeni* for each *tael* of gold, at which rate the tribute was to be paid. In the spring of 1569, according to articles appended to the laws governing the use of *zeni*, Nobunaga forbade the use of rice in making purchases and commanded that the sale of more than 10 *kattes* of silk thread or chemicals for medicines, more than 10 *tan* of damask, or more than 100 tea cups, should be transacted in gold or silver. Because deals involving such Chinese articles reached a suitably high price, and the price of other Chinese articles could be based on them, these provisions were included, but if gold and silver were not available, the original laws governing the use of *zeni* required the use of 'good' *zeni* rated at 15 *kan* per 10 *taels* of gold or 2 *kan* per 10 *taels* of silver. The rates of various kinds of *zeni* in terms of 'good' *zeni* were established by Nobunaga's government. When gold and silver were borrowed they were to be repaid in kind, and when gold and silver were not available, a rate in terms of 'good' *zeni* was to be fixed. The prescription of the use of gold and silver in transactions involving suitable quantities of Chinese goods, besides being dictated by the high prices involved, was probably due also to the fact that the values of imports were based on gold and silver. Though regulated temporarily, the rates of exchange of gold and silver for 'good' *zeni* were difficult to maintain uniformly in practice. Moreover, the rates of 'bad' *zeni* (poor quality copper coins) fixed by law were such that buyers sought to make all payments in these 'bad' *zeni*. For this reason debts incurred in gold and silver were to be repaid in gold and silver. The prohibition of secret sales of gold and silver was probably a result of fear that the exchange rates of gold and silver for copper coins would be disregarded. At the same time, requests by vendors for gold and silver, or conversion of gold and silver by ordinary merchants, was prohibited. Since the prohibition only applied to transactions of commodities by merchants, in which case the receipt or payment of sums stated in terms of gold could not be transacted in silver, and vice versa, the actual conversion of gold and silver in itself was not prohibited. Since under Nobunaga's regulations the places where exchange transactions involving gold and silver could be carried out were fixed, the free use of gold and silver in all transactions was not allowed.

At this time in Kyoto, Nara, Sakai, and other cities, gold and silver exchange houses (*ryô gae shô*) were numerous. The gold and silver received as donations or tribute by the nobility and temples was converted into *zeni* for use in payments, and to cover expenses and facilitate transport to distant places gold and silver were purchased. The establishments or merchants engaged in such conversion or sales of gold and silver were also called '*kin-ya*' or '*gin-ya*'. From the Tenshô (1573–92) and Bunroku (1592–96) periods, the '*gin-za*' (silver guilds) appeared as groups licensed by the regional lords to buy and sell gold and silver, convert mediums of exchange, refine precious metals, weigh and mark values, and cast *han-kin* and *han-gin* (gold and silver pieces with values stamped on them). In return for permission to form a *za*, dues were paid to the lords. Thus among the houses of gold- and silversmiths who could command great confidence were some *kin-ya* and *gin-ya*. In the Tamon-in Temple journal can be seen names such as those of Haramaki-ya Jinzaburo among the *kin-ya*, and Yotaro, Yosaburo, Kihichi, and Genzô among the *gin-ya* in Nara. At the *kin-ya* and *gin-ya* the prices of gold and silver were established. In the autumn of 1542, when Uchijima Hyôgo was given 500 *hiki* (5 *kan* of copper currency) as travel expenses from the temple Honganji, it was 'converted at the rate of 28 *kan* [for 10 *taels* of gold] in Kitashirakawa', which would have amounted to 1 *ryo* 3 *bu* 1 *shu* of gold. It seems that there was a *ryo-gae sho* in Kitashirakawa (Kyoto). The record of the Tamon-in for the seventh month of 1585 in which Kihichi is quoted as stating that, 'Recently 129 *monme* of silver (3 *mai*) are worth 15 *monme* 2 *bu* of gold', shows that the Tamon-in received information on the conversion rates from the Kihichi *gin-ya*.

Among the records in the archives of the Satake family, lords of the old Akita *han*, is one thin volume of twelve pages entitled, 'Record of Expenses Paid in Gold, 4th day of the 2nd month of the 5th year of Bunroku'. In the Hitachi (Ibaraki) domain of the Satake family there was a gold mine from which, as well as the levy of gold, the annual tribute and taxes were paid in gold sand. Gold sand was accumulated and used to pay the expenses of Satake Yoshinobu's period of residence in Kyoto. In the two and a half years up to November of 1598 all purchases, expenses of the mansion, and gifts to Hideyoshi and the officials Ishida and Masuda, cost Yoshinobu more than 1,014 *mai* of gold sand. Of this amount, much *han-kin* and *noshi-kin* were used as gifts. The *han-kin* was purchased at the rate of 44 *monme* and 5 or 6 *bu* of gold sand per 10 *taels*, and the *noshi-kin* seems to have been flattened pieces refined from the gold sand. The conversion or refining of the gold sand was done at a *kin-ya* or *ryo-gae ya*.

With the advent of the Bunroku period (from 1592), exchange houses for gold and silver spread to the castle towns (*jô-ka machi*) and post towns (*shuku eki*) of the provinces. In the autumn of 1593 Ôwada Shigekiyo, a retainer of the Satake's, left Nagoya in Hizen (Saga prefecture) and returned to Ôta in Hitachi by way of the Sanyo road (along the Inland Sea coast to Kyoto) and Tôsan road (through the mountains of central Honshu). According to the diary of this trip, the gold and silver carried were converted along the way into *zeni* for convenient use. In Gifu 10 *monme* 6 *bu* of silver were converted into 1 *kanmon* of *zeni*, and in Nojiri (Nagano prefecture) 1 *monme* of gold into 800 *mon* of *zeni*.

Since gold and silver were used in accordance with their weight, their measurement became a problem, but like that of volume, it was not unified. The differences corresponding to differences in area were large. In current accounts, there are even numerous statements of overpayments due to differences in weighing methods. For this reason, the use of scales to ensure confidence was agreed on by merchants engaging in transactions in gold and silver. The weighing became one of the functions of the *kin-ya* and *gin-ya* which collected a fee for doing it. The Gotô family, gold and silversmiths, which undertook the casting of *hankin* for Hideyoshi's use, were descendants of the famous artisan Gotô Yûjô. It was natural that scales having the stamp, '*Gotô han*', should command trust in the Kinai district around Kyoto and Osaka. In one article of the Rokuon Temple (Kyoto) records, dated the second month of 1592, 5 *koku* of rice and 4 *taels* of gold were converted into 3 *mai* of silver;

'scales marked with the Gotô seal' were used in the process. In the time of the Takeda's, as scales for gold those of the Shuzui family were provisionally approved for use in Kai (Yamanashi prefecture), and with the take over by the Tokugawa's in 1582, this privilege was confirmed. In the following year the Tokugawa's designated the scales of the Shuzui family as the ones to be used in sales in all their domains, and forbade the use of private scales. It can be concluded that Gotô Yôsuke's departure from Kyoto for Kaga, being about the same time as Gotô Mitsutsugu's departure for Kanto, was at the end of the Bunroku period (1592–1596). Even before this he had been given special permission to establish a *gin-za* in the Noto peninsula of the Kanazawa *han*. This *gin-za* was still called a *tenbin-za* (guild in charge of weights and measures) and had essentially the same constitution. In the time of the Niwa's, the Daimoji-ya of Komatsu (Kaga) was allowed to undertake weighing in the district of Nômi. After it became part of the Maeda domain, in 1600, the approval was confirmed and the Daimoji-ya was given a charter stating that it should continue to handle gold and silver as previously. The granting of special permission for the manufacture and exclusive use of scales where public confidence in their use had been demonstrated can be noted among the *han* at the beginning of the Tokugawa period. Afterwards the *Bakufu* divided the country into two halves, with Shuzui scales being recognized in the thirty-three eastern domains and the scales of Shinzen Shirô of Kyoto being approved for use in the thirty-three western provinces.

In the circulation of gold and silver, along with the measurement of weight, quality became a problem. It was the business of the *kin-ya* and *gin-ya* to appraise and certify quality. Refining was therefore also very important. According to the record of the Tamon-in for the twelfth month of 1581, the Tamon-in borrowed 1 *mai* of gold and in order to repay it in gold, bought 1 *mai* through the offices of the Yojirô *gin-ya* for 50 *koku* of rice. The refining and handling fee of 6 *tô* (1 *koku* = 10 *tô* = 100 *shô*) of rice was paid separately, but it was customary that, when borrowing, these fees (also called *suai*, or service fees) were paid by the creditor, and at the time of repayment their burden was born by the debtor. According to the Tamon-in record, however, when buying 1 *mai* of gold, the Tamon-in paid a service fee of 2 *tô*, refining fee of 2 *tô*, and marking fee of 2 *tô* in addition to the price of the gold in rice. In the second month of 1582, the Tamon-in paid a service fee of 1 *tô*, refining fee of 1 *tô*, and marking fee of 1 *tô* in addition to the rice price for half a *mai* of gold. The marking fee was a charge for stampting a seal on the gold and this seal's purpose was to guarantee the quality. Again, the same record, under the date of February 1589 concerning the Kihichi *gin-ya*, shows that it collected a marking fee of 500 *mon* per *mai* for stamping a total of more than 200 *mai* of gold in that year. At a rate of 500 *mon* for 1 *mai* of gold, that is, for 10 *taels*, the seal was affixed year after year.

According to the records of the family of Sueyoshi Kanbei who were put in charge of the Edo *Bakufu*'s *gin-za*, Ieyasu ordered Kanbei, and eleven other great establishments of Sakai and Hirano to cast *cho-gin* coins by combining copper with silver. Inspecting these *han-gin*, which had seals incorporating many patterns such as that of the chrysanthemum, Ieyasu found Daikoku Jyôzei's seal suitable and put Jyôzei in charge of minting in the *gin-za*. In the area of Sakai, from earlier times there had been a Nanryô-*za* whose several members had made *han-gin*. According to the 'Zono' inventory of the Toyotomi family, dated 1598, taxes amounting to 1,000 *mai* of gold for the use of the Gotô seal and 10,000 *mai* of silver from the Jyôzei-*za* are recorded. This was because the goldsmith Gotô and silversmith Jyôzei had been given the special rights of *kin-za* and *gin-za*. In the diary of Komai, lord of Sanuki (Shikoku), an entry states that about 1595 Jyôzei was made chief minter in Osaka, and twenty artisans were allowed to form a guild under him.

In the provinces also, *kin-ya* and *gin-ya* which could procure public confidence produced *han-kin* and *han-gin*, and some of them received special charters from the feudal lords. The Daimoji-ya of Komatsu was given the right to collect fees for wrapping and refining silver in the days of the Niwa family. The fee, collected in the form of silver, was a charge for wrapping a certain quality of *han-gin* and guaranteeing its weight. Such privileged *kin-za* and *gin-za* also cast *han-kin* and *han-gin* for use in the domains at the request of their lords. The '*kô-kin*' of Kai is said to be a legacy of the time of Takeda Shingen, but the four houses of Matsugi, Nonaka, Shimura, and Yamashita were all *kin-za* of Kai. According to the statement of the tax on land producing cotton under Asano Nagamasa in the third month of 1594, the tax contracted in terms of gold totalled 2 *mai* 7 *ryo* (*taels*) 4 *bu* 7 *rin*. This gold was marked with the seal of Noguchi Shinbei. It was Noguchi's *han-kin*, which had been cut into smaller pieces to be used as currency according to weight. In Edo, before the arrival of Gotô Mitsutsugi, it is recorded, 'There were three people who put their seals on gold: Shijô, Sanô, and Matsuta, who refined gold sand into *taels*, *bu*, and *shû* of good, fair, or poor quality, and put the weight and seal on the package, this being used from 1590 until 1595'.

In Japan, the forerunner of the gold and silver currency which first appeared in city markets was probably the product of the *kin-ya* and *gin-ya* of cities such as Kyoto, Sakai, and Nara which, with public confidence behind them, used their seals as guarantees of quality and established weights. Among such early currencies, some were designated for official use by the Toyotomi, Tokugawa, and others, and in the provinces such practices were carried on by the feudal lords.

The seal of the Gotô or the *han-kin* of the Kihichi *gin-ya* probably guaranteed that, for example, if the seal had been affixed to 1 *mai* and '10 *taels*' had been written on the gold with a brush in *sumi* (black ink), 1 *mai* was the actual weight. However, the *han-kin* of this period was not always used in transactions in the form of 1 *mai*, but was commonly used in the form of pieces of the original coins which were exchanged according to weight. In the year 1553 the gold sent from Shinano (Nagano) to the Hanya-in of Yamashiro (Kyoto) consisted of thirteen pieces amounting to 1 *mai* 2 *shu* by Kyoto measure, this being *han-kin*. Like silver, gold also was stamped many times so that even among the pieces cut from the original *han-kin* there are many in which the seal remains. Thus *han-kin* and *han-gin*, as guarantees of a certain quality, became the standards on which many transactions involving gold and silver refined at the mine were carried out at that time.

It is said that *ko-han kin* and *ko-han gin* were gold and silver pieces cast before the Tokugawa period, and with the advent of the Genroku period (1688–1704) the collection of earlier pieces of currency became popular along with the spread of counterfeiting.[1] The *Higashi-yama Hô-ô Maru* pieces (weighing 42 *monme*, and having a diameter of 3 *sun* 6 *bu* – about 11 cm.) were elaborate circular coins inscribed with the two characters for *Higashi-yama* (East Mountain) and a picture of the *Hô-ô* bird. Pieces such as the many *Taikô Fuku-ju* stamped with the three characters *tai*, *fuku*, and *ju*, or the *Taikô Ko-sakura* on which small *sakura*, or cherry blossoms, were engraved, seem to have been cast as toys or trinkets with which to conjure. The pieces of gold exported to Korea in the latter half of the fifteenth century weighed 42 *monme* and included some round pieces. These were pieces cast at 10 *taels* per piece. The many round pieces of gold transported by the *Gore* people of the Portuguese records were indubitably Japanese gold, and their seals which from the judgment of European people seemed to be those of the king, probably indicate that they were *han-kin* weighing 10 *taels*.

Thus the casting of *han-kin* to be used as a trading commodity is an early phenomenon, and even among the *daimyos* of the *Sengoku* period there were those who produced *han-kin*. The *Tensho Etsuza Ko-ban* (*han*) produced under Uesugi Kenshin weighed more than 4 *monme*, were round with a diameter of 1 *sun* 5 *bu* 5 *rin* – about 5 cm., and were stamped with the *Tensho Etsuza* seal. The *Ko-kin* of the time of Takeda Shingen were pieces weighing 4 *monme* 5 *bu* and labelled with the characters of the Matsugi name. The *han-kin* unearthed in a field of Shimotoyoura village near Azuchi in Omi (Shiga Prefecture) included one elliptical piece with a long axis of 8 *sun* (24 cm.), a short axis of 8 *bu* (2·5 cm.) weighing 4 *monme* 7 *bu*, and one with a long axis of 2 *sun* 4 *bu* (about 7 cm.), a short axis of 1 *sun* (3 cm.), and weighing 8 *monme* 7 *bu*, both pieces stamped with a seal showing three stars. It can be verified from records of the Ming dynasty that the *han-gin* (in the shape of chestnut-tree leaves entered in the records of 1593 and stamped as *Seki shu-gin* (from Iwami, now in Shimane prefecture)) was cast to be used as military funds for the expedition to Korea. The silver called *Hakata Gokuyo-gin* was another form of silver from Iwami. When Tokugawa Ieyasu visited Toyotomi Hidenaga at Nara in 1588, the gold presented by Ieyasu consisted of 100 *mai* all stamped with the Gotô seal. Hideyoshi ordered such large gold coins as the *Hishi Ôban* to be cast in the later years of the Tenshô period, and Gotô Tokujô was authorized to stamp them with the *Kiri* (paulownia) seal, but this was a means of employing the Gotô name belonging to the *kin-za* which had gained public confidence for its production of *han-kin*.

[1] Kondo Jūzō, *Kingin Zuroku* (Illustrated Record of Gold and Silver [Currency]), 1810, and Kusama Naokata, *Sanka Zui* (Illustrated Collection of Three Kinds of Currency), 1815. In these records the early *hankin* and *goku-in gin* are illustrated. For illustrations of the old coins preserved by the Japan Mint in Osaka, see T. Tsukamoto, *Nihon Kahei Shi* (History of Japanese Currency), 1923. and the Finance Ministry's *Dai Nihon Kahei Shi*, 1925.

The *han-kin* which the *daimyos* thus produced, like the three star *han-kin* of Nobunaga, was used as a currency which could be cut into pieces to be used according to their weight, but the real purpose of casting such coins was to provide funds for military campaigns or presents. Naturally such *han-kin* entered the ordinary channels of circulation at the same time. In the tenth month of 1595 Hideyoshi gave 4 *mai* of gold for repair of a bridge in Kyoto. This consisted of 3 *mai* of *Hishi-kin* and 1 *mai* of *Maru-kin*. About this time the gold exchanged in the Kinai area (Osaka, Kyoto) was mostly *han-kin* at 10 *taels* per *mai*; certainly that cast under Hideyoshi or privately by large merchants was of this value. *Han-kin* worth 10 *taels*, however, had too great a value for general use, and though its quality and weight were guaranteed, the areas in which such guarantees could command confidence were geographically limited. Though gold and silver were used in a wide sphere, coinage made from them formed only a nominal part, and *han-kin* was only used as a coinage which had to be weighed to establish its correct value. In order to bring about further progress in expanding the functions of a gold and silver coinage, a ruler with strong unified control would have to use that control to guarantee quality and weight and to cast a large number of gold and silver coins of small denominations.

It is said that Hideyoshi had *O-ban* and *Ko-ban* (kinds of large and small *han-kin*) made in 1588, and the names *Taiko Ichiryo Ko-ban* and *Ni-bu han* (very small coins) were used, but coins such as the *Ko-ban* and smaller are not recorded in authentic documents. The first small coins cast under the Tokugawa's were *Ko-ban* refined in Edo. They were cast by Gotô Shôzaburô Mitsutsugu, probably soon after 1595 when he went East from Kyoto in order to cast coins at the request of Ieyasu and at the order of Hideyoshi. His original family name was Yamazaki, but at the time of his trip to the East he was given the name Gotô by Gotô Tokujô, head of the Gotô house. In the third month of 1596, in a document presented to Tokujô, he promised to write the Gotô signature on the *O-ban*, not to use the *Kiri* seal on *O-ban*, *Ko-ban*, or *Ichi-bu han*, to keep the Gotô name only for himself, and as appreciation of the favour, to present 3 *mai* of gold every year in perpetuity. It is likely that *Ichi-bu han* was cast in 1596, but this was not rectangular, but probably in the shape of the circular *ko-ban*.[1]

At the request of Maeda Toshiie, a rival of the Tokugawa's about the end of the Bunroku period (1592–96), Gotô Yôsuke went to Kaga from Kyoto. There were *gin-za* and *tenbin-za* in Komatsu and Noto before this, but with the

[1] For research on the role of the Gotō House in the minting of *han-kin* through the sixteenth century and beginning of the seventeenth century, the most important document is the *Gotō Jotarō Monjo* (Records of Gotō Jotarō). This document was kept by the Kyoto house of Gotō Kanbeie, descendant of Gotō Tokujō's younger brother, Chōjō, and was given to Kyoto University History Dept. for safe-keeping. See the author's *Nihon no Kahei*, pp. 103–20. The family of Tokujō's descendant, Shirobeie, moved to Edo, where their house was burned in the great fire of 1657. The document kept in the Publications Section of Tokyo University is the *Gotō Monjo* (Records of Gotō) of the Shirobeie branch of the family and does not include many records of the pre-fire period. The Shirobeie branch, as the main branch of the family, minted the *ōban kin* for the Edo *Bakufu* and hence the importance of these two documents preserving the records of the Gotō.

arrival of Yôsuke, both a *gin-za* and *tenbin-za* were established in Kanazawa and gold and silver coins were cast.

The first casting of *O-ban*, *Ko-ban*, and *bu-han*, and *cho-gin* and *mame-ita-gin* in the third month of the sixth year of Keicho (1601) signified the inauguration of the currency system of the Tokugawa's. However, the spread of the system throughout the country in a short time and the achievement of uniformity were not possible. Refined but uncast gold and silver straight from the mines were widely used as weighable currency, and the *han-kin* and *han-gin* cast by each domain were used in an area centring in the domain. At the beginning of the Tokugawa period *cho-gin* was exported as such to foreign countries, but because of impurities it was discounted accordingly in foreign markets and those engaged in foreign and internal trade further refined it in Japan, or exported various kinds of pure refined silver such as the *Seda* (Sado Island?), *Nagites* (Nagato? Yamaguchi), *Somo* (Iwami?) and *Tagemon* (Tajima? Ikuno) silver noted in records of foreign traders. Such silver coins were also used widely within the country. In the domain of the *Bakufu*, the silver taken directly from the Sado mine, which was supplied as the main raw material for the gold and silver currency of the Keicho period (1596–1615), was even used originally in its form as refined but uncast pieces. After 1619 silver stamped with a seal was used in Sado for paying those working at the mines and for other purposes. In the Kanazawa *han*, silver cast and stamped by the Kanazawa *gin-za* and others was packeted as *Shuhô-gin* which was used along with refined but uncast silver from the mines and such silver from other domains as well as *cho-gin*. In 1633 such silver imported from other domains, recast and stamped with a new seal, was used along with *Shuhô-gin*, and in 1639 once again *Shuhô-gin* became the only medium of exchange. But about 1650 the use of stamped silver and imported silver was permitted and in transactions involving less than 1 *monme*, *Kanei zeni* were to be used. After the New Year of 1671 the system was converted to the exclusive use of *cho-gin*. In the Akita *han*, uncast silver and stamped silver were circulated along with Keicho gold and silver, and though not for general use, *zeni* made with gold and silver were also minted. In the spring of 1622 a 1 *ryo ko-ban* was worth 56 *monme* 5 *bu* of refined silver and 61 *monme* 6 *bu* of *cho-gin*. Such refined silver was known as *jo-gin* to distinguish it from *cho-gin*. The *Tenbin-ya* and *Fuki-ya* which undertook the weighing, refining, etc., of the silver were located in Kubôta (now Akita) and other important towns. The sealed silver coins cast in Yokobori in 1628 gained 5 per cent in weight, so that 105 *monme* were prepared from 100 *monme* of refined silver. In October of 1630 the *Tenbin-ya* and *Fuki-ya* of Kubôta were invited to cast silver coins which would gain up to 5 per cent in weight. At the order of Satake Yoshinobu, then in Edo, silver coins similar to *cho-gin* and *mame-ita gin*, and weighing from 20–30 *monme* uncut, were prepared. The conversion of the Akita *han* from the use of silver stamped with their own seals or simply refined silver to the use of *cho-gin* took place after the New Year of 1699.[1]

1 Concerning the appearance of *han-kin* and *han-gin* (*goku-in gin*) in various districts of Japan in the sixteenth century and beginning of the seventeenth century, see *Nihon no Kahei*, pp. 85–102, and concerning the Keicho *kin* and Keicho *gin*, pp. 103–24. For a study of gold and silver currency in the Kaga *han*, see S. Morita, *Kahan Kahei Roku* (Record of the Kaga *han's* Currency), 1901.

Among documents presented to the *Bakufu* in 1668 by Kano Shichirô Uemon, official in the Kyoto *gin-za*, was one entitled, 'Statement of the Relative Values of Silver from Each Domain'. According to this document, the four provinces of Harima (Hyogo), Inaba (Tottori), Tosa (Kôchi) and Mimasaka (Okayama) had previously used uncast silver but had recently changed to the use of *cho-gin*. For provinces which still used uncast silver, such as Aizu (Fukushima), Dewa and Yonezawa (Yamagata and Akita), Fukushima, Akita, Tsugaru (Aomori), Murakami and Takata (Niigata), Niigata, Sado Island, Kaga and Noto, Etchu (Toyama), Hyuga (Miyazaki), Tsushima Island, Tajima (Hyogo), Iwami, Shinano (Nagano), Hida (Gifu), and Bungo (Ôita), the qualities of their uncast silver from the mines are compared with that of *cho-gin*. These forms of silver are divided only into *jo-gin* and *cho-gin*, but the quality taken from each mine was quite uniform. Such unminted silver was usually stamped with a seal, or *goku-in*.[1] During the period between the Kanbun era (1661–) and the reminting of coinage undertaken in the Genroku era (1688–) the use of unminted and stamped forms of silver was gradually changed to use of *cho-gin*. These forms of silver had all been used simply according to their weight, and because refined silver and *han-kin* were originally used in large quantities as weighable currencies, there were *Tenbin-ya* established by special permission in each *han*. In 1653 the *Bakufu* extended the use of the Shuzui scales to the thirty-three eastern provinces, and established the exclusive use of the Shinzen scales in the thirty-three western provinces. This corresponded to the process of unification of the currency throughout Japan by the Tokugawa's.

[1] T. Tsukamoto's *Nihon Kahei Shi* includes many pictures of the *goku-in gin* thought to be that used in various districts at the beginning of the seventeenth century. Pictures of the *goku-in gin* are preserved in the Osaka Mint. Mukōyama Seisai, in the *Otsu-mi Zasshi* (Memoir of 1845) describes the *goku-in gin* from various districts in the Edo Ginza. Probably it was the same *goku-in gin* which the Osaka Mint collected from the Ginza.

4

Exports of Japan's Silver to China via Korea and Changes in the Tokugawa Monetary System During the Seventeenth and Eighteenth Centuries

Tashiro Kazui

Japan, which produced huge amounts of precious metals from the late 16th century, exported silver to China in order to purchase Chinese raw silk and silk fabrics.[1] In the beginning, the Portuguese had a monopoly of that trade through Macau. After the isolation policy of the 1630s enforced by the Tokugawa government, however, the Dutch, who had succeeded in excluding their Western «trade rivals», actively exported silver to China via the Nagasaki-Taiwan route. Despite this, Chinese ships were carrying many times more silver from Nagasaki than the Dutch. However, from the 1670s on, exports of Japanese silver from Nagasaki declined drastically because the *shogunate*, amazed at the huge flow, prohibited its exportation by the Dutch in 1668, and exported copper instead of silver. According to the popular view, exports of Japanese silver were negligible by the end of the 17th century.[2]

In fact, however, an alternative route for the export of Japanese silver remained open for 60 more years, and silver kept flowing to China through it. This route ran from Tsushima, the north-westernmost part of Japan, to China through Korea.[3] At that time, most silver carried to China was discharged into China's southern coastal provinces, such as Fukien or Kwangtung. Instead, the Tsushima-Korea route ran north, through several cities in Japan, Korea and China, and ended at Peking.

---

[1] W. ATWELL, *International Bullion Flows and the Chinese Economy (Circa 1530-1650)*, in: Past and Present 95 (1982); K. YAMAMURA and T. KAMIKI, *Silver Mines and Sung Coins. A Monetary History of Medieval and Modern Japan in International Perspective*, in: J.F. RICHARDS (ed.), Precious Metals in the Later Medieval and Early Modern Worlds, 1983.

[2] S. IWAO, *Kinsei Nisshi Boeki ni kansuru Sūryōteki Kōsatsu* [Sino-Japanese Trade in the Edo period from a Quantitative Perspective], in: Shigaku Zasshi 62-11 (1953), p. 28; YAMAWAKI TEIJIRŌ, *Nagasaki no Tōjin Bōeki* [Sino-Japanese Trade in Nagasaki], Tokyo, 1964.

[3] Added to this, there was the route from Satsuma, the southernmost province of Japan, to China via the Ryukyu Kingdom. This trade is not discussed in this paper, but the Tsushima-Korean route, almost totally neglected by historians, was much more important quantitatively than the Satsuma-Ryūkyū route.

Apart from trade at Nagasaki, the Tsushima-Korea trade route continued to carry exports of silver to China from the mid-17th century on, and in return flowed imports of Chinese raw silk, silk fabrics and Korean ginseng to Japan until the mid-18th century. Japanese traders were able to continue exports of silver to Korea in those days despite the shogunate controls because the trade system was completely different from that at Nagasaki. Since all silver exported from Tsushima was silver currency issued by the shogunate, however, the trade was seriously affected by the changing monetary system in Japan after the late 17th century.

In this paper, I will focus on the flow of silver in the Japanese-Korean trade, and will show quantitatively how this flow affected commerce between Korea and China and the monetary system in Japan.

1. Japanese-Korean Trade in the Tokugawa Period

In the pre-modern era, the island of Tsushima, in the Korea Strait (see Figure 1), was like a «bridge» connecting Japan with the Eurasian Continent. Taking advantage of its geographical location, the islanders made their living through trade with Korea. The So family, territorial lords of Tsushima, obtained political control over the island during the 14th century, and by the early 15th century they monopolized diplomatic and economic relations between Japan and Korea. This arrangement was interrupted by Japan's abortive invasion of Korea in the 1590s, but it resumed in the 17th century when the So restored relations with Korea for the Tokugawa shogun in 1607, and itself exchanged a diplomatic agreement with Korea in 1609. Thus, the So family's position as an intermediary was recognized by both the Tokugawa shogu nate and the Yi dynasty of Korea.[4]

The beginning of the Tokugawa isolation policy in the 1630s struck a serious blow at many Japanese engaged in overseas trade. The role of the So family, however, increased in importance. Basically, the Tokugawa policy of isolation had two roles to play with regard to the external world: the first, vis-a-vis the West, to expel Catholicism, and the second, vis-a-vis East Asia, to make clear Japan's independence from the Chinese world order - the Sinocentric community of pre-modern East Asia. As far as the latter was concerned it permitted only private trade with the Chinese in Nagasaki, and rejected

4 T. Tanaka, *Sakoku Seiritsuki Nitchō Kankei no Seikaku* [The Nature of Japan-Korea Relations in the Early Edo Period], in: Chōsen Gakuhō 34 (1965); H. Nakamura, *Nissen Kankeishi no Kenkyū* [A Study of the History of Japan-Korea Relations], 3 vols., Tokyo, 1969.

any diplomatic relations with the Chinese Government. Japan thus avoided joining the Chinese world order, while maintaining diplomatic relations with Korea as an equal partner and with the Ryūkyū Kingdom as a Japanese tributary.[5] The So family kept its unique position in East Asia throughout the Tokugawa era, and the shogunate regarded the Tsushima-Korea trade as important because it was one of the few trade routes for supplying Chinese goods to the Japanese domestic market.[6]

In contrast to the Japanese traders in Nagasaki who did not leave Japan, Tsushima islanders did travel for trade at the Japanese concession in Pusan, the southern tip of the Korean peninsula, in boats under the So's control. These were voyages specifically for overseas trading only during the Tokugawa era. Shipping from Tsushima was controlled by passports which the So issued, and was limited to 20 boats a year by the diplomatic agreement of 1609. However, over the years the Tsushima domain, with the acquiescence of the Yi government, managed to increase the number of boats beyond the quota so as to add to its profits. The domain developed a number of techniques to avoid the quota. For example, a number of small boats accompanied the officially approved mother ships on their two yearly trips; other ships and boats, on the pretext of carrying a «special mission» or a «letter to the concession,» made several trips a year. According to So documents, in 1709 90 boats came and went between Tsushima and Korea among which 30 or so of the smaller ships were carrying silver to the Japanese concession at Pusan.[7]

Tsushima-Korea trade took two forms: official trade, in which the Tsushima domain and the Korean government exchanged gifts and commodities of a kind and in amounts stipulated by the diplomatic provisions; and private trade, in which the Tsushima domain and merchants traded with the Korean officers and merchants at markets held 6 times monthly. Unlike the former, the latter was not subject to any restrictions, especially in exports of silver and imports of Chinese goods. This trade brought huge profits for the Tsushima domain.[8]

Although there is very little material concerning the private trade with Korea, an account book of the trade, recorded by Tsushima officials, does exist among the So family documents[9] and gives a record of the period from

[5] K. TASHIRO, *Foreign Relations During the Edo Period: Sakoku Reexamined*, in: Journal of Japanese Studies 8-2 (1982); R. TOBY, *State and Diplomacy in Early Modern Japan*, Princeton, 1984.

[6] K. TASHIRO, *Kinsei Nitchō Tsūkō Bōekishi no Kenkyū* [Studies on the History of the Diplomatic Relations and Trade between Japan and Korea], Tokyo, 1982, pp. 433-447.

[7] Ibid., pp. 202-218.

[8] K. TASHIRO, *Tsushima Han's Korean Trade (1684-1710)*, in: Acta Asiatica 30 (1976), pp. 85-86.

[9] There are 5 major collections of the So family documents as follows; National Diet Library (Tokyo), Historiographical Institute University of Tokyo (Tokyo), Keio Uni-

1684 to 1710. According to this, as is shown in Figure 2, silver comprised half of the annual total exports to Korea. In exchange for this silver, Tsushima islanders imported much Chinese white raw silk. As is shown in Table I, at the peak, in 1686, over 10 t. of silver currency was exported. The total amount exported from 1684 to 1710 amounted to about 188 t., an average of about 7 t. a year. During this same period, due to shogunate restrictions, the major commodity for export in Nagasaki was not silver but bar copper. Thus in 1686, for example, Chinese ships which were still permitted silver exports carried only 2 t. of silver from Nagasaki, one-fifth of the Tsushima-Korea trade. In 1708, moreover, the shogunate decided to fix an annual limit to silver exports of 0.6 t. in Nagasaki.[10] In other words, shogunate restrictions led to a economic reversal in the relative importance of the Nagasaki and Tsushima routes to the Japan-China silver trade.

2. The Changing Monetary System and Silver Exports

Tsushima's exports of silver to Korea created serious problems for the shogunate, both quantitatively and qualitatively, because all the silver was currency issued by the shogunate. The most important problem was caused by the recoinage in 1695 on, which had a great impact on the monetary system of the time.

Basically, Tokugawa Japan had a bimetallic standard system of gold and silver. In 1601, the central government, later to become the Tokugawa shogunate, issued gold coins (84% fine Keichō gold). These were circulated by *tael* (piece-by-piece) in East Japan. At the same time, silver currency (80% fine Keichō silver) was minted by the government and used in terms of «money by weight» in West Japan.[11] At first, the silver for trade was exported in the form of pure silver ingots. After the establishment of the monetary system, however, traders began to cast an eye on the high quality of silver currency. Adding to this, since the shogunate prohibited exports of pure silver ingots in 1609, a huge amount of silver currency flowed out of Japan.

About 100 years later, the then Finance Minister of the shogunate, Ogiwara Shigehide, reduced the quality of currency and increased the number in circulation. There were several reasons why he did this: a lack of

---

versity Library (Tokyo), Rekishi Minzoku Shiryōkan (Tsushima) and National History Compulation Committee of Korea (Seoul).

[10] YAMAWAKI, *op. cit.*, pp. 213-214.

[11] E. HONJŌ, *Economic Theory and History of Japan in the Tokugawa Period*, Tokyo, 1943, pp. 50-54; H. TAYA, *Kinsei Ginza no Kenkyū* [A Study of the Silver Mint in the Tokugawa Period], Tokyo, 1963, pp. 145-196.

currency resulting from the amount exported, an increase in the demand for the money which accompanied economic development, and a decline in mining productivity. Eventually, as Table 2 shows, his recoinages led to a drastic reduction in the quality of silver currency, which produced high inflation.[12]

The succession of debasements meant continual changes in the type of coin which the Tsushima domain was exporting. In 1697, Genroku silver (64% fine) in place of Keichō silver was minted by the shogunate, but the domain kept exporting Keichō silver for two more years for fear of introducing confusion into the private market. Meanwhile, Tsushima officials negotiated with Korea for the acceptance of Genroku silver since Keichō silver was in poor supply - as the saying goes, «bad money drives out good money». It took two years to make the Koreans accept Genroku silver because their poor melting techniques for precious metals made it difficult to fix an exchange rate between the old and the new.[13]

As already noted, Genroku silver had a purity rate of 64%. The exchange rate with Keichō silver should therefore have been at a premium of 25% (80/64). In the lack of any official announcement about the purity of Genroku silver, the Tsushima domain, which had been running some silver mines on the island for centuries, determined the exact purity by itself and imparted this information to Korea. However, because the Korean officials could not produce perfectly pure silver with their melting techniques, their first evaluation was that the silver content of Genroku silver was only between 60% fine and 62%, and thus should be at a 33% (80/60) to 29% (80/62) premium. After several negotiations and purity testings, it was finally decided that Genroku silver was «63% fine» and at a «premium of 27%» (80/63) compared with Keichō silver. Without doubt, this was a disadvantageous contract for the Tsushima domain. However, since long drawn out negotiations only stagnated the private market, the domain was obliged to make a contract with Korea as early as possible.[14] The Tsushima islanders began exporting Genroku silver to Korea, in 1699, but despite the high premium it had a bad reputation in the market. Moreover, as is shown in Table 2, after 1706, the Finance Minister again debased the silver currency, and finally, in 1711, issued Yotsuhō silver (20% fine), the worst silver currency in Tokugawa history. The Korean government and merchants were no longer interested in low value silver, such as the Yotsuhō silver. Inevitably, the Tsushima domain had to continue exporting Genroku silver, which

---

[12] TAYA, *op. cit.* pp. 168-169.

[13] SANG-UN CHUN, *Hanguk Kuahak Kishulea* [Science and Technical History in Korea], Seoul, 1983, pp. 270-273; K. TASHIRO, *Kinsei Tsushima-han ni okeru Nissen Bōeki no Ichi Kōsatsu* [A study for Japanese-Korean Trade by the Tsushima Domain], in: Nihon Rekishi 268 (1970), p. 94, p. 112.

[14] K. TASHIRO, *op. cit.*, pp. 94-95.

gradually became scarce as had Keichō silver. Thus, as Table 1 shows, total exports of silver in 1710 had dropped to about 2.3 t. of Genroku silver.

After negotiations that lasted for five years, in 1710 the Finance Minister finally permitted the minting of 5.3 t. per year of special quality silver (80% fine) and the exporting of it. Moreover, the Minister not only permitted the exchange of bad currency for the special quality silver with *no premium*, but also did not charge Tsushima for the expenses of minting it. The shogunate took these costs upon itself because it expected that the special silver would promote imports of ginseng, treasured as a medicine, from Korea. In fact, although Tokugawa Japan was experiencing a great boom in the demand for ginseng,[15] during the devaluation period it was in short supply due to the decline of silver exports. The special silver was even called the *ninjindai-ōkogin* (special silver to buy ginseng).

Although exports of the special silver did not begin until 1712 because of the time taken for diplomatic negotiations with Korea, its quality enabled the trade to survive the crisis caused by the successive debasements. Fortunately, in 1714 the shogunate restored the quality of money. Silver currency (Kyōho silver) remained 80% fine for 2 decades thereafter and trade gradually returned to the right lines. No account book of private trade survives for this era, but I can show clearly the quantity of silver carried to the Japanese concession in Pusan by the diary kept by head officials of the concession for over 184 years. The part of Figure 3 based on the period from 1711 to 1751 shows that the quantity of exports increased the year immediately following release of the special silver. Thereafter amounts again declined for a few years, only to rebound soon after. It also appears that the figures were far bigger than those for exports from Nagasaki, which totalled only 1.4. t. in the period 1714–1738.

During the period 1714–1738 when Kyōho silver was being issued, the amount of currency was reduced and there was a large reduction in the amount of silver currency in circulation. This caused a serious monetary shortage in West Japan, where silver currency was used and which was more developed economically, than East Japan, where gold coins were used. It is true that the exceptionally good quality of Kyōho silver helped cause the money shortage. In order to solve the problem, therefore, in 1730 the shogunate permitted the issue of paper money in several domains,[16] mainly in West Japan, and in 1738, the government again recoined to increase the amount of currency through debasement.

---

[15] T. Imanura, *Ninjin shi* [The History of Ginseng], 7 vols., Seoul, 1934–1940.

[16] The issue of paper money in domains was prohibited by the shogunate in 1707. The repeal in 1730 was because of the shortage of silver cash. See, A. Kohata, *Nihon no Kahei* [Money in Japan], Tokyo, 1958, p. 179; and Taya, *op. cit.*, p. 286.

However, I would insist that the high grade of silver currency was not the only reason for the money shortage. There was another reason as well; an increase in the exporting of silver cash. Table 3 shows the quantity of Kyōho silver carried to Korea by Tsushima while it has being issued by the shogunate. As is evident from this Table, the Tsushima domain exported, at the minimum, 86.4 t. of Kyōho silver, which is equivalent to 6.9% of all the silver minted during that time. Moreover, if one takes into account the years for which records have not survived, the figure would probably reach 8%; and if the amount of silver exported along the Satsuma-Ryūkyū-China routs is added (which is not discussed in this paper) it would reach about 10%. Given these facts, as discussed above, there seems little question that the export of great quantities of silver had a huge impact on the circulation system in periods of money shortage.

Setting aside the question of why exports of silver rapidly declined after the mid-1730s, let us turn to the movement of silver from Japan to Korea, and from Korea to China.

3. The Seasonal Cycle on the «Silver Road»

The Tsushima domain handled its transport of silver by the boats very carefully. These would start from Kyoto - the main marketing center for Chinese raw silk in Japan - where the Tsushima domain sold the silk it had bought in Korea and collected the silver it would export in return. The journey from Kyoto to Tsushima lasted about 7 days. At Tsushima the silver was unloaded and shortly after, carried to the Japanese concession in Pusan by other boats suitably reserved for the carrying of silver.

These special boats were sent to the Japanese concession in groups of 2-3, or 5-6, and the silver was passed on to the Korean merchants in private trade. There was a delay of several months after delivery of the silver before Tsushima received any commodities in return. As already noted, the diaries which the heads of the Japanese concession at Pusan kept over a long period of time provides us with valuable information on the operation of the boats. I would like to use these diaries, and to present some interesting data about the amounts of silver transported each month, according to the solar-lunar calendar from 1711 to 1750. Table 4 clearly shows that the amount fluctuated each month, and that there was a certain seasonal rhythm. For example, in the period 1711-1720, the amount of silver transported in the 8th, 10th and 11th months was more than that in other months, and over the whole period covered by the Table, the total ratio of silver transported annualy went up to 15.8% in the 8th month, 21.3% in the 10th and 13.0% in the 11th. Taken together with the 7th month (9.5%), the amount would equal about 60% of the totals.

This seasonal cycle was a result not of natural conditions, but was a phenomenon specific to silver transport. This becomes clearer if one looks at the following instructions given by the Tsushima domain to their officers in Kyoto in 1681[17]: first, send 2.6 t. of silver to Tsushima in the 6th month for the proffering ceremony of the Chinese calendar, second, send 5.6 t. of silver to Tsushima in the 8th month for the Winter Solstice ceremony. The former had to be carried to Korea during the 7th and 8th months, the latter during the 10th and 11th months. In other words, the seasonal cycle of silver transporting began in the supplying stage. Why was the domain so concerned to carry silver during these specific seasons?

«The proffering ceremony of the Chinese calendar» and «the Winter Solstice ceremony» which appear above were events which deeply affected diplomatic and economic relationships between Korea and China. Every year the Yi dynasty dispatched missions to the Ch'ing court in fulfillment of its tributary obligations under the Chinese world order. After the mid-17th century, annual missions went regularly to Peking twice a year, to attend both the Winter Solstice and the New Year's celebrations, and to receive the Chinese imperial calendar. Adding irregular missions, it is estimated that the Korean government dispatched missions on an average of 3-4 times a year. Although the official membership of each mission basically comprised less than 35 people, the total number expanded to about 300 people during the Winter Solstice mission. This was the largest one, because the retinue, stable keepers, and odd-jobbers joined in the party. Most of them were merchants, engaged in Sino-Korea trade, not only officially but privately as well. Although it took the form of a diplomatic mission, considering the constituent members and their economic activities, it only appeared to be a «caravan». This «caravan» brought goods which the Tsushima domain wanted from China and returned carrying mostly Japanese silver.[18]

While Japan began to produce a huge amount of precious metals after the late 16th century, thanks to a mining boom, Korea produced very little; rather she chose to close down her mines and to exclude precious metals from the list of the tributary goods sent to the Chinese emperor. However, her location sandwiched between China and Japan ensured that she was never indifferent to silver: on the one side, she had the world's largest importer of silver, on the other side, the greatest producer of silver in Asia. In addition, because Korea had no valuable goods other than ginseng, it was as a result of transit trade that silver imported from Tsushima flowed to China.

[17] *Nagasaki Kenshi* [The History of Nagasaki], II, Tokyo, 1964, pp. 615-616.

[18] M. Ejima, *Tensō Nenkan ni okeru Chōsen no Saihei ni tuite* [Sino-Korea Annual Tributary System During 1627-1643], in: Shien 101 (1969), pp. 39-62.

Indeed, a considerable amount of silver was re-exported from Korea to China in several ways: for example, through private trade by both Korean messengers and merchants, and through smuggling on the Sino-Korea border area, mostly at the time of the entrance and exit of the Korean missions. At first, official members of embassies were permitted to carry ginseng for their private trade, thereafter they were allowed to carry about 2,000–3,000 *taels* (75–113 kg.) of silver or ginseng per person. Therefore, in the case of the Winter Solstice mission, which had 35 official members, it was possible for them to carry a total of 74,000 *taels* (about 2.8 t.) of silver in all. Also, when private trade in silver as well as ginseng became allowed, most of the official members entrusted their privileges of carrying out silver to the trading merchants, and a fixed amount of silver export was no longer observed. It is said that one Winter Solstice mission carried a total of 80,000 *taels* (3 t.) or more of silver.[19] Moreover, silver was also exported by irregular missions and by smugglers, though it is impossible to make an estimate of the amount.

The movements of the regular missions were fixed: the mission for the acceptance of the Chinese calendar started from Seoul in the 8th month and returned from Peking in the 12th. The missions for the Winter Solstice and New Year's ceremony (a much larger group) left in the 11th month and did not come back until the 4th month of the next year. These undoubtedly had an effect on the seasonal cycle of silver transport from Kyoto, because silver transactions in the Japanese concession became more and more active as the start of a mission approached. To summarize, Japan's silver flowed to China via Korea on the «silver road» - Kyoto-Tsushima-Pusan-Seoul-Peking - where there was a twice yearly surge in conjunction with the movements of the Korean missions. We can call the 2 main as the «Summer flow» and the «Fall and Winter flow» respectably.

1. Summer flow:
   Kyoto → Tsushima in the 6th month; Tsushima → Korea between the 7th and 8th months; Korea → China after the 8th month.
2. Fall and Winter flow:
   Kyoto → Tsushima in the 8th month; Tsushima → Korea between the 10th and 11th months; Korea → China after the 11th month.

[19] SEUNG-JU LIU, *Choson Hugi tae Chong Muyok ui Chongae Kuajong* [Development of Trading with Ch'ing in the Latter Yi Dynasty Period], in: Paeksan Hakpo 8 (1970), pp. 327–398; TS'UN-WU CHANG, *Ch'ing Han Tsung-fan mao-yi* [Sino-Korean Tributary Trade, 1637–1894], Taiwan, 1978.

4. National Economy and Silver Export

Lastly, I would like to discuss the reasons for the rapid decline in the high level of silver exports for Tsushima after the mid 1730s.

The trade between Tsushima-Korea had 2 particular characteristics. The first was that the economy of the Tsushima domain depended on the export of silver because it lacked an agricultural base. Since the amount of trade was determined by the volume of silver exports, a decline in the latter meant at immediate fall in domain revenue. Moreover the Tsushima-Korea trade financed the diplomatic activities between Korea and the Tokugawa government for which the domain acted as an intermediator. This meant that any fall in trade also carried with it the possibility of a breakdown in the diplomatic activities between the two countries.

The second concerned the management of the Tsushima-Korea trade itself. As noted above, despite Japan's restrictive trade policy overall, trade took place on the Korean peninsula. The shogunate could not actually watch the activities of the Tsushima domain at the market, and could not monitor the exact amount of silver exported. In other words, the amount exported depended upon market business conditions. Even if the shogunate tried to limit the trade, unlike in Nagasaki it was not able to tell whether the domain had obeyed the law. It was because of these two characteristics that the shogunate allowed such exports of silver to continue for so long.

However, a Confucian scholar, Hakuseki Arai, who acted as adviser to the 7th Shogun eventually reversed the former policy toward silver exports in favor of hitherto unconsidered «national interest». The adviser first made estimates of the production, stock and exports of precious metals in Tokugawa Japan, and in 1714 restored the quality of the currency. Taking a broad view beyond the framework of the Tokugawa political and social system, he argued that Japan should reduce exports of valuable treasure, especially silver and copper, and should instead promote domestic industries and raise productivity. Although the currency was devaluated again later, the shogunate basically supported his ideas.

As a consequence of this, the shogunate decided that at some point they had to bring an end to Tsushima's high exports of silver. Eventually, in 1737, the shogunate had an opportunity to do so. The very year before, the 8th Shogun, Yoshimune Tokugawa, had debased the gold and silver currencies, reducing the purity of silver currency from 80% to 46%. According to precedent, the shogunate permitted Tsushima to mine and export 5.3 t. of special quality silver. But at the same time, it laid down the condition that Tsushima had to pay a premium of 99% in order to exchange currency for the special silver, except in the case of silver to be traded for ginseng imports which was at a premium of 50%. It is unlikely that in practice the special silver was ever used for the importing of ginseng because the Tsushima domain could never afford a premium of 99%. The shogunate thus took the power necessary to control the silver exports of the domain at the supply end, and was able to monitor the actual amount of ginseng imported because the amount of special silver permitted at a 50% premium was based on the amount imported in the preceding year. The shogunate moves had a definite effect and exports of silver currency from Tsushima began to fall as the result of a decline in Korea-China trade.[20] This is the reason for the rapid decline after the mid-1730s, as is shown in Figure 3.

Meanwhile, in order to continue diplomatic relations with Korea, the shogunate had to protect Tsushima from the consequences of its economic failure. It therefore gave financial help, in the form of grants or interest free loans. It is estimated this financial help reached a total of 1.5 million *ryō* of gold cash (corresponded to 8 million bushels of rice) from 1734 to 1841. This money was presumably in compensation for the closing down of the «silver road».[21]

---

20 K. Tashiro, *Kinsei Nitchō*, pp. 338-339.

21 *Ibid.*, pp. 394-395.

Table 1
Silver Export to Korea (1684-1710)

year	silver cash[a] (kg.)	pure silver (kg.)	per cent of total exports
1684	7,267	5,813	66%
85	7,527	6,021	64
86	10,827	8,662	69
87	7,665	6,132	59
88	9,327	7,461	63
89	7,480	5,984	50
90	8,366	6,693	47
91	10,239	8,191	47
92	9,139	7,311	52
93	8,528	6,822	41
94	9,671	7,737	46
95	9,185	7,348	47
96	9,149	7,319	49
97	9,018	6,230	40
98	5,250	3,360	53
99	7,425	4,752	68
1700	5,868	3,756	98
01	10,237	6,552	64
02	6,776	4,336	64
03	2,737	1,752	59
04	5,062	3,240	70
05	4,040	2,586	44
06	4,875	3,120	67
07	3,644	2,332	63
08	3,675	2,352	62
09	3,525	2,256	58
10	2,325	1,488	48
total	188,827	139,606	-
annual average	6,993	5,170	55

[a] the period 1684-1697: exports of Keicho silver (80%)
the period 1697-1710: exports of Genroku silver (64%)

Source: An account book of the private trade, in the So family documents, National Diet Library.

Table 2
The Kind of Silver Cash in Japan and Silver Exports to Korea, during the 17th and the mid-18th centuries

silver cash in Japan[a]			silver exports to Korea		
name	purity (%)	period of circulation	name	purity (%)	period of exports
Keicho silver	80	1601-1738	Keicho silver	80	1611-1697
Genroku silver	64	1695-1722	Genroku silver	64	1697[b]-1711
Hoei silver	50	1706-1722	Special silver (Ninjindai-okogin)	80	1712[c]-1715
Eiji silver	40	1710-1722			
Mitsuho silver	32	1710-1722			
Yotsuho silver	20	1711-1722			
Kyoho silver	80	1714-1738	Kyoho silver	80	1716-1738
Genbun silver	46	1736-1827	Special silver (Ninjindai-okogin)	80	1739-1750[d]

a Source : H. TAYA, Kinsei Ginza no Kenkyu [A Study of the Silver Mint in the Tokugawa Period], Tokyo, 1963.
b Genroku silver was recognized to export official in 1699.
c,d Special silver was minted from 1710 to 1715, and from 1737 to 1754.

Table 3
The Amount of Silver Currency in Japan and Silver Export to Korea

	α amount of issue (index)	β amount of export (index)	α/β x 100
Genroku silver	1,521 ton (100)	71.5 ton (100)	4.7%
Kyoho silver	1,242 ton (81)	minimum 86.4 ton (120)	6.9%

Table 4
The Amount of Monthly Export of Silver Korea:
The Seasonal Cycle of Silver Transported

month period	1st	2nd	3rd	4th	5th	6th	7th	8th	9th	10th	11th	12th	total (ton)
1711-20 T.	1.7	4.3	2.0	2.9	3.8	1.3	2.7	6.5	2.2	5.0	5.8	1.9	40.1
%	4.2	10.7	4.9	7.2	9.4	3.2	6.7	16.2	5.4	12.4	14.4	4.7	
1721-30[a] T.	1.4	0.8	0.2	0.3	0.2	0.2	6.0	5.4	4.7	15.2	0.8	0.4	35.6
%	3.9	2.2	0.5	0.8	0.5	0.5	16.8	15.1	13.2	42.6	2.2	1.1	
1731-40 T.	2.6	2.9	2.1	1.3	0.6	0.4	1.7	5.8	0.6	4.4	6.0	2.6	31.0
%	8.3	9.3	6.7	4.1	1.9	1.2	5.4	18.7	1.9	14.1	19.3	8.3	
1741-50 T.	0.5	1.4	0.5	0.6	0.6	0.0	1.0	1.1	1.2	1.1	3.0	0.9	11.9
%	4.2	11.7	4.3	5.0	5.0	0.0	8.4	9.2	10.0	9.2	25.2	7.5	
total T.	6.2	9.4	4.8	5.1	5.2	1.9	11.4	18.8	8.7	25.7	15.6	5.8	118.6
%	5.2	7.9	4.0	4.3	4.3	1.6	9.6	15.8	7.3	21.6	13.1	4.8	

[a] the period 1721-1730 is excepted the amount of the period 1724-1726 because the documents are not in order.
T. = ton.

Source : Mainikki (diary of the Japanese concession in Pusan), in the So family documents, National Diet Library.

FIGURE 1

Peking
KOREA
Seoul
JAPAN
Tsushima
Edo
Pusan
Kyōto
CHINA
Nagasaki
Satsuma
Fukien
Ryūkyū

FIGURE 2 Tsushima-Korea Trade by Commodities 1684-1688

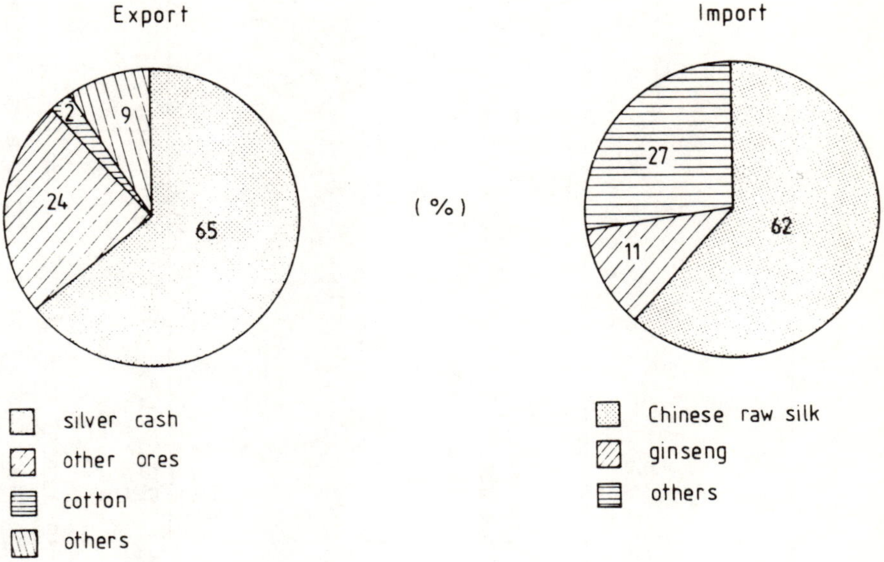

Source: An account book of the private trade, in the So family documents, National Diet Library.

FIGURE 3 : Annual Silver Export from Tsushima to Korea (1684 - 1751).

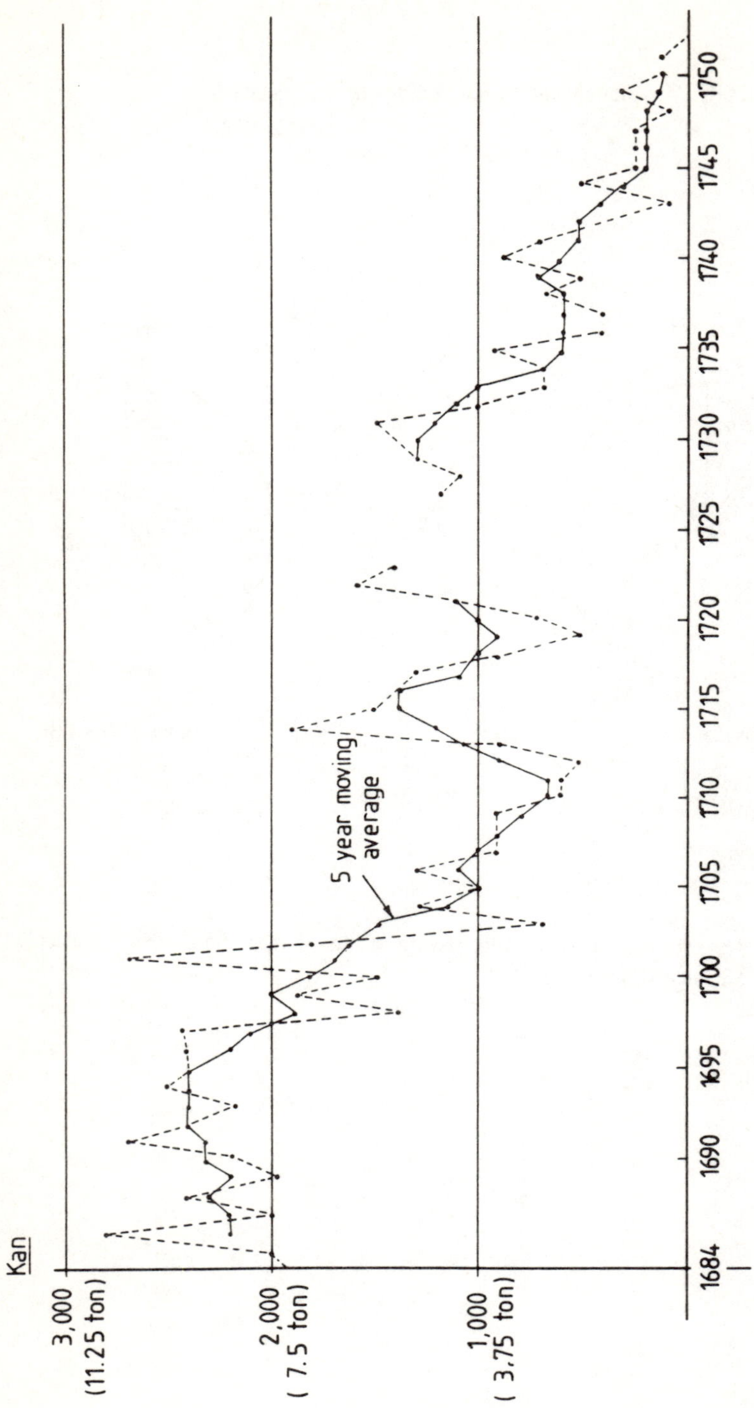

Source: the period 1684-1710 ; An account book of the private trade, the period 1711-1751 ; Mainikki (diary of the Japanese concession in Pusan), in the So family documents, National Diet Library.

5

The Dutch East India Company's Trade in Japanese Copper, 1645–1736

Kristof Glamann

I

Towards the end of the 1930's, Japanese copper appeared, like some strange bird in the northern skies, as an object of study in Scandinavian economic history. The occasion was the publication of the second volume of Professor Eli F. Heckscher's comprehensive work on Sweden's economic history,[1] in which the author amongst other things described the copper policy in the heyday of Swedish power. Making use of the abundant Swedish source material which, with few exceptions, contains the annual figures of Sweden's copper production since 1623, Professor Heckscher made a quantitative analysis of production and sale and discussed the question of currency policy, a subject which was, moreover, simultaneously discussed in a dissertation by his fellow-countryman, J. Wolontis.[2] Other aspects were also emphasized in Professor Heckscher's account which formed part of his description of Sweden's general economic history. As regards the extent of copper production the quantitative analysis led to the result that in the seventeenth century Sweden actually held a monopoly of copper production. As early as the year 1692, the Swedish metallurgist Erik Odhelius had come to the same conclusion after a four years' study tour to Germany, France, England, the Netherlands and Belgium. This was taken as a point of departure in the determination of the Swedish currency policy and the author concentrated especially on Gustavus Adolphus's decree of 1624 instituting a Swedish monetary standard of copper, or rather a double monetary standard of silver and copper. This event which is so interesting a trait in the history of currency had also, according to Heckscher, reference to the European copper market, the idea being

1 Eli F. Heckscher, *Sveriges ekonomiska historia från Gustav Vasa,* vol I, part 2, (Stockholm, 1936).

2 Josef Wolontis, *Kopparmyntningen i Sverige 1624—1714* (Helsingfors, 1936).

to create a development of prices favourable to the producer, Sweden, by restricting the export of Swedish copper without reducing the production. Gustavus Adolphus's policy failed. Heckscher attributed this to interior weaknesses in the scheme, but the late Professor Albert Olsen, of Copenhagen, maintained that the failure was rather to be ascribed to a Swedish underestimation of the competition in the European copper market, where Sweden's position as a monopolist could by no means be taken for granted.

Centering on the decree of 1624 a comprehensive discussion of the seventeenth century European copper market as a whole gradually developed.[3] Amongst the factors outside Sweden which might have influenced the status of Swedish copper Albert Olsen attributed a decisive importance to Japanese copper, imported by the Dutch East India Company. From information contained in Pieter van Dam's description of the Dutch company[4] Professor Olsen showed that the capacity of the Japanese copper industry was as great as that of the Swedish, and for certain decades exceeded it; through the medium of the Dutch East India Company there existed a connection and interaction between the European and the Asiatic markets, a European boom producing a Dutch import which in its turn influenced the prices in the market in Amsterdam. This occurred, for example, in the 1620's when Swedish sources mention that the immediate cause of the slump was the import of Japanese copper. This establishment of the connection between the two copper markets was further elaborated by another Danish historian, Professor Astrid Friis,[5] who pointed out that copper was amongst the goods with which, in the sixteenth century, the Portuguese in Antwerp provided themselves for the East India trade. The same author also connected the enormous rise in prices of copper in Amsterdam after 1606 with the extension of the Dutch East India trade and the colonization in the East, in contrast to Professor Heckscher, who connected the rise in price mainly with the Spanish coinage policy (vellon). In the first decades of the seventeenth century the Dutch were supposed to have used European copper. This was, however, soon abandoned, it being superseded by Japanese copper, which was cheaper and not inferior in quality. This had its counterpart in the eighteenth

[3] Albert Olsen in *Scandia* X : 1—2 (1937), pp. 38—75, pp. 297—305, and XI : 1—2 (1938), pp. 157—8, pp. 307—8. Eli F. Heckscher in *Svensk Historisk Tidskrift* 1937, pp. 287—314, 1938, pp. 130—6, and *Scandia* XI : 2 (1938), pp. 214—79.

[4] Pieter van Dam, *Beschryvinge van de Oostindische Compagnie, II : 1*, ed. F. W. Stapel ('s-Gravenhage, 1931), *R. G. P.* No. 74.

[5] *Scandia* XII : 1 (1939), pp. 151—80.

century. Then it was European copper that superseded Japanese copper. It was now the English East India Company which found exportation of European copper to India to be a profitable transaction.

II

With the object of estimating Swedish copper's status in relation to Japanese copper in the seventeenth century European market Heckscher as well as Olsen endeavoured to make a quantitative analysis of the imports of the Dutch company. Both naturally utilized the material at hand, i. e. chiefly the description by Pieter van Dam and Oskar Nachod's work on the connection between Japan and Europe in the seventeenth century,[6] but neither of these works gave information on the supplies direct to Europe. Shortly before his death Albert Olsen found sporadic evidence of the company's import in the 1690's in the reports from Nicolas Abo, the Danish consul in Amsterdam. But although this material was valuable, it was too defective to be made the basis of a renewed discussion.[7]

The following investigation will show that, for a considerable part of the period in question, it is possible to determine the quantities of Japanese copper imported into Europe by the Dutch company. Although it has thus been possible to undertake the analysis that Heckscher and Olsen desired, it must, however, be pointed out that a revival of the old discussion is not intended, it being scarcely possible to treat the subject exhaustively simply by means of a quantitative juxtaposition of the two kinds of copper. The primary object of this investigation has been to see Japanese copper in its true position as one amongst the many commodities in which the company dealt, and the greatest importance is attached to determining the place held by copper in the trade of the company. The possibility that the Dutch company's importation of copper into Europe was also determined by factors other than Swedish or European copper should be held open. The following investigation therefore is fully as much a contribution to the commercial history of the Dutch East India Company illustrated by its trade in Japanese copper as an account of the relation between copper originating from various countries.

[6] Oskar Nachod, *Die Beziehungen der Niederländischen Ostindischen Kompagnie zu Japan im siebzehnten Jahrhundert* (Leipzig, 1897).

[7] *Dansk Historisk Tidsskrift* 11 : II (1949), pp. 533—40.

III

W. H. Moreland, an authority on the economic history of India, has shown how the Dutch East India Company, in its first years in the East, had already obtained a firm footing within the Asiatic trade. "Beginning by the direct purchase of spices for Europe, the Dutch were led gradually to engage in nearly every important line of commerce throughout Asiatic waters." India especially quickly came into "their scheme of operations. The necessity arose from the fact that pepper and other spices which they sought were ordinarily obtained from the producers by bartering Indian-made cotton cloth".[8] It was especially the cloths from the Coromandel coast which met the demand of the Indonesian market. This coast therefore became of vital importance to the Dutch trade, so much so that it was named "Den slincken arm" of the company (the company's left arm).[9] Cash was needed for carrying on the Indian trade. To the regret of the adherents of mercantilism this was sent out from Europe. The Asiatic trade, however, was also capable of bringing in cash, if only carried on in the proper way. In the first part of the seventeenth century it was possible through commercial intercourse with Persia, and especially with China and Japan, to get precious metals by barter. This is at the root of the activity which sprang up in these extreme regions of the Asiatic trade and which led to the Dutch superseding the Portuguese in the dominion of the Eastern seas. They became "de beurtschippers van het Oosten", as they were already in the West.[10] This fact makes it necessary to investigate the Dutch company's trade in its entirety. This applies to the goods as well as to the factories. There can be no division between East and West. The actual importance of each individual article of trade can only be decided by following it on its route through this vast commercial machine.

As early as 1608 the Dutch received permission from the emperor to carry on trade from Firando in Japan, where the first factory was established. The English also established a factory, but for various reasons they had to leave Japan in 1623. The Dutch, however, succeeded in holding their ground. It is true that Firando had to be abandoned, but in Nagasaki they found a stronghold for the next two

8 W. H. Moreland, *From Akbar to Aurangzeb, A Study in Indian Economic History* (London, 1923), pp. 66 and 30.

9 H. Terpstra, *De Nederlanders in Voor-Indië* (Amsterdam, 1947), p. 40, the same author, *De vestiging van de Nederlanders aan de kust van Koromandel* (Groningen, 1911), *passim.*

10 H. Terpstra, *De Nederlanders in Voor-Indië*, p. 10.

centuries. During this period the company, unlike its European competitors, enjoyed the privilege of trading in Japan. The administration of the Nagasaki factory was placed under the factory at Taiwan (Formosa), which until the conquest of Formosa by Koxinga, the Great Mandarin, in 1665, was the centre of the Dutch trade in Japan and China. The governor at Taiwan was responsible for the supply of the Indian markets with gold and silver.[11]

In a letter dated September, 1640, to the governor-general and council of Batavia the court of directors of the Dutch company, the so-called Heeren XVII, described the Japanese and Chinese trade as the principal trade in the entire Indies.[12] The occasion was the arrival, the year before, of 6 ships carrying a total of 3 million fl. in silver from Japan to Formosa. The flow of silver from Japan is a distinctive feature of the Nagasaki trade up to 1668. As part of their currency policy the Japanese in that year issued an ordinance against the export of silver. After this the Dutch turned their attention to coubangs, the Japanese gold coins. As early as 1665 the company had by the emperor's permission started an export of coubangs, which could be profitably sold on the Coromandel coast. The loss of Taiwan in 1665 had interrupted the regular connection with China, which had until that time supplied most of the gold.[13] At the close of the 1660's an abundant supply of gold was found in Japan, and the company exported large quantities at a cheap price. In 1668 the export amounted to 2.99 million fl., of this 74.72% were in gold. The following year the export totalled 2.55 million fl., 77.68% of this were in gold. Most of this was carried by way of Malacca to the Coromandel coast.[14] In 1670 the Japanese raised the price of the coubang from 56 to 58 maas, and in 1672 to 68 maas.[15] This slowed down the export of gold and the quota of gold in the export figures decreased considerably in the years that followed. Another aspect of the Japanese policy was a temporary ordinance against exporting gold, 1685–86.[16] The raising of the price of gold and the ordinance against exporting it are explained by the Japanese' fear of the country's being depleted of cash. In 1696 the Dutch company sustained a severe loss through the general devaluation of the silver and

11 W. H. Moreland, *op.cit.*, p. 66.

12 Algemeen Rijksarchief, 's-Gravenhage, Kol. Arch. No. 488, 11/9 1640.

13 W. H. Moreland, *op.cit.*, p. 65 note 1. Kol. Arch. No. 579, 6/5 1664. *Dagh-Register* Batavia 1665, 25/4 1665. Oskar Nachod, *op.cit.*, tables relating the exportation of silver and gold.

14 Pieter van Dam, *op.cit.*, p. 436. Kol. Arch. No. 457, 5/9 1670.

15 Pieter van Dam, *op.cit.*, pp. 446 and 469.

16 Kol. Arch. No. 460, 25/10 1686 and 16/11 1687.

gold coins, which was effected in Japan. The new gold coubangs contained less gold than the old ones, but retained the same weight – and the same price in silver, 68 maas. While the coubangs had hitherto been resold at a profit in India, they were now sold at a loss of 15 %.[17] A further devaluation of the gold coin in 1720 increased this figure to 37 or 38 %.[18] Notwithstanding this the company persisted in exporting these coins. In spite of the decifit, they brought a supply of cash to the Coromandel coast. Moreover, the general trade restrictions helped to keep up the export of gold, which continued right up to 1752, although it fell considerably in the 1730's and was completely at a standstill, 1739–42.[19] During the years 1701–24 a total of 291,585 pieces of coubangs were exported from Japan to Coromandel. This brought the company a total loss of 1,566,600 fl.[20]

The two periods of silver and gold export (in the years up to 1668 and the years 1669–72) were, in the last quarter of the seventeenth century – as well as in the eighteenth century – superseded by a third metal for coinage, namely copper. As early as in the 1620's Japanese copper was exported. In 1638–1645, a shortage of fractional currency led to an export prohibition. In 1646, however, a regular export set in, which in 1656 was stimulated by a vigorous demand from Europe.[21] Towards the close of the 1660's the export was checked by a considerable scarcity of copper in Japan brought about by the minting of copper coins. A temporary ordinance was issued against export in 1668.[22] But in 1670 the export re-commenced on a large scale and reached a preliminary peak in the early 1680's; thus the total value of the export for 1680 amounted to 1.30 million fl., 80.62 % of which were in copper.[23] The diminishing interest in gold coins was succeeded by an increasing interest in copper, which became the prevailing feature during the remainder of the existence of the Japan factory.

As compared with the metals, the other goods which the Dutch company exported were of inferior importance and will not be further considered. An appreciable quantity of camphor was purchased, various textiles, china, tea, etc.

---

17 Pieter van Dam, *op.cit.*, pp. 529, 534, 537 and 540.

18 J. Feenstra Kuiper, *Japan en de buitenwereld in de achttiende eeuw* ('s-Gravenhage, 1921), p. 115.

19 J. Feenstra Kuiper, *op.cit.*, table II.

20 Kol. Arch. No. 1931, fol. 2178 ff.

21 Oskar Nachod, *op.cit.*, table F. Pieter van Dam, *op.cit.*, pp. 403 ff.

22 Pieter van Dam, *op.cit.*, pp. 435 ff. Oskar Nachod, *op.cit.*, p. 357.

23 Provided that the price has been 11.9 tael per picol copper, cf. Oskar Nachod, *op.cit.*, table F. Pieter van Dam, *op.cit.*, p. 468.

Among the commodities imported into Japan, silk must above all be mentioned, both silk fabrics and raw silk. The Chinese and Bengalese silks ranked first, but Persian, Siamese and Toncquinese silks could also be sold. Chinese silk was, according to ancient custom, sold 'en bloc' to the governor of Nagasaki, the so-called 'pancado'-sale. Owing to scarcity of Chinese silk the 'pancado'-sale was abolished in 1655, but it was re-introduced in 1685 and extended to include silk from all the other countries. From the same date the Japanese demanded that one third of the entire trade should be in silk.[24] This stipulation caused considerable difficulty towards the end of the period as the increasing Japanese production of silk and the expanding Japanese textile industry brought about a fall in the demand. European textiles were also in demand in Nagasaki. There was in addition a considerable demand for hides and leather, which were sent direct from Toncquin. From the factory in Batavia pepper was sent and, from the 1670's, considerable quantities of sugar.

Until 1672 trade was free and was done by the ordinary auction system, apart from the special practice in the sale of silk mentioned above. In this year the Japanese, however, introduced a regulated trade. By this system free prices were superseded by a fixed price for each kind of commodity, regulated by a joint Japanese-Dutch-Chinese council.[25] This system involved a decline in the prices of the goods imported. The company retaliated by restricting its imports into Japan. This ultimately resulted in a considerable decline in the value of the total export of the company. As seen from the figures given above, the Dutch exported goods at a value of 2.5–3 million fl. in the 1660's and the beginning of the 1670's. In 1674 there was a fall in the export to a little above 2.1 million fl. and in the following years it varied between slightly more than 1 and 1.5 million fl. In the eighteenth century there was a further decline in the export, and from 1728 it never reached 1 mill. fl.[26] In spite of repeated Dutch complaints the regulated trade was maintained until 1685, when the prices became once more free. A maximum limit for the total trade was, however, fixed.[27] The Dutch company was now allowed to sell goods at a total of 300,000 Japanese tael, and the Chinese at a total of 600,000 tael. With a short interruption in 1698–99 the restriction lasted up to 1715, when the regulation system was re-introduced.[28]

24 Pieter van Dam, *op.cit.*, p. 828.
25 Pieter van Dam, *op.cit.*, p. 446.
26 Pieter van Dam, *op.cit.*, *passim*. J. Feenstra Kuiper, *op.cit.*, table II.
27 Pieter van Dam, *op.cit.*, pp. 487 ff.
28 Pieter van Dam, *op.cit.*, pp. 541—2. J. Feenstra Kuiper, *op.cit.*, pp. 106 ff.

The volume of the total trade was kept within the old limits. In 1698 the exportation of copper was for the first time restricted to a quota of 25,000 picols a year.[29] When the regulation system was re-introduced, in 1715, this was further reduced to 15,000 picols p. a. In 1720 a new regulation for the exportation of metal was made and the quota for copper was reduced to 10,000 picols. This quota continued till the next regulation in 1744.[30]

In the beginning the Dutch company complained of the Japanese trade restrictions which set a limit to their activities. In a letter to Batavia of December, 1688, the court of directors expressed their regret that a capital of 342,490 fl. had been left unemployed in Nagasaki.[31] The complaints, however, soon ceased. They were followed by bursts of anger, because the Dutch did not succeed in exporting the full amount of the quotas. The decline of their trade had two causes: a diminishing demand for the principal import commodity brought about by the incipient Japanese culture of silk and textile industry, and a greater consumption of copper in Japan itself owing to the increasing money economy.

IV

The system which the factory at Nagasaki used in computing its profits was a simple one. The cost price of the imported commodities – at the place where they had been bought – was deducted from the total of the selling price in Nagasaki. The surplus constituted the profit. The winnings were calculated at so many per cent of the cost price; sometimes, in addition, a 'clear profit' was calculated. In this the expenses in Japan of customs duties, factory rent, representation, etc. were deducted.[32] This, from a modern point of view, primitive method of book-keeping is – as shown by the Dutch historian Dr. W. M. F. Mansvelt[33] – characteristic of the entire financial administration of the company. In the ledgers of the company all expenses were regarded as indirect acquisition-costs and entered into a general costs account without any attempt being made to

[29] Kol. Arch. No. 488, 23/6 1700. Pieter van Dam, *op.cit.*, p. 542.

[30] J. Feenstra Kuiper, *op.cit.*, p. 115.

[31] Kol. Arch. No. 460, 1/12 1688.

[32] For illustration of the calculation of gross and net profit cf. Pieter van Dam, *op.cit.*, *passim*. The yearly net profit percentages are given in the works of Oskar Nachod and J. Feenstra Kuiper.

[33] W. M. F. Mansvelt, *Rechtsvorm en geldelijk beheer bij de Oost-Indische Compagnie* (Amsterdam, 1922).

calculate the costs of the individual commodities. Thus, for instance, the costs of freight were never calculated. This is explained by the company itself building and owning its ships. The working expenses of the shipyard were simply added to the costs involved in the equipment of the ships. This total was subtracted from the total income of the returns, when sold at the auction sales. For each individual commodity profit and loss were always computed gross, as the difference between cost price and selling price.[34]

Another difficulty in assessing the net return of the trade in a single commodity is the difference in the value of the rixdollar in the Netherlands and in the East. This dated from a decree made by the court of directors in 1665 (April 23). By this the Dutch rixdollar was authorized to be counted as 60 stuivers (or 3 fl.) in all the overseas possessions of the company. All foreign coins were in accordance with this converted into stuivers or florins. This created a regularity in the accounts of the company, which had long been wanting. In the Netherlands, however, the rixdollar was only calculated at 48 stuivers at this time. In other words, there arose from this the so-called 'swaar' money in the Netherlands and the 'ligt' money in the East. Thus an imaginary gain of 25 % on all overseas capitals was created. As far as the silver exported from Europe was concerned, the profit was, however, some real gain, silver – in contrast to gold – being of higher value in India than in Europe. The factories were informed of the order of regulation by the court in a circular of April, 1666, and it was maintained until 1742. It was then abolished at the request of G. W. van Imhoff, the energetic governor-general.[35] In Japan the company had up to 1666 calculated 1 Japanese silver tael at 57 stuivers, now it was calculated at 70 stuivers 'ligt' coin. In similar manner the rate of exchange of the other principal Asiatic coins, the rupee, the pagoda, etc. was adjusted.[36] In spite of the Japanese devaluation of the silver and gold coins in 1696 and the violent inflation in the beginning of the eighteenth century the tael remained equal to 70 stuivers in the accounts of the company though the rate of exchange for the rupee and the pagoda was, as mentioned below, altered again in 1683.

---

34 W. M. F. Mansvelt, *op.cit.*, p. 108.

35 Oskar Nachod, *op.cit.*, p. 341. G. W. van Imhoff, *Consideratiën over den tegenwoordigen staat van de Nederlandsche Oost Indische Maatschappij* (Amsterdam, 24/11 1741), ed. J. E. Heeres, *Bijdragen tot de Taal-, Land- en Volkenkunde van Nederlandsch-Indie,* deel 66 ('s-Gravenhage, 1912), pp. 485 ff.

36 G. C. Klerk de Reus, *Geschichtlicher Ueberblick der administrativen, rechtlichen und finanziellen Entwicklung der Niederländisch-Ostindischen Compagnie* (Batavia, 1894), p. 201, *Verhandelingen van het Bataviaasch Genootschap van Kunsten en Wetenschappen,* deel XLVII.

V

Copper was a marketable commodity nearly everywhere in the East. It was produced not only in Japan, but also in China, Toncquin, Siam, India and Persia. While the output of the mines in Northern India seems to have been constantly decreasing[37] and China consumed her full output and more, the Japanese copper mining industry was in constant growth throughout the seventeenth century. The production of copper in Further India and Persia did not come into full force until the eighteenth century and it did not become of importance in international trade relations in the period dealt with here. The copper was used for many purposes. In the first place it was a coinage metal. The fractional currency of Japan, the sen, and that of India, the nevel, were both copper coins. The metal was also used in the building industry, for the making of roofs, in the decoration of temples and for statues of Buddha. Numerous household goods and containers were manufactured of copper, e. g. boilers and vats for the making of arrack, refining of salpetre, etc. Copper was employed, too, in the shipbuilding industry. The quality of the metal varied. Thus the Dutch East India Company regularly sent small lots of copper to its workshops in Batavia to be used for specific purposes, e. g. assorted red copper plates for the sheathing of the bottoms of ships, yellow copper or brass for mounting and for compasses, and copper-wire.[38] These sorts of copper were bought and sold in the Indian trade, especially copper plates for ships.[39] The Japanese, too, manufactured brass and also plates of red copper. In 1669 the price of the latter commodity was the same as that of Swedish copper, counting freight, interest and risk.[40] It was, however, the so-called 'Staaf-koper' or copper bars, which predominated in the export from Japan. They were named after the moulds in which they were cast after refinement, and were sold by weight. The unit of weight was a pikol or chest, which was, up to 1698, equal to 125 Dutch lbs., later to 120 lbs.[41] The copper bars were the cheapest of all the various qualities mentioned above. The Dutch also occasionally exported copper coins from Japan, the so-called 'kopere casjes', e. g. a little less than 27 million pieces in 1665 and nearly 24 million

[37] W. H. Moreland, *op.cit.*, pp. 183 ff.

[38] Kol. Arch. No. 10.061, *Petitien en Eyschen uÿt India, passim.*

[39] Kol. Arch. No. 1204, fol. 501. Kol. Arch. No. 10.813—22, *Bataviaasche Negotie Grootboeken* 1713—33, the account *Koper in Platen en Bodems.*

[40] Pieter van Dam, *op.cit.*, p. 438.

[41] Pieter van Dam, *op.cit.*, pp. 541 f. Algemeen Rijksarchief, 's-Gravenhage, Collectie Radermacher, No. 431.

pieces in 1666.[42] By weight and value, however, they only constituted a fraction of the copper export in spite of the great number of pieces exported. They were, moreover, used mainly in the Toncquin trade. For this reason, this special form of copper export is left out of consideration in the following pages. Copper bars were, on the other hand, a commodity of international importance.

Table I shows the company's annual exports from Japan of copper bars for the years 1646–1734 and the imports into Europe for the same period in Dutch lbs. The export figures are taken from Oskar Nachod's and J. Feenstra Kuiper's works, which in their turn build on Pieter van Dam's information, the court minutes, and accounts from the factory in Japan. The European import figures have been taken from the invoice lists of the company up to 1692, from lists of lots offered for sale at auction by the six chambers of the company in the years 1693–1713, and finally, for the years after 1713, from a list of the actual sales that has been checked by examination of the ledgers at the company's head office in Amsterdam.

This quantitative analysis shows clearly that the principal part of the Japanese copper remained in the East. A smaller part was carried to Europe in very varying quantities. Of the further distribution to the Indian trading stations our knowledge is complete for the first 25 years of the eighteenth century, as will be seen from Table II, p. 54, which shows the annual sale of copper from each factory. Of the total of 29,048,244 lbs. of Japanese copper bars that the company sold in the years 1701–24, the factories in Bengal took 27.03 %, the Coromandel coast 24.57 %, Surat 23.99 %, Ceylon 11.59 %, the Malabar coast 8.37 %, Mocha 3.46 % and Persia 0.99 %.

It is not possible to draw up a similar survey for the seventeenth century, but from much desultory information in the archives of the company it appears that the quota for the Coromandel coast was larger in this period and that of Bengal smaller than in the years 1701–24. It can, for instance, be mentioned that a total of 14,290,755 lbs. of copper bars were sold in Surat during the years 1645–84[43] (the annual figures are found in table IV below). Of this a sale of 4,972,204 lbs. took place in the years 1679–1684, while the factories in Bengal (Hooghly, Balasore, Kasimbazar, Dacca and Patna) sold only 2,483,472 lbs. in the same period.[44] The sale on the Coromandel coast for one of these years, 1684, is known. In this year the factories at Palliacata,

[42] Pieter van Dam, *op.cit.*, pp. 431 ff.
[43] Kol. Arch. No. 1297, fol. 806—22.
[44] Kol. Arch. No. 1286, fol. 429—37.

Table I *

The Dutch Company's Export from Japan and Import into (or Sale in) Europe of Japanese Copper, 1646–1734. (Dutch lbs.)

Year	Japan Export	Europe Import or sale	Year	Japan Export	Europe Import or sale
1646	520,004	—	1681	3,000,000	— [4]
1647	577,881	—	1682	3,125,000	— [4]
1648	?	—	1683	2,000,000	— [4]
1649	381,875	143,238	1684	2,850,000	— [4]
1650	399,375	—	1685	2,625,000	— [4]
1651	550,000	306,250	1686	2,500,000	— [4]
1652	487,500	418,125	1687	1,875,000	— [4]
1653	377,998	—	1688	1,953,125	—
1654	412,500	—	1689	2,450,000	193,750
1655	401,625	—	1690	1,812,500	274,375
1656	1,111,625	—	1691	1,125,000	71,375
1657	1,762,500	39,140	1692	2,250,000	68,750[6]
1658	1,512,500	1,007,160[1]	1693	1,500,000	353,750
1659	1,030,531	367,080	1694	2,000,000	618,750
1660	1,424,919	480,700	1695	2,125,000	353,325
1661	1,899,500	336,000	1696	2,062,500	—
1662	2,325,000	116,400	1697	3,125,000	348,842[7]
1663	1,920,250	528,000	1698	3,480,000	492,364
1664	3,024,375	60,000	1699	2,500,000	414,950
1665	1,135,500	639,600[2]	1700	1,796,280	650,331
1666	1,573,438	156,000[3]	1701	1,990,200	379,250
1667	500,000	662,500	1702	1,758,000	46,875
1668	1,130,000	506,250	1703	1,956,000	30,250
1669	1,206,250	400,500	1704	2,195,280	75,000
1670	2,828,875	192,000	1705	2,196,000	523,120
1671	1,999,375	195,000	1706	1,800,000	968,625
1672	2,808,250	679,625[4]	1707	1,800,000	532,250
1673	1,880,500	568,750[4]	1708	992,640	306,250
1674	2,240,000	833,125	1709	1,800,000	698,000
1675	1,275,875	817,265	1710	1,800,000	231,026
1676	2,570,125	458,125[4]	1711	1,200,000	118,750
1677	2,129,375	— [4]	1712	997,920	—
1678	2,011,000	— [4]	1713	1,200,000	93,750
1679	2,937,500	814,211[4]	1714	1,260,000	127,277
1680	3,125,000	17,275[4a, 5]	1715	1,380,000	95,949

Table I (cont.)

Year	Japan Export	Europe Import or sale	Year	Japan Export	Europe Import or sale
1716	1,560,000	41,853	1726	1,032,000	—
1717	1,560,000	41,512	1727	1,200,000	—
1718	1,560,000	—	1728	1,152,000	—
1719	—	219,918	1729	1,200,000	—
1720	1,560,000	396,922	1730	996,000	68
1721	2,340,000	396,212	1731	532,800	—
1722	780,000	93,249	1732	1,188,960	—
1723	1,200,000	561,094	1733	972,480	417,246
1724	720,240	574	1734	1,224,000	453,694
1725	1,200,000	470			

* Japanese figures: Oskar Nachod, *op.cit.*, table F, J. Feenstra Kuiper, *op.cit.*, table II. European figures: 1) cargo lists 1646—92, *Gen. Miss.* and *Sommaria* in the collection *Overgekomen Papieren,* Kol. Arch., suppl. by printed lists in *Hollantse Mercurius* and *Amsterdamsche Courant,* 2) quantities for sale 1693—1713, Kol. Arch. No. 10.228 checked by the collection *Resolutiën Heeren XVII,* Kol. Arch., 3) auction sale figures 1714—34, Kol. Arch. No. 10.241 checked by *Negotieboeken Kamer Amsterdam,* Kol. Arch. The converting of Japanese picols into Dutch lbs.: 1646—97, 1 picol = 125 lbs., 1698 and later, 1 picol = 120 lbs. — 1 *kist* = 1 picol, and 1 *catti* = 1¹/₈ lbs.

[1] 275,160 lbs. of this quantity, loaded in two ships, were destined for return voyages in the year 1657, but the ships were delayed and arrived in Europe in 1658.

[2] 7,995 *caskens* reckoned at 80 lbs. per *casken* according to information in Gen. Miss. Batavia 18/1 1649, Kol. Arch. No. 1066, fol. 140-1.

[3] Cargoes of two ships unknown.

[4] Cargoes of the Ceylon ships unknown. The company only occasionally sent copper home from Ceylon.

[5] Cargo of one ship from Batavia unknown.

[6] Cargoes of five ships from Ceylon and Batavia unknown.

[7] Figures from an auction sale, spring 1698. Generally the copper was sold at the autumn auction sales.

Golconda, Bimlipatam, Nagapatam, Tegnapatam, Porto Novo and Sadrangapatam sold a total of 2,279,528 lbs.[45] (against a total for the same year of 1,265,310 lbs. in Surat and 393,986 lbs. in the whole of Bengal). In the early 1680's copper sales were extraordinarily large on the Coromandel coast. This will be considered in greater detail later.

Although, in the eighteenth century, nearly all the copper exported from Japan

[45] Kol. Arch. No. 1287, fol. 387—8.

Table II *

Quantities of Copper Sold at the Company's Indian Factories 1701–24 (Dutch lbs.)

Year	Ceylon	Coromandel	Malabar	Bengal	Surat	Mocha	Persia	Total
1701—02	164,243	456,939	126,741	291,312	816,904	—	—	1,856,139
1702—03	91,562	406,066	31,424	516,256	156,042	—	—	1,201,350
1703—04	108,160	184,588	88,556	—	158,185	—	—	539,489
1704—05	93,774	158,403	90,709	—	101,248	30,000	—	474,134
1705—06	51,331	269,532	109,860	356,514	—	62,500	75,000	924,737
1706—07	101,047	351,747	143,415	297,500	365,000	62,500	—	1,321,209
1707—08	101,705	429,026	141,042	499,664	650,000	—	—	1,821,437
1708—09	125,900	414,567	124,815	561,000	473,063	75,000	37,500	1,811,845
1709—10	59,847	414,134	123,261	635,868	525,000	124,369	62,500	1,944,979
1710—11	40,965	378,312	89,927	676,098	520,239	49,945	—	1,755,486
1711—12	67,590	406,203	105,800	143	400,000	—	—	979,736
1712—13	106,032	499,166	50,974	335,920	325,000	79,431	—	1,396,524
1713—14	278,603	99,995	94,702	387,784	293,750	69,575	—	1,224,409
1714—15	145,616	346,156	76,379	547,536	34,375	62,292	—	1,212,354
1715—16	136,476	158,397	70,598	650,934	187,500	70,000	25,000	1,298,905
1716—17	180,254	364,554	96,210	148,512	125,000	50,000	25,000	989,530
1717—18	208,403	245,594	129,310	433,160	187,500	40,000	31,250	1,275,217
1718—19	159,846	132,740	135,623	340,000	165,625	50,000	31,250	1,015,084
1719—20	227,353	148,273	128,126	176,800	606,516	50,000	—	1,337,068
1720—21	176,858	212,263	105,530	234,600	73,166	—	—	802,417
1721—22	262,355	427,553	—	182,455	328,662	59,562	—	1,260,587
1722—23	265,744	366,306	155,812	399,221	292,393	69,795	—	1,549,271
1723—24	215,667	265,368	213,409	179,157	182,749	—	—	1,056,350

* Kol. Arch. No. 1931, fol. 2178-85, *Berigt aantonende de behaalde Winsten sedert den jare 1700 tot 1724 in Japan, alsmede op de goederen uÿt dat Rÿk op andere plaatsen omgezet, nevens en summiere aanwysing van de geledene schade in dien tÿd, gedateert 5 July 1726.*

was shipped to Batavia and from there distributed to the various Indian factories, in the seventeenth century only a small quantity was carried thence, the bulk of it, up to 1666, being taken to Taiwan (Formosa) for distribution. Later it was sent to Malacca and distributed there. To this must be added the copper which the factories bought occasionally from private traders. From the quotas which are occasionally found mentioned in the court minutes from Batavia it appears that in the 1670's the Malabar coast was a considerable customer for copper. In 1677 for example, the coast received as much as one third of all the quotas from Malacca and Batavia. This was, however, exceptional, and was balanced by nothing being sent the following year.[46]

VI

Having thus given a quantitative account of the Dutch company's exports of Japanese copper bars to Asiatic and European markets we have to consider the question of the importance of copper to the company in these markets.

In view of the facts mentioned above it is not surprising that, throughout the period, Heeren XVII, i.e. the court of directors of the company, regarded the Asiatic market as the most important one. A systematic examination of the so-called Eysch van Retouren[47] – the demands for return goods which the court of directors in Amsterdam sent to the court in Batavia every year – shows that Amsterdam did sometimes demand that great quantities of copper should be sent to Europe. The demands were, however, always made with the important reservation that the sale in India should not be curtailed thereby. The Japanese copper bars were in the first place reserved for the company's important Asiatic markets.

During the export prohibition of 1638–45, copper had to be imported to India from Europe. Thus Batavia demanded 100,000 lbs. of copper plates in 1641–42 and no less than 600,000 lbs. of Swedish or Hungarian copper plates for its trade in 1642–43. With the re-opening of the export from Japan, copper bars came again into the market and on December 31, 1646, the court in Batavia wrote to the court of directors in the Netherlands that European copper was no more needed "now that the exportation of Japanese copper is permitted to 8 and 9 tael per picol and Taiwan is able to send us the necessary supplies."[48] The quantitative analysis of the export from Japan shows that the first consignment of copper bars after the cancelling of the export prohibition came to Europe in 1649 and the next consignment in 1651. In the next four years nothing was sent, but towards the end of 1655 the court of directors for the first time mention copper in the list of goods demanded for 1657. Stating that the price of copper in Holland had risen from 36 to 56 fl. (a 100 lbs.) a demand was made for as

[46] *Dagh-Register* Batavia 1677 and 1678, *passim.*

[47] In the series *Resolutiën Heeren XVII,* Kol. Arch.

[48] Kol. Arch. No. 10.061. »Synde de uÿtvoer van Japans Coper tot 8 ende 9 tail t picol toegestaen ende van solxcher can ons Taÿuan genoech provideren.«

much copper as India could spare. Accordingly 39,140 lbs. were sent to Holland in 1657 and in 1658 the vast quantity of 1,007,160 lbs.[49] Up to and including 1676 Japanese copper in varying quantities was sent to Europe every year – the supplies being especially plentiful in 1658, 1663, 1665, 1667 and 1672–75.

In October 1657 the court of directors stated that the price of copper in Holland was still 56 fl. and seemed to show no tendency to fall.[50] In the following year, for the first time, the court of directors made a demand for a definite quantity, namely 600,000 lbs.; in 1659 the demand was increased to 800,000 lbs., in 1663 to 900,000 lbs., and in 1668 it reached the very high figure of 2 million lbs. A postcript was added saying that the quantity of copper sent to Europe could not be too large.[51] In 1674 the demand for copper fell to 1.2 million lbs. and in November, 1679, it was desired that no Japanese copper should be sent except such quantities as were needed for ballast in ships or as India might want to part with.[52]

In the late 1670's and the early 1680's the European prices of copper appear to have been low,[53] and in the years 1681–1688 no copper was carried to Europe. In other words, the Asiatic market absorbed the whole of the company's very large exports of copper from Japan, which reached a climax during these years. As will be mentioned in another connection (infra pp. 75–8), considerable quantities of Japanese copper exported by Chinese merchants also circulated in the East during this period. This caused the Dutch company much grievance and led to various projects for buying up the Chinese exports which interfered with the Dutch monopoly of exportation of Japanese copper. In reference to some such project the court of directors wrote, in December, 1684, to the court in Batavia that they would be very pleased if the company succeeded in monopolizing the copper trade and added that Batavia might count on the quantities left unsold in India being sold in Europe. In these circumstances salpetre as ballast would be superseded by copper.[54] In the demands made in October, 1686, and November, 1687, the court of directors re-asserted that they were ready to accept the copper that could not be disposed of in India, adding, in a letter of 1686, that the price in Holland was 55 fl.[55] Europe was thus used as a

[49] Kol. Arch. No. 252, 1/11 1655, and table I above.
[50] Kol. Arch. No. 253, 4/10 1657.
[51] Kol. Arch. No. 255, 21/8 1668.
[52] Kol. Arch. No. 257, 14/11 1679.
[53] N. W. Posthumus, *Nederlandsche Prijsgeschiedenis* (Leiden, 1943), tables 173—4.
[54] Kol. Arch. No. 459, 11/12 1684.
[55] Kol. Arch. No. 259, 25/10 1686 and 5/11 1687.

supplementary market. In 1688 a demand for definite quantities from the East was made again, so far a modest 150,000 to 200,000 lbs. It was stated that the price in Holland had risen to 58–60 fl.[56] For the next years we are able to check the prices of Japanese copper at the auction sales of the company in Holland, as seen in table III p. 58. Here the prices of Swedish and Norwegian gar-copper are also seen as are the Japanese cost prices (calculated at so many florins a hundred lbs.).

In its demands for 1692, sent in December, 1690, the court of directors stated that in 1690 copper only fetched a price of 54 fl. at the auction sales, but the court did not therefore find reason to cancel further consignments as the commodity took up no room in the ships but was only used as ballast. As usual it was emphasized that the copper should not be sent to Europe if it could be sold more profitably in India.[57] In 1692 the auction prices reached an average of 61 fl. a 100 lbs. and the court increased its annual demand from 200,000 to 300,000 lbs.[58] In 1694, 618,750 lbs. came to Holland from Batavia. The court in Batavia explained that this was necessitated by the scarcity of salpetre as alternative ballast goods. This quantity proved excessive, however, and the court in Amsterdam cut down its demand for copper to 200,000 lbs.[59] The question of ballast weighed heavily in determining the demands for these years. The court of directors repeatedly complained that on their return voyages the ships were filled with scrap iron as ballast instead of more merchantable goods, such as salpetre, sugar, tin and copper. Usually salpetre was considered the most profitable ballast goods, next to that, sugar, copper and tin.[60]

The extremely high prices fetched by copper at the auction sales in the late 1690's caused another increase in the import of this commodity. This can be seen from the figures in table I of the quantities annually offered for sale. In consequence of this, the directors' demands again increased to 300,000 lbs. In 1699 they reached 5–600,000 lbs[61] and remained at that figure until the spring of 1705. Among the copper that Holland received in these years was also Japanese copper bought in Japan by the Chinese and re-sold by them to the Dutch company. The court of directors complained of its inferior quality and

56 Kol. Arch. No. 259, 26/11 1688.
57 Kol. Arch. No. 259, 14/12 1690.
58 Kol. Arch. No. 260, 6/12 1692 and 18/11 1693.
59 Kol. Arch. No. 260, 8/11 1694.
60 Kol. Arch. No. 488, 10/12 1692, 27/8 1694, 7/9 1696, 18/9 1702 and 24/7 1704.
61 Kol. Arch. No. 261, 28/10 1699.

Table III *

European and Japanese Copper Prices 1688–1735 in Florins per 100 Dutch lbs.

Year	Swedish gar copper Sale	Norwegian gar copper Sale	Japanese copper bars Sale	Japanese copper bars Cost	Year	Swedish gar copper Sale	Norwegian gar copper Sale	Japanese copper bars Sale	Japanese copper bars Cost
1688	60.00	59.50	59.00[1]	33.32	1712	60.57	59.93	—	?
1689	?	?	57.50[1]	33.32	1713	59.25	58.93	58.07	?
1690	?	?	54.00[1]	33.32	1714	62.08	62.18	65.57	40.14
1691	57.50	57.00	?	33.18	1715	74.50	73.50	71.97	?
1692	59.00	58.00	61.00[1]	33.18	1716	74.68	73.70	67.17	40.39
1693	?	?	58.00	33.18	1717	73.11	71.39	74.98	?
1694	59.00	58.00	53.88	33.18	1718	73.08	72.56	—	?
1695	?	?	55.75	33.18	1719	59.92	65.50	56.91	45.18
1696	?	?	—	33.18	1720	56.00	59.00	50.61	?
1697	?	?	60.13	33.18	1721	56.00	54.25	51.91	?
1698	?	?	63.13	?	1722	57.50	57.00	57.51	45.23
1699	?	?	66.30	?	1723	—	62.00	55.87	?
1700	?	?	62.47	36.05	1724	66.00	65.50	59.41[2]	45.33
1701	64.00	62.50	62.32	37.33	1725	—	69.25	43.83[2]	45.47
1702	?	?	60.96	37.33	1726	—	63.75	—	?
1703	57.17	56.75	55.31	?	1727	—	64.00	—	45.25
1704	?	?	55.66	?	1728	—	64.75	—	45.19
1705	55.50	55.00	51.93	?	1729	—	63.50	—	?
1706	51.60	51.10	49.35	?	1730	63.25	66.50	54.41[2]	?
1707	?	?	52.75	?	1731	63.13	65.50	—	45.26
1708	56.75	56.25	55.97	?	1732	—	65.63	—	?
1709	55.08	55.50	49.05	?	1733	67.50	61.83	57.70	?
1710	54.70	53.90	54.26	?	1734	62.46	60.17	58.03	?
1711	58.88	58.38	61.38	?	1735	61.45	61.55	55.87	?

* The prices of Swedish and Norwegian gar copper, according to N. W. Posthumus, *op.cit.*, tables 173-4, suppl. by 147 original *Cours der Koopmanschappen tot Amsterdam* found in *Amsterdamske Faktorregnskaber,* Rigsarkivet, Copenhagen, covering the years 1708—32. Auction sale prices of the Dutch East India Company, according to Kol. Arch. No. 10.228, 1693—99, and *Negotieboeken Kamer Amsterdam,* Kol. Arch. No. 10.406-22, 1700—35. Japanese cost prices, according to Oskar Nachod, *op.cit.*, table F, and J. Feenstra Kuiper, *op.cit.*, table II, suppl. by *Bataviaasche Negotie Grootboeken,* Kol. Arch. No. 10.813-22, 1713—33, under the account *Japans Staafkoper.*

[1] This is an approximate price compiled from the demands of the court of directors.

[2] Average based on a very small quantity of copper sold this year, cf. table I.

was of the opinion that it had been bought at too high a price.[62] Owing to a fall in prices, the demand was reduced to 4–500,000 lbs. in 1705, and again reduced to 3–400,000 lbs. in the spring of 1710, a note being added that copper was now at so low a price in Holland that the sale of it brought in very slight profits.[63] The court of directors therefore urged Batavia to try to increase the sale in India. About 1712 the state of the market changed again, and the prices of Swedish and Norwegian as well as Japanese copper rose once more. In the beginning of 1715, the court of directors increased its demand for copper to 5–600,000 lbs., giving as a reason for this the state of war in Sweden, which was expected to cause a decline in the Dutch imports and a corresponding rise in prices.[64] But already in the following year the demand had fallen to 3–400,000 lbs., and it remained at this figure during the rest of the period. In spite of the extremely high prices of copper in the years up to and including 1718 – the highest prices throughout the seventeenth and eighteenth centuries – only small quantities were sent to Holland from the East. Consideration of the Indian market – in which copper fetched very high prices in these years too – still came first, and the decreasing copper exports from Japan left a constantly smaller surplus for Europe. In the beginning of 1717, the court of the company proposed that the consignments destined for Europe should be supplemented by copper from Toncquin,[65] and from 1719 to 1723 the sale increased considerably in proportion to that of the year immediately before. Prices, however, were rapidly declining, and while, formerly, the price of Asiatic copper was nearly equal to those of Swedish and Norwegian gar-copper, it now seems to have been lower. This may possibly be accounted for by the inferior quality of the Toncquin copper. From 1724 up to 1734 hardly any copper was sold at the auction sales of the Dutch company, though the European market price seems to have remained at a comparatively high level. The court of directors persistently demanded the usual 3–400,000 lbs. of Japanese or Toncquin copper bars, but Batavia did not comply with the demands. Not until 1734 did the sale of copper at the auction sales of the company again become brisk.

In the eighteenth century, Japanese copper bars were of subordinate importance in the European trade of the Dutch company. In the period 1700–1736, goods were annually sold from the company's chamber in Amsterdam at a

[62] Kol. Arch. No. 262, 7/3 1704.

[63] Kol. Arch. No. 263, 27/2 1705, and Kol. Arch. No. 264, 4/3 1710.

[64] Kol. Arch. No. 198, 4/3 1715.

[65] Kol. Arch. No. 198, 22/2 1717.

total varying between a minimum of about 6.2 million fl. (1734–35) to a maximum of 10.4 million fl. (1719–20). The percentage of copper bars was generally only about 1 %, with the extremes 0.13–3.46 %.[66] In some years no copper at all was sold. For all the six chambers of the company the percentages were even lower, Japanese copper being generally sold in Amsterdam. In the years of extraordinarily large supplies in the seventeenth century, copper constituted at a rough estimate about 3–4 % of the total sale.[67] It will be seen that from the point of view of the company copper cannot be claimed to belong to the more important commodities sold in Europe.

It is another question what was the percentage of Japanese copper in the European copper market. As to this Professor Heckscher has shown that in 1669 Japanese copper appears in the price lists in Amsterdam – in the same year as Norwegian copper – and remains there up to 1688.[68] This seems to point to Japanese copper bars having been a factor of some importance in the European market in these years. This impression is corroborated by the fact that the Dutch company imported large quantities of Japanese copper during this period (see table I p. 52); in 1672–75 these imports amounted to $^1/_3$ or $^1/_2$ of the simultaneous copper and brass exports from Sweden.[69] It is also remarkable how the peak in the Dutch copper imports from Japan coincided with the wars in Europe, a fact that is, of course, also reflected in the development of prices. The large import of copper at the close of the 1650's coincides with the war between Denmark and Sweden 1658–60. Similarly, the record import of the 1670's coincides with the great wars from 1672 to the peace at Nijmegen in 1678. Conversely there was no import in the peaceful decennium, 1678–88, – apart from the years 1679 and 1680. As was mentioned above the court of directors gave as a reason for their demands for larger quantities in 1715 the Great Northern War and the risk that it would cause a scarcity of copper.

It will thus be seen that the influx of Japanese copper to Europe was decided by various factors, the most important one being the state of the Asiatic market. Compared with this the European market was of subordinate importance. The secondary character of the European market is apparent in the projects for

66 Kol. Arch. No. 10.406—22, *Negotieboeken Kamer Amsterdam.*

67 The yearly average of the total sales by all chambers in the period 1656—60 is, for instance, about 8.5 mill. fl. The average quantity of copper taken home in the same years was about 0.55 mill. lbs. At an average selling price of 55 fl. per 100 lbs. this amounts to 302,500 fl. or 3.5 % of the whole sales amount.

68 *Scandia* XI: 2 (1938), p. 255.

69 *Scandia* XI: 2 (1938), p. 242.

monopolization set forth in the 1680's, it being here regarded as a mere supplementary market. On the other hand, both the list of goods demanded – the figures of which should be regarded as merely indicative – and the import figures show that the importance of Europe as a buyer of copper in the 1650's and 1670's was considerably greater than in later years. As will be mentioned in another connection there is reason to assume that the great European demand from the middle of the 1650's gave an impetus to the company's copper exports from Japan.

In the second place, the import was determined by the difference between the selling prices in the Dutch market and the cost prices in Japan. With too small a margin between these prices an import did not pay. It is true that the court of directors did not reckon with a definite profit; as was mentioned above they were incapable of calculating the net profits. But an examination of the company's trade in Japanese copper in Europe shows that the directors wanted to sell at a gross profit of some 50 %; if the profit was less it did not pay to take the goods home. When the distinction between 'ligt' money and 'swaar' money was introduced in 1666 it might be expected that, in the calculation of profits by sale at the company's auctions, the cost prices were converted into the European monetary standard. This seems, however, not to have been the case. As far as can be judged, the amounts in the invoices of goods of the returning ships were not converted into 'swaar' money, but the profits in Europe were calculated as if the company's Indian and European monetary units were one. In actual fact the gross profits became in this way reduced by 25 %. The difference in value was, however, not unheeded. It was felt to be an impediment in calculating the exact profit. This is clearly seen in a letter of October, 1686, to the court in Batavia, in which the court of directors in Amsterdam writes that Japanese copper can be sold profitably in Europe "when it is purchased at 11:9:– tael and the tael is calculated at 57 stuivers only, as has formerly been customary in the ledgers".[70]

In addition to prices of copper in Europe and Japan a third factor was of moment, namely, the question of ballast. As mentioned above, the Dutch company classed Japanese copper with ballast goods. In the importation of this it therefore considered the prices and profits obtainable for other ballast goods,

---

[70] Kol. Arch. No. 460, 25/10 1686. »Te weten als wanneer het voor Ts 11 : 9 : — daer te krÿgen is, de teÿl gereeckent tegens 57 Stuÿvs :, gelyck die in voorgaende tÿden niet hogher in de boecken, is gewaerdeert en uyt getrocken geweest.«

in the first place salpetre. It is remarkable that the company's importation of copper was large exactly in those years when the prices of salpetre in Europe showed a falling tendency.[71]

VII

Turning our attention from the European to the Asiatic market we shall first consider the Dutch factory in Surat, which was one of the company's most important centres of commerce in the beginning of the seventeenth century and which remained one of the corner stones of the Dutch company's Asiatic trade. Moreland has proved that, in the course of the 1620's, there was a distinct rise in the prices of copper in Surat which occasioned Dutch imports of copper into Surat, for the first time in 1635.[72] In the following years the import of copper bars to Surat was prevented by the ordinance against export from Japan (cf. *supra* p. 46). Instead the Dutch company sold Swedish and other European copper in the 1640's. From 1645 to 1684 the quantities of copper sold are noted in the Dutch company's factory, as well as the prices in rupees per man (34 $^1/_2$ Dutch lbs.) and stuivers per 1 Dutch lb. respectively. These are shown in table IV (p. 64) where the prices in stuivers per lb. are also converted into fl. per 100 lbs. and compared with the company's cost prices in Japan, which have been converted into fl. per 100 lbs.

It appears from table IV that, apart from 1649 and 1669, there was a sale of copper in Surat every year. Seen over the vista of years the sale reveals an upward tendency, and in 1653 and 1680 there was a virtual boom. The price of copper which in 1645–46, the first years of the period mentioned above, was 15 rupees per man fell in the next years to an average of 14.22 rupees. But in 1654 a rise in prices set in, which in the course of the next 17 years gradually raised the price to the high figure of 22.50 rupees in 1670. In 1669 the price made a sudden leap rising from 17 to 20 rupees. This was explained in a letter from Surat to Batavia (copied in the court minutes of that year) by a shortage of copper caused partly by insufficient Dutch import, partly by disturbances at the copper mines in Northern India brought about by the civil war.[73]

---

[71] N. W. Posthumus, *op.cit.*, table 209.

[72] W. H. Moreland, *op.cit.*, p. 184. Moreland says that the export of Japanese copper was prohibited during the years 1640 to 1648. As has been stated above the suspension of the Dutch copper export from Japan extended over the period 1638—45, cf. Oskar Nachod, *op.cit.*

[73] W. H. Moreland, *op.cit.*, p. 184.

In judging the Dutch prices in the table the changes in the valuation of the rupee should be taken into consideration. The converting of rupees into stuivers shows that the company up to the issuing of the decree of April 23, 1665, reckoned the rupee at 24 stuivers, from 1666 up to and including 1682 at 28 stuivers, and from 1683 at 30 stuivers. This rise in the valuation of the rupee involved a considerable though illusory rise in the price of copper in 1667. The writing up of the rupee in 1683 likewise involves a partial obliteration of the downward tendency in the price of copper (as expressed in stuivers) which made itself felt after 1671. The prices were especially low during the later years of the period. This may possibly be explained by the fact that very considerable quantities were sold in the beginning of the 1680's, possibly also by the political situation in India, where the centre of unrest was moved from the North to the South when, in 1681, Aurangzeb, the Great Mogul, attacked the kingdoms of Bijapur and Golconda and the Marāthas.[74]

The occasionally vehement fluctuations in the prices of copper were presumably due to conditions of a local character (famine, etc.). A detailed explanation of these conditions may be found in the writings of the experts on the economic history of India. It will here suffice to ascertain what was the status of copper in proportion to the other goods sold by the company in Surat. In the course of the period, which extended over 43 years, 40 different sorts of commodities were dealt in. The sale of the following commodities was fairly regular: clove, mace, cinnamon, pepper, arrack, tea, cinnabar, mercury, benzole, alum, Radix China, camphor, sapan-wood, sandal-wood, ivory, tortoise-shell, zinc, tin, lead, copper and European cloth. By far the largest sales were of clove and copper, followed by the so-called 'Rompen', i. e. dried mace, while the rest brought in comparatively small sums. By trade in these commodities the factory in Surat made a total gross profit of 20 million fl. during this period, to which should be added 0.78 million fl. brought in by the carrying trade. On the other hand the costs amounted to a little more than 4 million fl. Unfortunately we have no means of ascertaining the profit realized by the individual commodities in the whole of the period, but from a general knowledge of cost prices it may with certainty be said that the sale of clove brought in the largest proportion of the profit, while copper came second, at any rate during the later years. The court in Batavia mentioned in its general report sent to Holland in Decem-

[74] W. H. Moreland & Atul Chandra Chatterjee, *A Short History of India* (London, 1936), p. 251.

Table IV *

Quantities and Prices of Copper Sold at the Surat Factory and Cost Prices in Japan, 1645–84.

Year	Quantities lbs.	Sale prices			Cost prices Japan Fl. per 100 lbs.
		Rupees per man (or 34½ lbs.)	Stuivers per lb.	Fl. per 100 lbs.	
1645	34,378	15.00	10.43	52.15	—
1646	16,540	15.00	10.43	52.15	18.24—20.52
1647	37,939	14.22	9.88	49.40	18.24—20.52—20.98
1648	50,376	14.22	9.88	49.40	?
1649	—	—	—	—	26.45
1650	95,017	11.50	8.00	40.00	26.45
1651	53,084	14.22	9.88	49.40	23.94—25.08
1652	60,465	14.22	9.88	49.40	22.80—25.08
1653	149,467	12.00—14.22	8.42— 9.88	42.10—49.40	23.48
1654	233,528	15.11	10.52	52,60	23.48—23.94
1655	259,721	15.56	10.81	54.05	24.85
1656	159,373	15.56	10.81	54.05	25.99
1657	197,379	17.78	12.37	61.85	25.99
1658	327,036	17.78	12.37	61.85	25.99
1659	336,862	18.00	12.52	62.60	22.34—26.22—27.36
1660	340,192	18.00	12.52	62.60	22.34—25.08—26.90
1661	200,130	17.00—20.00	11.83—13.91	59.15—69.55	26.90
1662	616,317	20.00	13.91	69.55	27.13
1663	382,201	20.00	13.91	69.55	26.90
1664	482,874	19.00	13.22	66.10	28.04
1665	360,944	20.25	14.09	70.45	28.27
1666	296,228	20.00	13.91	69.55	28.27
1667	527,816	20.25	16.43	82.15	?
1668	285,261	21.00—21.25	17.04—17.25	85.20—86.25	?
1669	—	—	—	—	?
1670	525,344	22.50—20.50	18.26—16.64	91.30—83.20	35.56
1671	602,378	20.00—19.00	16.23—15.33	81.15—76.65	35.28
1672	392,703	19.13	15.52	77.60	35.00
1673	658,224	17.00—21.00	13.80—17.04	69.00—85.20	34.58
1674	474,198	19.00	15.33	76.65	34.16
1675	600,169	17.00	13.80	69.00	33.88
1676	599,459	16.50	13.39	66.95	33.32
1677	743,108	17.00	13.80	69.00	33.32
1678	219,844	15.50	13.00	65.00	33.32

Table IV (cont.)

Year	Quantities lbs.	Sale prices			Cost prices Japan Fl. per 100 lbs.
		Rupees per man (or 34½ lbs.)	Stuivers per lb.	Fl. per 100 lbs.	
1679	782,195	16.75—16.00	13.59—12.99	67.95—64.95	33.32
1680	956,731	17.25	14.00	70.00	?
1681	613,757	16.00	12.99	64.95	33.32
1682	417,665	16.63	13.49	67.45	33.32
1683	936,546	16.63—15.50	14.46—13.48	72.30—67.40	33.32
1684	1,265,311	15.88—15.00	13.80—13.04	69.00—65.20	33.32

* The Surat figures: Kol. Arch. No. 1297, fol. 806-22, *Authentique copie prÿscourant der coopmansⁿ sedert aº 1643 tot 1684 incluÿs, door de Compᵉ in Suratta verkoght, met de winsten en lasten in deselve tijd en ieder jaar specificerende.* The Japanese cost prices: Oskar Nachod, *op.cit.*, table F.

ber, 1679, that a total profit of 677,512 fl. had been made in the latest financial year.[75] Of this amount 'Garioffel Nagelen' (the first quality of clove) brought in 386,203 fl. and copper bars 199,880 fl. These were the two chief items. The sale of mace brought in the sum of 22,823 fl. The general report also gave the characteristic computation of the gross profit in connection with the sale. Of this mace brought in a profit of 1960 ¾ %, a dazzling figure, while 912 ⅝ % was gained by the sale of clove and only 69 1/16 % by the sale of Japanese copper. The example shows clearly the status of monopoly held by the Dutch company in the spice trade. But it also shows how necessary it is to supplement the knowledge of the gross profit by information of the quantities sold. This also appears from the following example from the Coromandel coast.[76] In the factories on this coast the gross profit on copper varied between 93 and 116 % in the financial year 1683–84, while the gross profits on clove and mace were 1026–1381 % and 556–2503 % respectively; the profit on other spices was similar. The quantity of mace sold was, however, only 251½ lbs., the profit totalling 556 fl. A considerably larger quantity was sold of clove, namely 25,943 lbs., this yielded a profit of 116,463 fl. But of Japanese copper bars no less than 2,279,528 lbs. were sold; this brought in an income of 881,040 fl. or 60.30 % of the total profit of the Coromandel coast, which was 1,457,547 fl.

[75] Kol. Arch. No. 1230, fol. 478.

[76] Kol. Arch. No. 1287, fol. 387—8.

In the seventeenth century, the Coromandel coast was, as has been mentioned above, the largest buyer of copper from the company, and, at least from the 1670's, copper became one of the chief, if not the chief commodity sold by the company. The above example is from one of the most profitable years. In estimating the profit it should be taken into consideration that in 1683 the value of the pagoda was raised from 105 to 120 stuivers. The profit thus became to a certain degree illusory, a fact to which attention was later called by the court in Amsterdam.[77] Thus in the financial year 1680–81 the profit was 816,250 fl. Of this the copper sold in Golconda and the northern factories alone brought in 369,954 fl.[78] In 1681–82 the profit rose to 1,054,222 fl. More than two thirds of this – 711,046 fl. — came from the Golconda factory. Of copper 1,178,916 lbs. were sold in that year and the payment for this constituted a considerable proportion of the above profit.[79] The high prices obtained for copper should be seen in relation to the disturbed and warlike conditions during these years, when Aurangzeb, in 1686, captured the town of Golconda. This put a sudden stop to the Dutch East India Company's adventure in copper.[80] The principal customer of the Dutch in Golconda in 1684, a native merchant by the name of Zierum Chodenda, had made a contract for the purchase of 12,000 chests or 1.5 million lbs. annually chiefly for coinage of the nevel, the Indian copper coin. The company had delivered a large quantity of copper on credit. When Chodenda went bankrupt at the fall of Golconda, the Dutch sustained a loss of more than 900,000 fl. This led to the dismissal of the 'Opperhoofden' (the managers) and the bringing of a lawsuit against them.[81]

In the wake of the wars followed famine and disease, and trade suffered consequently. The first mark of decline was seen in the textile sales of the Coromandel coast. This was one of the most decisive events in the history of the Dutch company. As regards Japanese copper, however, it was expected that there would still be a large market for it. In 1688 a letter was sent by the court of directors in Amsterdam to the court in Batavia saying that it was expected that, in the next two years, it would be possible to sell 14–15,000 chests of copper a year in Golconda for the coinage of copper coins, now that conditions

77 Kol. Arch. No. 459, 8/10 1685.

78 Kol. Arch. No. 459, 8/6 1682.

79 Kol. Arch. No. 459, 22/6 1683.

80 Kol. Arch. No. 460, 25/10 1686.

81 Kol. Arch. No. 459, 8/10 1685, Kol. Arch. No. 460, 25/10 1686, and Kol. Arch. No. 461, 8/7 1693.

were again settled after the wars.[82] In the beginning of the 1690's the supplies on the Coromandel coast seem, however, to have been very large, a fact which presumably indicates that the sale scarcely reached the heights expected.[83] This impression is also corroborated by a statement from the factory at Nagasaki that the smallness of the exports was due to a diminishing sale of copper in the Indian markets, and that the rise in the export at the close of the 1690's was due to an increased demand from Europe.[84] The decline in the Asiatic consumption can also be traced in the prices.

From 1677 Japanese copper bars appear in the statistics of prices which were once a year made in Batavia from the ledgers of the various factories – except Ceylon – and which were sent to the head factory. These statistics show clearly in which places Japanese copper was sold. Specimens of the prices have been given in table V below converted into fl. per 100 lbs. and compared with the company's cost prices in Japan, similarly converted.

This table exhibits considerable divergences between the individual sale localities. Thus high market prices on the Coromandel coast in the beginning of the 1680's occur simultaneously with a downward tendency in the prices in Bengal. In a letter written in the summer of 1684 the court of directors in Amsterdam reprimanded the factory in Bengal urging it to increase the sale in this densely populated area.[85] In 1685 an improvement in the prices in Bengal set in, and in the course of the 1690's they came up to the prices in Masulipatam on the Coromandel coast. Usually, however, the prices in Bengal and Surat were lower than those on the Coromandel and Malabar coasts. The decreasing demand in the 1690's is reflected in low prices in nearly all factories. At the close of the century a temporary rise in prices occurred. In 1698–99 the extraordinarily high price of 100.31 fl. per 100 lbs. was obtained by sale of copper in Mocha, which was, however, not one of the large markets. The years 1703–06 were disastrous to the company. It appears from table II that the sale of copper failed completely in Bengal in 1703–05 and in Surat 1705–06, while on the Coromandel coast very small quantities were sold. The price statistics show, too, that prices were low. About 1710 things began to look brighter and an upward tendency can be perceived. Especially towards the close of the 1720's this tendency gained impetus raising the Asiatic copper prices to a figure which

82 Kol. Arch. No. 460, 1/12 1688.
83 Kol. Arch. No. 461, 27/8 1692.
84 Pieter van Dam, *op.cit.*, p. 526 f.
85 Kol. Arch. No. 459, 24/6 1684.

Table V *

Asiatic Copper Prices from the Dutch East India Company's Factories, 1678–1739. Florins per 100 lbs.

Year	Japan Cost	All factories: Lowest Sale	All factories: Highest Sale	Porto Novo Sale	Masulipatam Sale	Bengal[1] Sale	Cochin Sale	Surat Sale
1678—79	33.32	65.63	87.50		65.63	69.14		? [2]
1679—80	33.32	63.44	92.81		74.38	68.75		63.44
1680—81	?	65.31	88.44		71.25	65.94		65.31
1681—82	33.32	? [3]	? [3]	? [3]	? [3]	? [3]	? [3]	? [3]
1682—83	33.32	55.63	82.50	75.00	—	55.63	70.63	66.56
1683—84	33.32	59.69	94.69	75.00	—	59.69	72.50	63.75
1684—85	33.32	55.00	87.50	75.00	87.50	62.81	71.88	59.06
1685—86	33.32	52.81	81.25	75.00	—	70.63	70.31	—
1686—87	30.94	59.38	76.88	75.00	—	72.50	65.00[4]	59.38
1687—88	33.32	62.81	75.00	75.00	—	72.81	? [5]	62.81
1688—89	33.32	59.22	81.25	? [5]	74.77	—	? [5]	59.22
1689—90	33.32	59.90	81.25	75.00	75.00	68.96	? [5]	59.90
1690—91	33.32	66.18	75.00	75.00	75.00	66.18	? [5]	70.00
1691—92	33.18	61.09	76.25	—	? [5]	66.18	? [5]	61.09
1692—93	33.18	64.31	78.75	75.00	75.00	70.28	? [5]	64.31
1693—94	33.18	60.00	78.75	77.53	75.00	70.63	? [5]	61.15
1694—95	33.18	47.34	77.31	75.00	75.00	71.25	? [5]	60.74
1695—96	33.18	64.38	79.18	75.00	64.38	66.29	74.68	68.80
1696—97	33.18	64.81	78.75	75.00	64.81	66.17	73.83	68.13
1697—98	33.18	60.94	76.88	75.00	60.94	66.17	—	66.48
1698—99	?	63.75	100.31	75.00	65.63	66.04	75.00	72.29
1699—1700	?	65.63	76.88	75.00	65.63	66.25	75.00	67.50
1700—01	36.05	? [3]	? [3]	? [3]	? [3]	? [3]	? [3]	? [3]
1701—02	37.33	62.19	82.50	75.00	65.63	71.88	75.00	62.19
1702—03	37.33	65.00	80.31	76.25	65.63	69.22	75.00	65.00
1703—04	?	65.31	76.88	75.00	65.63	—	75.00	65.31
1704—05	?	63.43	76.88	75.00	63.75	—	75.00	65.00
1705—06	?	? [3]	? [3]	? [3]	? [3]	? [3]	? [3]	? [3]
1706—07	?	57.50	77.50	66.25	61.88	57.50	75.00	66.68
1707—08	?	57.50	77.50	66.25	61.88	57.50	75.00	62.81
1708—09	?	56.56	77.50	66.25	61.88	57.19	75.00	65.00
1709—10	?	63.75	80.00	66.25	73.75	66.25	75.00	63.75
1710—11	?	63.75	77.50	66.25	70.63	66.25	75.00	63.75
1711—12	?	65.00	77.50	66.25	70.63	66.25	75.00	65.00

Table V (cont.)

Year	Japan Cost	All factories		Porto Novo Sale	Masu-lipatam Sale	Bengal[1] Sale	Cochin Sale	Surat Sale
		Lowest Sale	Highest Sale					
1712—13	?	65.63	77.50	66.25	70.00	66.25	75.00	65.63
1713—14	?	66.25	77.50	67.50	71.88	66.25	75.00	66.68
1714—15	40.14	67.50	80.00	75.00	75.00	67.50	75.00	67.50
1715—16	?	67.50	80.00	75.31	75.31	77.50	75.00	68.75
1716—17	40.39	60.94	80.00	77.50	79.38	72.19	75.00	68.44
1717—18	?	61.56	83.75	80.00	82.19	75.62	75.00	68.75
1718—19	?	60.31	85.00	85.00	80.00	68.44	75.00	68.75
1719—20	45.18	62.81	85.00	82.50	80.00	66.25	75.00	63.75
1720—21	?	? [3]	? [3]	? [3]	? [3]	? [3]	? [3]	? [3]
1721—22	?	65.31	85.00	82.50	80.00	68.44	—	65.31
1722—23	45.23	61.56	85.00	80.00	80.00	68.44	75.00	65.31
1723—24	?	68.75	82.50	82.50	81.25	68.75	75.00	68.75
1724—25	45.33	70.63	86.88	85.94	80.63	70.63	75.00	—
1725—26	45.47	68.75	94.68	90.94	80.00	77.81	75.00	68.75
1726—27	?	83.44	97.19	96.56	—	83.44	—	—
1727—28	45,25	69.07	100.63	100.00	83.13	88.13	95.93	69.07
1728—29	45.19	80.31	107.81	107.50	80.31	89.38	102.19	83.13
1729—30	?	80.31	107.81	107.81	80.31	—	102.19	—
1730—31	?	86.88	107.81	107.81	86.88	88.13	102.50	—
1731—32	45.26	82.19	107.81	106.25	88.13	88.13	102.50	82.19
1732—33	?	82.19	102.50	102.50	88.13	82.19	102.50	82.50
1733—34	?	78.13	102.50	102.50	88.75	82.81	102.50	78.13
1734—35	?	79.38	102.50	102.50	88.75	86.88	102.50	79.38
1735—36	?	78.13	102.50	102.50	89.69	83.75	102.50	78.13
1736—37	?	83.75	102.50	—	89.38	83.75	102.50	—
1737—38	?	78.75	102.50	102.50	91.88	83.75	102.50	78.75
1738—39	?	? [6]	102.50	102.50	—	83.75	102.19	? [6]

* Indian figures: The yearly *Indische Prijscourant van verkoghte Coopmanschappen* in the collection *Overgekomen Papieren,* Kol. Arch. Japanese figures: See sources for table III.

[1] Later on Hooghly when the Bengal factories are specified.
[2] The price of 62.89 fl. is to be found in the price-list of the financial year, 1677—8.
[3] The price-list of this year missing.
[4] The price will be found in the price-list of the following year.
[5] In these years the ledgers did not arrive in Batavia. For this reason the figures are missing from the price-list.
[6] The price-list of this year fragmentary.

Table VI *

Profit and Loss of the Copper and Total Trade of Indian Factories, 1701–24. (Florins).

Year	Ceylon		Coromandel		Malabar		Bengal		Surat	
	Copper	Total	Copper	Total	Copper	Total	Copper	Total	Copper	Total
1701—02	+ 63,330	÷ 169,257	+ 137,433	+ 318,913	+ 46,384	÷ 68,472	+ 96,094	+ 256,884	+ 178,309	+ 524,820
1702—03	+ 34,040	÷ 349,921	+ 125,688	+ 26,876	+ 10,768	÷ 93,185	+ 147,442	+ 150,245	+ 34,546	+ 212,576
1703—04	+ 40,214	÷ 278,601	+ 45,310	+ 124,227	+ 30,704	÷ 422,220	—	+ 120,171	+ 23,292	÷ 57,846
1704—05	+ 35,395	÷ 315,022	+ 31,905	÷ 98,864	+ 31,118	÷ 113,286	—	÷ 312,000	+ 20,633	÷ 277,817
1705—06	+ 18,658	÷ 171,010	+ 57,092	+ 20,218	+ 39,888	÷ 75,710	+ 56,471	÷ 180,656	—	+ 363,427
1706—07	+ 36,478	÷ 200,075	+ 76,918	÷ 69,258	+ 50,663	÷ 104,665	+ 50,168	÷ 42,302	+ 88,635	+ 177,626
1707—08	+ 34,402	+ 192,695	+ 102,889	+ 16,076	+ 49,157	÷ 28,475	+ 83,863	+ 47,757	+ 147,336	+ 600,973
1708—09	+ 42,520	+ 77,609	+ 105,862	+ 54,229	+ 40,860	÷ 89,745	+ 99,989	+ 194,588	+ 114,951	+ 694,217
1709—10	+ 20,584	+ 32,340	+ 118,801	+ 224,816	+ 45,257	÷ 143,828	+ 169,854	+ 54,909	+ 125,542	+ 288,563
1710—11	+ 14,183	+ 41,157	+ 106,610	+ 13,985	+ 31,692	÷ 232,646	+ 177,689	+ 258,053	+ 111,751	+ 425,359
1711—12	+ 23,054	+ 66,920	+ 97,627	+ 30,439	+ 37,913	÷ 139,246	+ 38	+ 203,866	+ 98,465	+ 450,610
1712—13	+ 35,731	÷ 85,992	+ 126,467	+ 50,425	+ 18,450	÷ 129,221	+ 81,875	÷ 184,844	+ 81,124	+ 451,033
1713—14	+ 92,691	+ 236,329	+ 29,983	+ 224,790	+ 34,003	÷ 149,857	+ 92,997	+ 3,613	+ 75,960	+ 408,643
1714—15	+ 49,807	+ 277,179	+ 120,293	+ 140,679	+ 25,850	÷ 136,557	+ 150,519	+ 105,001	+ 9,088	+ 582,920
1715—16	+ 46,341	+ 514,137	+ 56,171	+ 217,573	+ 23,464	÷ 171,171	+ 197,152	+ 52,867	+ 52,491	+ 434,442
1716—17	+ 62,357	+ 437,658	+ 136,139	+ 130,760	+ 32,040	÷ 498,952	+ 43,782	+ 280,828	+ 34,645	+ 372,724
1717—18	+ 67,463	÷ 53,152	+ 95,530	+ 211,202	+ 37,030	÷ 373,039	+ 130,190	÷ 9,556	+ 52,616	+ 387,437
1718—19	+ 58,637	÷ 29,326	+ 46,356	+ 64,703	+ 36,991	÷ 486,097	+ 77,362	+ 85,700	+ 46,712	+ 413,643
1719—20	+ 87,568	+ 31,182	+ 51,312	+ 205,536	+ 34,626	÷ 501,245	+ 36,148	÷ 57,200	+ 130,148	+ 427,610
1720—21	+ 69,468	+ 80,080	+ 74,863	+ 293,888	+ 29,738	÷ 378,201	+ 52,353	÷ 59,162	+ 15,543	+ 448,357
1721—22	+ 102,513	+ 187,078	+ 145,203	+ 336,515	—	÷ 196,347	+ 40,527	÷ 20,694	+ 62,918	+ 338,173
1722—23	+ 105,225	+ 249,554	+ 121,245	+ 476,725	+ 44,304	÷ 228,300	+ 87,936	÷ 42,397	+ 52,591	+ 422,476
1723—24	+ 84,181	+ 87,483	+ 98,368	+ 294,604	+ 60,173	÷ 191,947	+ 40,078	÷ 42,829	+ 39,546	+ 182,588

* Copper figures: See sources for Table II. Total trade figures: G. C. Klerk de Reus, *op.cit.*, Table IX.

– the difference in value between 'ligt gelt' and 'swaar gelt' being taken into account – is considerably above that obtained in Europe. Now that conditions were favourable, it was a profitable transaction to carry European copper to the East. The English did so after 1731, and on several occasions the Dutch planned to do so.[86] The cause of the steep rise in prices is in the first place to be found in the decreasing export from Japan. The rising prices of copper in the 1720's also explain the fact that the Dutch company sent no copper to Europe at that time.

The general importance of copper to the Asiatic trade of the Dutch is also apparent on several occasions in the eighteenth century when the profitableness of the Japan trade came under discussion, for instance in 1724 when a thorough examination was made in order to elucidate the importance of the Japan trade in its most comprehensive sense, i. e. also counting the profits gained by the re-sale in India of Japanese products. The result of the examination, in respect of copper, is seen in table VI in which the gross profits of the individual factories are compared with the total annual net profit of the factories' entire trade, this being the usual procedure of the court of directors in Holland when they wanted to form an impression of the proportion held by copper in the total trade.

The result of the examination was that, in the course of the 25 years mentioned above, a gross profit of 7,932,855 fl. had been made by the sale of copper in India. The sale of Japanese camphor had yielded a gross profit of 87,977 fl., while the exportation of gold coubangs had involved a loss of 1,566,600 fl. Taken altogether a total profit of 6,454,231 fl. had been gained in Asia by the sale of these Japanese products. In the same period, the profit in Japan itself was 9,461,326 fl. From this sum must be deducted the value of lost cargoes, i. e. 1,725,625 fl. Counting both the profit from trade in Japanese products in India and the profit from trade in Japan itself the final results is a total profit of 14,189,932 fl. In this statement the profits gained by the sale of Japanese goods in Europe have been left out. These profits constitute a sum of 2,433,347 fl. for copper only. – Demands for the closing of the factory in Nagasaki gave rise to this examination which, however, resulted in a proposal to continue the trade on Japan. The proposal was accepted.[87]

---

[86] H. Hamilton, *The English Brass & Copper Industries to 1800* (London, 1926), p. 280. For the Dutch schemes in 1735 for copper exportation from Malacca, Persia and Sweden see J. Feenstra Kuiper, *op.cit.*, p. 152.

[87] Kol. Arch. No. 1931, fol. 2178 ff. Kol. Arch. No. 201, 21/8 1727 and 4/8 1728.

The decisive rôle of copper in the company's Asiatic trade becomes still more obvious when seen in relation to the ever increasing losses incurred in the trade of several Indian factories. This point was particularly stressed by van Imhoff, the enterprising governor-general, who in his attempts to make the Asiatic trade profitable attached most importance precisely to copper and spices, these two principal commodities having always yielded large profits.[88] Van Imhoff was of the opinion that the main point was again to raise the export of copper. This would improve the situation, and to this end he examined the business methods in Asia, including Japan, more closely. He was especially amazed at the cost prices of copper bars. Since the 1720's, these had been a point of contention between the Japanese and the Dutch. In the following section a brief outline of the price development of Japanese copper will be given as the background against which van Imhoff's reform proposal should be seen.

VIII

Influenced by the increasing Dutch demand from 1656 the cost prices of copper in Japan tended to rise.[89] Furthermore, a similar increase in the domestic demand for copper made itself felt there in the middle of the 1660's by reason of the coinage of copper coins. The result was a steep rise in the price of copper bars. In the beginning of the 1650's the company bought its copper at 10–11 tael per pikol, in 1665–66 the price had risen to 12.4 tael and in 1670 it was 12.7 tael. In the intermediate years, it had been still higher although it is not possible to give the exact figures.[90] The rise in prices caused the court of directors in Amsterdam some anxiety.[91] The Japanese production of copper, however, soon became large enough to satisfy the considerable demand of both the Dutch and, more particularly, of the Chinese.

When an ordinance against the export of silver was issued in 1668, the Dutch company began to use gold coins in its trade in Nagasaki. The appreciation in 1672 of the coubang from 58 to 68 maas became, as has been shown, disastrous to the company's trade in gold. Conditions were favourable, however, for the export of copper, the Japanese merchants being also bound to accept coubangs

---

88 G. W. van Imhoff, *op.cit.*, p. 494 ff., and his *Consideratiën over den Japanschen handel,* published in *Tijdschrift v. Nederlandsch-Indië* 1853.

89 Pieter van Dam, *op.cit.*, p. 408.

90 Oskar Nachod, *op.cit.*, table F. Pieter van Dam, *op.cit.*, pp. 438 and 441.

91 Kol. Arch. No. 256, 2/9 1670.

at the new high price of 68 maas though they had still to deliver copper at the old price.[92] In view of this it is in fact surprising that prices, in the course of the following years, fell to 11.9 tael per pikol (1676) and remained at that level up to 1691. The company had expected that the change of coin would involve a rise, but the abundant output had evidently kept the prices down. In the 1690's, the Japanese copper merchants in Nagasaki began to complain of the decreasing output of the copper mines.[93] Later the complaints grew ever more frequent. There can thus be no doubt that the rise in the price of copper in the beginning of the eighteenth century was partly due to a shortness of supply.

To this must be added that the change of currency in 1696 was so radical that it could not avoid influencing the price of copper, which rose to 12.36 tael per pikol in 1700 and to 12.80 tael per pikol in 1701. The violent inflation which harried Japan in the following years made the prices rise further. One of the reasons for this was that the inflation involved a shortness of metal. In 1716 the company's cost price was as much as 14.84 tael per pikol. With this figure, however, the upward tendency reached its climax, and after 1721 prices in Japan tended to decline.[94] During the last years of the period the company watched jealously over its cost prices for Japanese copper. The Japanese asserted in 1729 that they virtually delivered copper to the company at a figure which was below the market price and proposed that the prices should be raised. In this way the company would get a higher price for its imported goods, but would on the other hand have to pay more for copper.[95] The court in Batavia directed the Nagasaki factory firmly to decline this proposition. In drastic terms the decreasing profits of the Indian factories were described, copper being nearly the only commodity which brought in a profit. If this profit were cut down, owing to a higher cost price, the factory in Japan might as well be closed.[96]

When at intervals a closing of the Nagasaki factory was proposed in the 1720's and 1730's, this was due to the decreasing profits realized by the trade in Japan. As mentioned above the profit was reckoned as the difference between the selling price less current expenses and the invoice price of the imported goods.[97] Right up to 1743 the factory in Nagasaki brought in a profit, but the profit on the imported goods was veiled by the fact that the export of gold

[92] Pieter van Dam, *op.cit.*, p. 446 f.
[93] Pieter van Dam, *op.cit.*, *p.* 524 f. J. Feenstra Kuiper, *op.cit.*, p. 100.
[94] J. Feenstra Kuiper, *op.cit.*, pp. 102—3 and 110 together with table II.
[95] J. Feenstra Kuiper, *op.cit.*, p. 118.
[96] Kol. Arch. No. 646, 31/3 1730.
[97] J. Feenstra Kuiper, *op.cit.*, table II.

coubangs had involved continuous losses since 1696. In 1730 the company began to deduct these losses from the gross profits of the Japanese factory. The result was that the net profit fell from 64 % in 1729 to 9 % in 1730. It would have been consistent with this procedure also to include the profit on copper and other products re-sold in India and Europe – in 1724 an attempt to do so was in fact made, as shown above – but this would have necessitated a complete re-organization of the method of keeping accounts. So far as that the company was not ready to go. Already in 1734 it gave up deducting the loss on gold from the profits. The net profits rose accordingly, from 19½ % in 1733 to 45 % in 1734. In spite of this the profits on the sale of imported goods continued to decrease in later years.

In these circumstances van Imhoff's project for a revision of the value of the company's coins in relation to the Asiatic coins must act as a radical break with the past. For if the Japanese tael was reduced from 70 to 40 stuivers in the company's accounts this would be equal to the loss of all profit in the trade of the Nagasaki factory. In van Imhoff's view, however, the Japan trade was identical with the copper trade, and the only way in which larger quantities of this commodity, the company's principal one, could be secured was by meeting the demands of the Japanese for higher prices. The dominant concern of the court of directors – acting in accordance with Amsterdam's way of conducting business – was with the sale of imported goods in Japan. According to van Imhoff's project, however, the determining factors would be the cost price of copper in Japan and the prices obtainable by re-sale in Asia and Europe.

IX

The figures given above are the figures of the company: those of the company's exports from Japan and imports into Europe, the sales and the profits of the company's factories in India. Although the Dutch company held a privilege among its European competitors on the Japan trade itself, it did not, however, hold a complete monopoly of Japanese copper in Asia, partly owing to the Chinese trade, partly owing to the activity of private traders and interlopers. We here touch a sphere where information as to quantity is, by the nature of the matter, scarce and uncertain. Nevertheless it claims our attention as at intervals it occupied the thoughts of the court of directors in Amsterdam and of the governor-general and court in Batavia.

There is much evidence that points to the Chinese trade in Nagasaki having been very considerable compared to that of the Dutch. In their silk the Chinese possessed an excellent key with which to open up the Japanese market, and by the introduction of the trade limitation in 1685 their maximum sale was fixed at 600,000 tael annually while the Dutch trade was fixed at only 300,000 tael annually. At the same time the activity of the Dutch private traders was sanctioned by the Japanese and fixed at 40,000 tael annually. Against this an ineffectual protest was made by the court of directors in Amsterdam.[98] In the following years the Chinese quota became, in fact, still larger. About the year 1700 Pieter van Dam made a note to the effect that the Chinese junks in Nagasaki sold three times as much as the Dutch company.[99] Contributing to bringing this about was the fact that the Chinese carried on business as private merchants in contrast to the company, which, in relation to the Japanese, held the position of one big business house. After the ordinance against the export of silver in 1668 the Dutch company was obliged to use Japanese gold as a means of payment, whereas the Chinese merchants were still allowed to use silver in the retail trade, and this was a more profitable form of tender.[100]

Like the Dutch the Chinese merchants were interested in copper. The greater part of their export was presumably shipped to China but a certain part of the copper was carried to farther regions of Asia and sold there. This was done, partly through other European nations' trade with China, partly by Chinese junks which traded to the Fillipines, Further India and Java. Throughout the period – apart from a short interruption at the beginning of the eighteenth century – there was a regular trade between China and Batavia. The Chinese junks also frequently called at Bantam, the centre of the pepper trade and one of the strongholds of the independent interlopers' trade in the seventeenth century. Thus there were many channels by which Japanese copper could reach India without passing through the Dutch company. It was even possible for the supplies to Batavia to come outside the company's trade. If the court did not want to buy the copper at the public auction sales, the copper went into the private trade. The Chinese trade based on Batavia was in other ways of great importance to the Dutch company, which was provided thereby with many commodities that the company, after the conquest of Formosa, was debarred from buying in direct trade with Chinese territory. For certain years, too, the

98 Pieter van Dam, *op.cit.*, p. 487. Kol. Arch. No. 461, 14/7 1695.

99 Pieter van Dam, *op.cit.*, pp. 534 and 543.

100 J. Feenstra Kuiper, *op.cit.*, p. 99 f.

Chinese supplies of Japanese copper were a welcome addition to the company's own export from Japan.[101] In 1684 the company gave permission to the Chinese merchants, domiciled in Batavia and Bantam, to trade with Japan, the permission being given in view of the fact that they had for many years been carrying on this trade.[102] As early as in 1664, the company had likewise given permission to Chinese, residing in Japan, to trade in Malacca which at this time became the centre from which the company distributed the products of the Japan trade to the various Indian factories.[103] The Dutch looked less kindly at the Chinese traffic based on Cambodia, where in the 1650's and 1660's the company tried to monopolize the export of hides and leather to Japan. In a report of 1664 mention is made of an agreement with the king of Cambodia concerning this question. The agreement provided that no Chinaman should be allowed to sail to North Cambodia, although he might freely sail south, to Siam, Malacca, Borneo and Batavia. The terms of this agreement could, however, not be enforced.[104]

About the middle of the 1650's, as a result of the European demand for copper, the Dutch company took measures to secure larger quantities of Japanese copper bars, and realized then that it might expect competition from the Chinese. In 1657 the court in Batavia wrote to the factory in Japan that it should try to put obstacles in the way of the Chinese in the buying up of copper, Chinese merchants having the year before conveyed 17–18,000 pikols of copper to Batavia. This copper had been sold to private traders at 20–21 reals per pikol and from there carried to Coromandel, Surat and Persia – in the company's own ships.[105] The Dutch company therefore recommended that the factory in Japan should bid up the price of copper somewhat in order to secure the required quantities.

The endeavour – through the buying up of larger quantities of copper in Nagasaki – to prevent the Chinese from conveying copper to India and the necessity of buying up 'free' copper at high prices in Malacca, Siam or Batavia were the two problems that the Dutch company had to cope with in the last quarter of the seventeenth century. According to information which is given by

---

101 Kol. Arch. No. 461, 7/9 1696.

102 Kol. Arch. No. 459, 8/10 1686.

103 Kol. Arch. No. 579, 1/4 1664.

104 *Dagh-Register* Batavia 1656, p. 36 ff. Hendrik P. N. Muller, *De Oost-Indische Compagnie in Cambodja en Laos* ('s-Gravenhage, 1917), pp. 386—7, *Werken uitg. door de Linschoten-Vereeniging,* deel XIII.

105 Pieter van Dam, *op.cit.,* p. 403 ff.

Pieter van Dam, the Chinese copper purchases nearly equalled those of the Dutch in the course of the 1670's. In the year 1679 the company exported 23,500 pikols of copper bars, while the Chinese purchased 18,477 pikols, which were partly carried to Taiwan, partly to Siam, where it was sold to the English, the Danes and others, and carried on to Coromandel and Bengal "involving a great loss by putting obstacles in the way of the Company's sale there".[106] In the same year Chinese craft carried at least 4,358 pikols of copper bars to Batavia,[107] and the court formed a plan for buying up the total quantity. In accordance with this plan the company, in March, 1680, secured 3,972 pikols of copper arriving in a Chinese junk, at 20 rixdollars per pikol.[108] A similar quantity arrived in Batavia later in the summer, and the same year the Dutch resident in Bantam reported the arrival of an English hooker from Amoy, carrying 1,000 pikols, as well as a Chinese junk from Japan, with 2,000 pikols of Japanese copper bars.[109] Later in the same year it was reported from the Coromandel coast that part of the copper from Bantam had been carried to the coast. Two Danish ships, for instance, carried 900 pikols to the colony at Tranquebar.[110] From Malacca, too, came mention of private traders' interest in copper. In April, for instance, an English ship had sailed by, carrying amongst other goods 300 pikols of copper from Macao to Bengal.[111]

The interlopers' activity, which was rendered possible by an increasing Chinese export in connection with a simultaneous rise in the demand for copper in India, is the background against which the great project formed by the Dutch company, in the years 1683–84, should be seen. The project aimed at buying up all the Japanese copper intended for export which was estimated at between 50,000 and 60,000 pikols. A proposition was made to the Japanese for the taking over of this quantity, at a price of 11.9 tael per pikol to be paid for by supplying Japan with such commodities from outside the country as might be desired. The terms of exchange were to be governed by the cost prices of the imported commodities and the contract should, in the first instance, be binding for a period from 3 to 4 years. As mentioned above, the court of directors in Amsterdam undertook to take over for sale in Europe the part of the copper which could

106 Pieter van Dam, *op.cit.*, p. 464, »tot een groot hinder en nadeel in den vertier van 't koper van de Compagnie aldaar.«

107 *Dagh-Register* Batavia 1679, pp. 222—3 and 343—5.

108 *Dagh-Register* Batavia 1680, p. 101.

109 *Dagh-Register* Batavia 1680, pp. 406—8 and 57.

110 *Dagh-Register* Batavia 1680, p. 728.

111 *Dagh-Register* Batavia 1680, pp. 366—7.

not be disposed of in India.[112] The Japanese, however, did not accede to the proposition. They maintained that the copper must be paid for in cash, asserting, moreover, that they were unable to supply such large quantities. After this the factory in Nagasaki was commissioned to buy as much as possible on the previous terms, and in Batavia attempts were made, in various ways, to obstruct the trade of the Chinese, living there, with Japan.[113]

According to Pieter van Dam, the Chinese – who exported not only copper bars but also scrap copper and copper plates – seem to have taken the lead, superseding the Dutch, at the close of the 1680's. In 1687 they exported nearly 34,000 pikols, in 1689 no less than 57,352 pikols and again 34,000 pikols in 1691. The Chinese export for 1696 is estimated at 90,000 pikols, compared to which the Dutch export was moderate, namely 16,500 pikols.[114] The buying up of Japanese copper from the Chinese junks in Batavia continued in the 1690's. The expensiveness of this procedure was pointed out in a letter sent by the court of directors in Amsterdam to the governor-general and court in Batavia in the autumn of 1696.[115] The year before, the court in Batavia had bought 46,604 lbs. of copper bars at 63 fl. per pikol (of 122 lbs.), while the price in Nagasaki was 41 fl. 7 stuivers for 1 pikol (of 125 lbs.). An increase of the purchases in Japan was therefore recommended and the possibility of loading the ships in a more competent manner was discussed. The court of directors also suspected the employees of the company of carrying on an interloping trade and thought that this was at the root of the low export figures. As mentioned above the Dutch exports of copper from Japan rose rapidly in the years 1697–99, one of the reasons for this being the temporary demand from Europe. Later, however, the figures fell. The Chinese export also, no doubt, decreased considerably, the Japanese measures for regulating the foreign trade applying to both parties.

As mentioned above the prices at which the company bought Japanese copper in Batavia were considerably higher than the cost prices in Nagasaki. Part of the difference in price was due to carriage expenses. When the court in Batavia introduced a fixed price for the copper which it compelled the Chinese merchants to hand over, this price was also a fairly high one. In 1701 the price was, for example, 24 rixdollars or 72 fl. per pikol, and in 1711 from 20–21 rixdollars or 60–63 fl. per pikol. As was only natural, this method of trade, involving a

112 Kol. Arch. No. 459, 11/12 1684 and 8/10 1685. Pieter van Dam, *op.cit.*, p. 480.
113 Kol. Arch. No. 460, 25/10 1686.
114 Pieter van Dam, *op.cit.*, pp. 493, 500, 509, 518 and 530.
115 Kol. Arch. No. 461, 7/9 1696.

compulsory sale at fixed prices, provoked a good deal of discontent among the Chinese merchants.[116] As late as 1713–14, however, copper was bought from the Chinese in Batavia, though only 44,125 lbs. As is seen from the ledgers of the Batavia factory the purchase stopped after this year though it is not impossible that the Indian factories bought copper from interlopers as was at times the case in the seventeenth century.[117]

The English, who in the beginning of the eighteenth century carried on an extensive trade with China and since 1715 owned a factory in Canton, caused the Dutch company constant anxiety, also in respect of copper. The company realized that the English might gain access to the Japanese market and thus rob the company of the profit that copper still yielded in India. Fear of the English company was not the least reason for the tenacious upkeep of the Japan factory. The future was to prove hat it was the English East India Company that ruined the Dutch trade in copper in Asia. In 1731 the English began to carry English copper into the East. This marks the beginning of the struggle which ended in the defeat of Japanese copper.[118]

---

116 Kol. Arch. No. 732, 25/1 1701, 3/2 and 17/4 1711. J. Feenstra Kuiper, *op.cit.*, p. 112.

117 Kol. Arch. No. 10.813 ff.

118 H. Hamilton, *op.cit.*, p. 280. Holden Furber, *John Company at Work, A Study of European Expansion in India in the Late Eighteenth Century* (Cambridge, Mass., 1948), p. 286, *Harvard Historical Studies*, vol. LV.

6

A Note on the Relative Importance of Slaves and Gold in West African Exports

Richard Bean

WHEN one thinks of European trade with sub-Saharan Africa before the nineteenth century, the trade in slaves immediately comes to mind. Slaves made up a part of African exports from the beginning of European contact, but slaves did not at first dominate the trade. In West Africa, the export of cotton cloth, wax, hides, wood, gum, pepper, ivory, and most especially gold, were together much more important in terms of value than were slaves for quite some time. The biggest part of West African gold was exported through the Gold Coast, a fact which goes far toward explaining why this area was consistently the scene of the most intense rivalries between European traders in Africa.

It is not always appreciated how late was the date when slaves became the most important African export.[1] Nevertheless, there does exist sufficient data, already published, which can be combined to throw considerable light upon the makeup of the foreign trade of West Africa through the last five centuries. The data at hand are crude and incomplete, but there is quite enough when properly combined to indicate that our generally held views on the relative importance of slaves in West African exports are substantially incorrect. Quantitative methods are certainly not the whole key to understanding the past, but even such elementary arithmetic as that in this note can be useful.

There are no really reliable statistics on pre-colonial African trade although Philip Curtin has provided aggregate figures for the slave trade which future researchers are unlikely to substantively alter.[2] Much less work has been done on other African exports, but it is possible to delineate at least the general dimensions of the trade in gold. In the late fifteenth and early sixteenth century, Europeans (almost all Portuguese) were carrying away about £100,000 worth of gold per year.[3] By the mid-sixteenth

[1] For contrasting statements, see Paul Bohannan and Philip Curtin, *Africa and Africans* (Garden City, New York, revised ed., 1971), 261; K. O. Dike, *Trade and Politics in the Niger Delta, 1830–1885*, Oxford, 1956), 1; Marion Kilson, 'West African Society and the Atlantic Slave Trade, 1441–1865', in *Key Issues in the Afro American Experience*, 1 (New York, 1971, edited by Huggins, Kilson and Fox), 39; and J. D. Fage, 'Slavery and the Slave Trade in the Context of West African History', *J. Afr. Hist.*, x (1969), 397.

[2] Philip Curtin, *The Atlantic Slave Trade: A Census* (Madison, 1969).

[3] K. Y. Daaku, *Trade and Politics on the Gold Coast, 1600–1720* (Oxford, 1970), 8; and J. W. Blake, *European Beginnings in West Africa, 1454–1578* (New York, 1937), 83. Both cite Duarte Pacheco Pereira, *Esmeraldo de situ orbis* (edited by G. H. Kimble, Hakluyt Society, 2nd ser., no. 69: New York, 1937 but written about 1507), in estimating 170,000 dobras of gold per year moving from West Africa to Portugal after 1481. This amounts to 3,400 marks of gold or £108,000 sterling (1601 standard) if we use the rate of 50 dobras per mark found in W. A. Shaw, *The History of Currency, 1252 to 1894* (New York, 1895),

century, French and English interlopers had cut into the Portuguese monopoly and were diverting perhaps £50,000 worth of African gold to northern Europe.[4] A century later, the Dutch West India Company (as successor of the Portuguese) was averaging almost exactly the same amount per year—£99,584 per year between 1621 and 1634.[5] Although the Dutch West India Company was by far the largest gold trader at this time, Dutch interlopers and the traders of other nationalities may have together accounted for an equivalent amount of gold. Nicholas Crispe claimed that he and his English associates had taken out of Africa a half million pounds worth of gold in the eight-year period 1636 to 1644, but this claim of £62,500 per year for the Guinea Company was probably an exaggeration.[6]

In 1665, the Company of Royal Adventurers of England Trading into Africa made the anticipatory estimate that its gold trade should be worth £200,000 per year.[7] The expectation was exaggerated and surely could only have been realized if the Company had ousted all other Europeans from the trade. The Company of Royal Adventurers failed, and the much more powerful successor to the English monopoly over African trade, the Royal African Company, managed to produce gold shipments which ran at only about £25,000 per year in the decade before the Company's trade was reduced by the Glorious Revolution and war after 1689.[8] Estimates of total West African gold exports run at about £250,000 per year at the beginning of the eighteenth century.[9] This should be taken as an upper

---

323 ff. However, a careful reading of Pereira indicates that the 170,000 dobras figure was probably for Elmina alone, while Axim produced another 30,000 to 40,000 dobras per year (p. 117) and the Sierra Leone area yielded about 9,000 dobras per year (p. 107). These additions would add another £30,000 sterling to the total of West African exports. This is confirmed by Vitorino Magalhães-Godinho, *L'économie de l'empire Portugais aux XVe et XVIe siècles* (*Paris, 1969*), 219.

[4] Ibid. 216, documents the decline of the Portuguese gold traffic. The dimensions of the leakage are indicated by Thomas Astley, ed., *A New General Collection of Voyages and Travels* . . . (London, 1968, first published 1745), I, 142, 147, 169; and James A. Williamson, *Sir John Hawkins, The Time and the Man* (Westport, Conn., 1970), 41 ff.

[5] K. K. Daaku, *Trade and Politics on the Gold Coast, 1600–1720* (Oxford, 1970), 11, 14; 40,461 marks of gold in thirteen years is reported.

[6] R. Porter, 'The Crispe Family and the African Trade in the Seventeenth Century', *J. Afr. Hist.*, IX (1968), 64.

[7] Great Britain. *Calendar of State Papers, Colonial Series*, VI, 226.

[8] K. G. Davies, *The Royal African Company* (London, 1957), 360. He has annual data, 1673–1713, for the number of guineas coined by the Mint for the Royal African Company which includes nearly all the gold imported by the Royal African Company.

[9] W. Bosman, *A New and Accurate Description of the Coast of Guinea* (New York, 1967), 89, says that in peacetime the Gold Coast should supply a total of 7,000 marks (£224,000). It seems unlikely that this flow could have been sustained over a long period of peace as the previously accumulated stocks declined. Bosman considerably overestimated the size of the annual Royal African Company exports as 1,200 marks (£38,400) while we know that the Royal African Company actually only reached this level in two years, 1685 and 1688. To balance against Bosman's probable overstatements for the Gold Coast, one must add in the relatively small exports of the Senegambia which he presumably was not including. J. D. Fage, *A History of West Africa: An Introductory Survey* (Cambridge, 1969), 74, guesses the total for West Africa at this time to have been £250,000 per year.

limit figure at a time when Brazilian gold was actually flowing into West Africa in exchange for slaves.[10] Through the course of two centuries, 1500 to 1700, West African gold exports seem to have risen from something over £100,000 per year to perhaps as high as £250,000 per year.[11] An average of £200,000 worth of gold being carried away each year by the Europeans would probably not be far off.[12]

The figure of £200,000 in gold per year is, of course, an educated guess. The only 'hard' numbers for total West African gold exports are Pacheco Pereira's figure of £130,000 for about 1500, and William Bosman's figure of £225,000 plus for about 1700. In both cases these were informed contemporaries with access to official data. The more fragmentary data mentioned are all consistent with an overall annual average of £200,000 for the period 1500 to 1700, but do not rule out £150,000 or £250,000 being the true figure. Neither figure would contradict the conclusions reached later in this paper. There can be little doubt that the true export figure varied substantially from year to year with the vagaries of war and peace in Africa and in Europe. The only reason that we have *any* reasonably hard figures for total gold exports before the modern era is because the trade was a tightly held monopoly of the Portuguese crown at first, and was then struggled over by a Dutch and a British monopoly company.

After 1700 the number of separate traders was greatly expanded and no one had either the incentive or the means to keep track of the aggregate trade. As a consequence, information on aggregate gold exports during the eighteenth century is even more sparse than for the earlier period. Nevertheless, there are enough hints to indicate that gold exports were much more likely to have fallen below the level of the previous century than to have risen above it. English ship captains testified to importing gold into the Gold Coast to trade for slaves in the 1760s and 1770s, so it is even possible that West Africa stopped being a net exporter of gold for a time.[13] West African gold exports may well have declined in the eighteenth century as the increased wealth stemming from rising slave exports and rising slave prices induced the Africans to retain more of their gold for domestic use.[14] A decline in gold production may also have occurred.

[10] A. F. C. Ryder, 'The Re-establishment of Portuguese Factories on the Costa de Mina to the Mid-eighteenth Century', *Journal of the Historical Society of Nigeria*, I, no. 3 (1958), 158.

[11] Two modern estimates based on the same data are E. W. Bovill, *The Golden Trade of the Moors* (London, 2nd ed., 1968), 118; and A. F. C. Ryder, 'Portuguese and Dutch in West Africa before 1800', in *A Thousand Years of West African History* (Ibadan, 1969, edited by J. F. Ade Ajayi and Ian Espie), 222.

[12] Raymond Mauny's estimate of nine tons per year, inclusive of the gold going north over the Sahara, is almost certainly two or three times too large. Raymond Mauny, *Tableau Géographique de L'Quest Africain au Moyen Age.* (Mémoires de L'Institut Francais d'Afrique Noire, no. 61, Dakar, 1961), 300.

[13] Marion Johnson, 'The Ounce in Eighteenth-Century West African Trade', *J. Afr. Hist.*, VII (1966), 204.

[14] J. D. Fage, *A History of West Africa: An Introductory Survey* (Cambridge, 1969), 74; W. Rodney, 'Gold and Slaves on the Gold Coast', *Transactions of the Historical Society of Ghana*, X (1969), 17.

Higher export prices for slaves should have driven up the labour costs of gold extraction in Africa and the depletion of the most easily accessible deposits would have had a similar effect.

Slaves may be compared in importance with gold as an African export by multiplying the annual number shipped by their price in America. This C.I.F. valuation is comparable with the gold trade valuations which were C.I.F. in Europe. If we use £20 sterling as the average price of a slave in America, it would take an annual export of 10,000 slaves to match in value a £200,000 annual gold trade.[15] Total annual slave exports did not average 10,000 until the third quarter of the seventeenth century, and if we exclude the Angola–Brazil shipments, not until the last quarter.[16] This relative unimportance of slaves is reflected in the 1665 estimate by the Company of Royal Adventurers that, of their gross annual yearly income in the African trade, one-fourth would be derived from slave sales and three-fourths from gold and other African commodities.[17] Even the Royal African Company was to derive 40 per cent of its gross income from the sale of African commodities other than slaves.[18]

From the perspective of the Africans, slaves must have seemed less important in the seventeenth century than the comparison of the C.I.F. values of slaves and gold implies. The high transport and transaction costs of the trans-Atlantic slave trade caused a much larger difference between the African and American prices of slaves than was the case for gold. The F.O.B. price of a slave in Africa in the seventeenth century was less than £4 in European goods (valuing the goods at their prices in Europe) in contrast to a sale price in America nearly £20 sterling.[19] Gold worth £20 in Europe would command more than twice the amount of European goods in African as would a slave worth £20 in America.[20] By this criterion, it would require an export of 25,000 slaves per year to generate as much value to the Africans as was produced by the annual export of £200,000 in gold (C.I.F. value). Curtin estimated that the average annual exports of slaves from all of Africa in the last quarter of the seventeenth century was 24,100.[21] Gold exports and slave exports were thus worth about the same to Africans between 1676 and 1700. Both the price of slaves in Africa and the number of slaves exported doubled after the Peace

[15] Davies, *The Royal African Company*, 364. For a more complete time series of slave prices, see R. N. Bean, 'The British Trans-Atlantic Slave Trade, 1650–1775', unpublished Ph.D. dissertation, University of Washington, 1971, Appendix B.

[16] Curtin, *The Atlantic Slave Trade*, 119.

[17] Great Britain. *Calendar of State Papers, Colonial Series*, VI, 1661–8, 226.

[18] Davies, *The Royal African Company*, 179–80.

[19] Bean, 'The British Trans-Atlantic Slave Trade', Appendices A and B.

[20] An ounce of gold dust (containing substantial adulteration or impurities) usually brought in trade 40 shillings or a little less, worth of European goods at 'prime' cost. Karl Polanyi, 'Sortings and "Ounce Trade" in the West African Slave Trade.' *J. Afr. Hist.*, V (1964), 381 ff.; Johnson, 'The Ounce in Eighteenth-Century West African Trade', 197–214.

[21] Curtin, *The Atlantic Slave Trade*, 119.

of Ryswick in 1697.[22] From then until the reductions in the trade following 1807 slaves were clearly the most important export of Africa.

For sub-Saharan Africa, as a whole, slaves were the most important export for not much more than one century. For the other four centuries of European contact the Africans earned less from the export of slaves than they did from the export of more conventional commodities. There were important regional variations. The bulk of the gold exports emanated from the Gold Coast. Only in the single decade of the 1780s did the Gold Coast earn substantively more than £100,000 per year from the export of slaves.[23] Thus, it was only for a ten-year span in the eighteenth century that slaves clearly earned more foreign exchange annually for the Gold Coast than gold exports had earned per year in the seventeenth century. It is, however, likely that gold exports declined in the eighteenth century so that slaves may have actually dominated Gold Coast exports for a longer period than just the 1780s.

Other areas, such as those now in eastern Nigeria and in Angola, produced no gold. There, slaves may have been more important than the total of all other exports through almost the whole period of European contact. It is certainly possible that if we had statistics for trade in commodities other than slaves and gold, we would find that slaves were less important than they seem in the exports of those areas as well. After 1807, the decline in the number of slaves exported and the rise in the amount of 'legitimate' trade rapidly eclipsed the importance of the slave trade for most of Africa. Between 1820 and 1850, the whole of Africa probably received an average of a little under £300,000 worth of European goods per year in exchange for slaves exported to the Americas.[24] Slaves ceased to be an important source of export earnings for West Africa north of the equator by the 1820s—earning less than £15,000 per year. In contrast, by 1830 Africans were receiving £100,000 per year just for the palm oil that was shipped to England. By 1850, English palm oil imports were earning Africans over £450,000 per year.[25]

In summary, the combination of data on slave prices with Curtin's series on the number of slaves exported yields an estimate for the foreign

[22] Ibid. 216, for the number of slaves. Bean, 'The British Trans-Atlantic Slave Trade', 158, for slave prices in Africa up to 1775. Later prices are shown in Archibald Dalzel, *The History of Dahomy* (London, 1967: reprint of 1793 ed.), 191 n.; John R. Spears, *The American Slave Trade* (New York, 1900), 51; Karl Polanyi, *Dahomey and the Slave Trade* (Seattle, 1966), 167; and W. W. Claridge, *A History of the Gold Coast and Ashanti* (New York, 1964, 2nd ed.), 261.

[23] At least 6,000 slaves per year were exported through the Gold Coast in that decade. The Europeans paid for them an average of a little less than £20 in goods. Curtin, *The Atlantic Slave Trade*, 221; and Bean, 'The British Trans-Atlantic Slave Trade', p. 158.

[24] Ibid. 234 and 258 for the volume of the nineteenth-century slave trade. For nineteenth-century slave prices in Africa, see E. Phillip LeVeen, 'British Slave Trade Suppression Policies, 1821–1865, Impact and Implications', unpublished Ph.D. dissertation, University of Chicago, 1971, 15.

[25] Thomas DeGregori, *Technology and the Economic Development of the Tropical African Frontier* (Cleveland, 1969), 341; and LeVeen, 'British Slave Trade Suppression Policies', 95, give prices and quantities for the palm oil trade.

exchange earnings over time accruing to West Africa as a result of the Atlantic slave trade. This figure was clearly exceeded by West Africa's earnings from gold exports from 1450 until late in the seventeenth century. Not much more than a century later, in the 1830s, the earnings from the export of palm oil surpassed the earnings of the slave trade. If we had sufficient data on price and quantity for the lesser exports, such as spices, ivory, wax, etc., for the earlier period and for those commodities plus gold for the eighteenth and nineteenth centuries, there can be no doubt that the overall dominance of the slave trade would be seen to have been even less.

Concealed within these aggregates figures are very large regional variations. The relative values of slaves and other commodities exported varied between regions, and within regions over time. The slave trade began early and lasted longest in Angola, and it is there that the popular image of slavery dominating African trade probably has the most validity. For aggregate African trade, the dominance of slaves was for a much more brief period than is generally recognized.

7

International Bullion Flows and the Chinese Economy, *circa* 1530–1650

William S. Atwell

DURING MANY OF THE YEARS ENCOMPASSED BY THIS STUDY, MASSIVE amounts of foreign silver flowed into China in exchange for raw silk, silk and cotton textiles, porcelain, copper cash, gold, precious stones, mercury and a wide variety of other products.[1] Students of European, Ottoman and South Asian history during this period perhaps will not be surprised to learn that by facilitating rapid monetary growth and greatly enhancing the efficiency of exchange, these imports of silver played an important role in determining the pace and, to a certain lesser extent, the direction of the country's economic development. As was true elsewhere, however, those imports ultimately proved to be a mixed blessing and, for various reasons to be discussed below, did much to undermine the economic and political stability of the Ming (1368-1644) empire during the last few decades of that dynasty's existence.[2]

THE JAPANESE CONNECTION

From the 1530s until at least the 1570s, China's primary source of foreign silver was western Japan, where substantial deposits of the

* Some of the research for this article was made possible by a study leave from the School of Oriental and African Studies, University of London, and by an award from the National Endowment for the Humanities administered by the American Council of Learned Societies. Although neither the School, nor the Endowment, nor the Council assumes any responsibility for my conclusions, I would like to take this opportunity to thank them for their assistance. An earlier draft of this article was discussed at the Ming History Workshop which was held at Princeton University during the summer of 1979, and I would like to express my appreciation to the participants at the workshop and particularly to Dr. James R. Geiss for their many useful comments at that time. I am of course solely responsible for the errors and shortcomings which remain.

[1] For an exhaustive list of items regularly exported from China to Manila during the late sixteenth and early seventeenth centuries, see Antonio de Morga, *The Philippine Islands, Moluccas, Siam, Cambodia, Japan and China at the Close of the Sixteenth Century*, trans. Henry E. J. Stanley (London, 1868), pp. 337-8. For a list of ninety-six predominently Chinese goods carried by Portuguese merchants from Macao to Nagasaki in 1637, see C. R. Boxer, *The Great Ship from Amacon: Annals of Macao and the Old Japan Trade, 1555-1640* (Lisbon, 1959), pp. 191-7. See also Iwao Seiichi, *Shuinsen to Nihon-machi* [Red-Seal Ships and Japanese Trading Communities Overseas] (Tokyo, 1978), pp. 66-8; Katō Eiichi, "The Japanese-Dutch Trade in the Formative Period of the Seclusion Policy", *Acta Asiatica*, xxx (1976), p. 64.

[2] These issues are discussed in a preliminary fashion in my "Notes on Silver, Foreign Trade and the Late Ming Economy", *Ch'ing-shih wen-t'i*, iii no. 8 (1977), pp. 1-33. Parts of the present study draw upon that article, and I would like to thank the former editor of *Ch'ing-shih wen-t'i*, Dr. Ramon H. Myers, for permission to use that material here.

metal were discovered throughout the sixteenth century.[3] As Kobata Atsushi, Sakuma Shigeo and others have demonstrated, some of that silver was soon being used by Japanese merchants in a flourishing but often illegal commerce with China. Indeed, despite repeated attempts by the Ming government first to control and then to prohibit trade between the two countries, smuggling was widespread along the Chinese coast during the middle decades of the sixteenth century, and by the 1540s, if not earlier, merchants from Chekiang, Fukien and Kwangtung provinces were exchanging Chinese goods for silver at ports in southern Japan.[4]

They were joined there in 1542 or 1543 by the Portuguese, who quickly became middlemen in Sino-Japanese trade, a role which proved particularly lucrative when Chinese merchants temporarily stopped going to Japan at the height of the so-called *wakō* ("Japanese pirate") disturbances during the 1550s and 1560s.[5] Once those disturbances had ended, however, economics prevailed over politics, and the partial relaxation of restrictions on maritime trade by the Ming government in 1567 constituted official recognition of an established fact: foreign silver had become so important to the Chinese economy that the country's merchants would do almost anything to procure it. To be sure, even after 1567 those merchants were prohibited from trading directly with the dreaded Japanese, but that does not seem to have stopped them.[6] Two years before the restrictions were eased, for example, the Portuguese encountered a number of Chinese merchants engaged in business at the port of Fukuda in western Kyūshū.[7]

In part because so much of the trade between the two countries was conducted illegally, it is impossible to determine just how much silver flowed from Japan to China during the middle decades of the sixteenth century. Drawing upon the research of Kobata, C. R. Boxer and others, Evelyn S. Rawski has written:

> We have only one item of information on this question, derived from the capture of three merchant ships in 1542. The ships, which were on their way to [Chang-chou] from Japan, carried 80,000 [Chinese] ounces [3,000 kilograms] of silver. With per-

[3] The most important work on mining in pre-modern Japan is by Kobata Atsushi, whose major findings have been collected in his *Nihon kōzanshi kenkyū* [Studies on the History of Japanese Mining] (Tokyo, 1968). In English, see Delmer M. Brown, *Money Economy in Medieval Japan* (New Haven, Conn., 1951), pp. 56-61; Tetsuo Kamiki and Kozo Yamamura, "Silver Mines and Sung Coins: A Monetary History of Medieval and Early Modern Japan in International Perspective" (paper presented at the Workshop in Pre-Modern World Monetary History, Univ. of Wisconsin-Madison, 28 Aug.-1 Sept. 1977; hereafter P.-M.W.M.H. Workshop, 1977).

[4] Sakuma Shigeo, "Mindai kaigai shibōeki no rekishiteki haikei", [The Historical Background of Private Maritime Trade during the Ming Period], *Shigaku zasshi*, lxiii no. 1 (1953), pp. 19-21; Kobata Atsushi, *Kingin bōekishi no kenkyū* [Studies on the History of the Bullion Trade] (Kyoto, 1976), pp. 2-3, 243-4, 284-5.

[5] For a recent discussion of these disturbances in English, see So Kwan-wai, *Japanese Piracy in Ming China during the Sixteenth Century* ([East Lansing, Mich.], 1975).

[6] See, for example, Kobata, *Kingin bōekishi*, pp. 284 ff.

[7] Boxer, *Great Ship from Amacon*, pp. 30-1.

haps ten to twenty ships plying the Japan-China trade during this period, silver imports into Chekiang and Fukien may have been as much as 530,000 ounces [19,875 kilograms].[8]

Although there is no way to verify this estimate, it would appear to be well within the realm of possibility. For example, the distinguished Japanese authority Iwao Seiichi has calculated that during the opening years of the seventeenth century, Chinese merchant vessels were carrying an average of approximately 23,500 taels (883 kilograms) of silver from Japan to China, a figure which is quite close to the average of 26,666 taels (1,000 kilograms) seized from each of the three ships mentioned in the above quotation.[9] Moreover, there is considerable evidence to suggest that there were many more than "ten to twenty ships" involved in Sino-Japanese trade at any one time during the period in question. Indeed, individual Chinese entrepreneurs are said to have controlled entire fleets of trading vessels, and when the Ming authorities tried to stamp out smuggling along the Chekiang and Fukien coasts during the late 1540s and early 1550s, they encountered literally hundreds of "pirate" ships operating out of illicit ports such as Shuang-hsü-kang on Liu-heng Island near Ning-po.[10]

As impressive as these figures are, however, it was not until the late sixteenth and early seventeenth centuries that trade between Japan and China reached its peak. This occurred as mining technology in Japan was improved[11] and as military and political leaders there sought to bolster their economic positions through foreign trade. The most successful of those leaders were Toyotomi Hideyoshi (1536-1598) and the first Tokugawa shogun, Ieyasu (1542-1616), whose commercial and foreign policies have been the subjects of numerous studies.[12] For these men and their contemporaries, foreign trade essentially meant trade in Chinese raw silk and silk textiles, items which proved extraordinarily lucrative for those who dominated the rapidly expanding Japanese domestic market for luxury goods. Both Hideyoshi and Ieyasu amassed immense personal fortunes, as did some of

[8] Evelyn Sakakida Rawski, *Agricultural Change and the Peasant Economy of South China* (Cambridge, Mass., 1972), p. 76. See also Kobata, *Kingin bōekishi*, p. 59.

[9] Iwao, *Shuin-sen to Nihon-machi*, p. 78.

[10] There is considerable literature on this subject in Chinese and Japanese. For brief but informative surveys in English, see So, *Japanese Piracy*, pp. 41-79; John E. Wills Jr., "Maritime History from Wang Chih to Shih Lang: Themes in Peripheral History", in Jonathan D. Spence and John E. Wills Jr. (eds.), *From Ming to Ch'ing: Conquest, Region, and Continuity in Seventeenth-Century China* (New Haven, Conn., 1979), pp. 204-13.

[11] Kobata, *Kingin bōekishi*, pp. 221-8; Brown, *Money Economy in Medieval Japan*, pp. 56-61.

[12] See, for example, Nakada Yosunao, "Ieyasu seiken to gaikoku bōeki no kichō" [The Administration of Tokugawa Ieyasu and the Underlying Tone of Japanese Foreign Trade], *Rekishi kyōiku*, viii no. 10 (1960), pp. 22-36; Izumi Chōichi, *Sakai to Hakata: Sengoku no gōshō* [The Cities of Sakai and Hakata: The Great Merchants of the Sengoku Period] (Osaka, 1979), pp. 99-266.

the merchants and officials upon whom they depended for help in administering the trade.

In part because of the chaotic political situation which existed in Japan during much of this period, reliable mining and trading figures for the country as a whole are not available. Nevertheless, after considering the fragmentary information which does survive, Tetsuo Kamiki and Kozo Yamamura have recently suggested that between 1560 and 1600, exports of silver from Japan probably averaged between 33,750 and 48,750 kilograms per year.[13] This estimate may be high, but it is not impossible. Japanese and Chinese vessels were exporting substantial quantities of the metal at this time, and an English traveller in Asia during the 1580s later wrote the following about another important aspect of this trade:

> When the Portuguese go from Macao in China to Japan, they carry much white silk, gold, musk, and porcelain; and they bring from thence nothing but silver. They have a great carrack which goes there every year and she brings from thence every year above six hundred thousand cruzados [approximately 17,000 kilograms]. And all this silver . . . they employ to their great advantage in China.[14]

Moreover, as C. R. Boxer has shown in his classic study on this subject, between 1557 and 1590 the Portuguese often employed more than one ship per year in their trade with Japan.[15] Under the circumstances, then, it is neither surprising that Japan became known in Europe during the late sixteenth century as the "Silver Islands" nor that the profits made there by the Portuguese were the envy of sailors, merchants and adventurers from Amsterdam to Manila.[16]

That envy is likely to have increased following the establishment of the Tokugawa shogunate in 1603. Perhaps the world's leading authority on the subject, Kobata Atsushi, believes that in some years during the early seventeenth century, exports of silver from Japan on Japanese, Chinese, Portuguese and Dutch ships may have reached annual totals of between 150,000 and 187,500 kilograms.[17] Although

[13] Kamiki and Yamamura, "Silver Mines and Sung Coins", p. 24.

[14] J. Horton Ryley (ed.), *Ralph Fitch: England's Pioneer to India* (London, 1899), p. 179. The spelling and punctuation of the original have been modernized. Cf. Michael Edwards, *Ralph Fitch, Elizabethan in the Indies* (London, 1972), pp. 125-41. Despite the existence of Kobata Atsushi's important article, "The Production and Uses of Gold and Silver in Sixteenth- and Seventeenth-Century Japan", trans. W. D. Burton, *Econ. Hist. Rev.*, 2nd ser., xviii (1965), pp. 245-66, scholars in the west have generally not appreciated the magnitude of Japaneses bullion production and exports during this period. Among the exceptions to this statement are Kamiki and Yamamura, "Silver Mines and Sung Coins"; Kristof Glamann, "The Changing Pattern of Trade", in E. E. Rich and C. H. Wilson (eds.), *The Cambridge Economic History of Europe*, v (1977), pp. 257-8.

[15] Boxer, *Great Ship from Amacon*, pp. 24-55.

[16] Kobata Atsushi, *Nihon to Kingintō* [Japan and the Gold and Silver Islands] (Osaka, 1943), pp. 54 ff.

[17] Kobata Atsushi, "Jūroku, jūnana seiki ni okeru Kyokutō no gin no ryūtsū" [The Circulation of Silver in the Far East during the Sixteenth and Seventeenth Centuries], in *Kobata Atsushi Kyōju taikan kinen kokushi ronshū* [A Collection of Essays on Japanese History in Honour of the Retirement of Professor Kobata Atsushi] (Kyoto, 1970), p.

(cont. on p. 72)

trading restrictions imposed by the Ming government meant that much of this silver first had to go to Macao, Taiwan, the Ryūkyū (Liu-ch'iu) Islands, Korea and South-East Asia, most of it ultimately reached China where it was used to purchase goods for the Japanese market. Moreover, this trade was so lucrative that the great Japanese merchant houses vied for shogunal permission to send their ships overseas, and sizeable Japanese trading communities were established in most of the above locations. There Japanese merchants were able to trade without interference from Tokugawa or Ming officials, and, like those of the Portuguese in Macao, their profits are said to have been enormous.[18]

EXPORTS OF SPANISH-AMERICAN SILVER TO EAST ASIA DURING THE SIXTEENTH AND SEVENTEENTH CENTURIES

China's other major source of silver at this time was the New World, where fabulously rich mines had been discovered by the Spanish during the 1540s. As D. A. Brading, Harry E. Cross and others have demonstrated, however, it was not until the mercury amalgamation process of refining was disseminated throughout Spanish-America several decades later that bullion production there began to soar to the heights which would transform world monetary history.[19] At Potosí in the viceroyalty of Peru this occurred during the early 1570s, and silver output there rose from an annual average of 41,048 kilograms during the 1571-5 quinquennium to 124,050 kilograms in 1576-80, 187,591 kilograms in 1581-5, 202,453 kilograms in 1586-90, and 218,506 kilograms in 1591-5.[20] A prime beneficiary of this Potosí-led boom was Ming China, to which Spanish-American silver flowed along several routes.

The most important of these ran directly across the Pacific from Acapulco on the west coast of modern Mexico to the Philippine Islands, the latter having been conquered by the Spanish during the late 1560s and early 1570s. Almost as soon as they arrived in the area of Manila in the spring of 1571, the conquistadores began a brisk

(n. 17 cont.)
8. A partial English translation of this article is appended to the same volume. Kobata's estimate is slightly higher than those given in Iwao Seiichi, *Shuin-sen bōekishi no kenkyū* [Studies on the History of Trade by Red-Seal Ships] (Tokyo, 1958), pp. 328-30; Iwao, *Shuin-sen to Nihon-machi*, pp. 78-9.

[18] The classic study of these overseas trading communities is Iwao Seiichi, *Nanyō Nihon-machi no kenkyū* [Studies on Japanese Trading Communities in the "South Seas"] (Tokyo, 1978 edn.).

[19] D. A. Brading and Harry E. Cross, "Colonial Silver Mining: Mexico and Peru", *Hispanic Amer. Hist. Rev.*, lii (1972), pp. 545-79.

[20] Harry E. Cross, "South American Bullion Production and Export, 1550-1750" (paper presented at P.-M.W.M.H. Workshop, 1977), appendix 2. See also P. J. Bakewell, "Registered Silver Production in the Potosí District, 1550-1735", *Jahrbuch für Geschichte von Staat, Wirtschaft und Gesellschaft Lateinamerikas*, xii (1975), pp. 92-7; Pierre Vilar, *A History of Gold and Money, 1450-1920*, trans. Judith White (London, 1976), pp. 119-33.

trade with merchants from China's Fukien and Kwangtung provinces in which they used Peruvian and Mexican silver to purchase large quantities of Chinese goods.[21] In 1573, for example, two galleons sailed from the Philippines to Acapulco with a cargo which included, among other things, "712 pieces of Chinese silk and 22,300 pieces of 'fine gilt china and other porcelain ware' ".[22] Six years later Francis Drake seized Chinese silk, porcelain and linen during his famous raid along the Pacific coast of North and South America,[23] and when Thomas Cavendish took the Acapulco-bound galleon *Santa Ana* off California in 1587, he is said to have acquired silks and other items worth more than 2,000,000 pesos.[24] Another indication of the growth and extent of Sino-Spanish trade during this period is the fact that between 1570 and about 1600, the number of Chinese living in the Philippines increased from approximately forty to more than fifteen thousand.[25] Moreover, the latter figure does not include the large number of merchants from Fukien and Kwangtung who travelled across the South China Sea every year to trade.

Despite the vigorous efforts of the Spanish government to control the amount of bullion shipped from the New World to East Asia, it is clear that much of the commerce between Acapulco and Manila at this time was conducted illegally.[26] It is thus difficult to say with any

[21] The classic study of the Manila-Acapulco trade is William L. Schurz, *The Manila Galleon* (New York, 1939). See also Pierre Chaunu, *Les Philippines et le Pacifique des Ibériques, XVIe, XVIIe, XVIIIe siècles: introduction méthodolique et indices d'activité* (Paris, 1960); C. R. Boxer, "*Plata es Sangre*: Sidelights on the Drain of Spanish-American Silver in the Far East, 1530-1750", *Philippine Studies*, xviii (1970), pp. 457-68; O. H. K. Spate, *The Spanish Lake* (London, 1979), pp. 176-291. There are three excellent articles on Sino-Spanish trade in Manila by the distinguished Chinese economic historian Ch'üan Han-sheng, all of which have been reprinted in his *Chung-kuo ching-chi shih lun-ts'ung* [Studies on Chinese Economic History], 2 vols. (Hong Kong, 1972), i, pp. 417-73.

[22] Schurz, *Manila Galleon*, p. 27.

[23] Derek Wilson, *The World Encompassed: Drake's Great Voyage, 1577-1580* (London, 1977), p. 148. The recent discovery of nearly six hundred shards of late sixteenth-century Chinese porcelain on the coast of California near San Francisco has led to renewed speculation about where Drake landed in 1579 after plundering Spanish shipping in the eastern Pacific. Whether or not these particular shards came from Drake's *Golden Hind*, they are further proof of the extensive trade which existed between China and the New World at that time. See Christopher Reed, "Porcelain Clue to Drake's Landing", *Guardian*, 13 May 1980, p. 7. The "Drake's Landing" controversy is far from over, however. See Roy Perrott, "Hindsight on the Hind Sight", *Sunday Times*, 2 Aug. 1981, p. 14; David Cressy, "The Mystery of Francis Drake's Californian Voyage", *History Today*, xxxi (Aug. 1981), pp. 10-14.

[24] Schurz, *Manila Galleon*, pp. 305-8.

[25] Victor Purcell, *The Chinese in Southeast Asia*, 2nd edn. (London, 1966), pp. 500-1. For further information on the Chinese in the Philippines at this time, see Ch'en Ching-ho, *Shih-liu shih-chi chih Fei-lü-pin hua-ch'iao* [Chinese in the Philippines during the Sixteenth Century] (Hong Kong, 1963); Albert Chan, "Chinese-Philippine Relations in the Late Sixteenth Century and to 1603", *Philippine Studies*, xxvi (1978), pp. 51-82.

[26] On this point, see Woodrow Borah, *Early Colonial Trade between Mexico and Peru* (Berkeley and Los Angeles, 1954), pp. 124-5; Schurz, *Manila Galleon* , pp. 77, 184-6, 194; Cross, "South American Bullion Production and Export".

assurance just how much silver flowed through the Philippines into China, although current evidence suggests that in a good year during the early seventeenth century it probably amounted to between two and three million pesos (57,500-86,250 kilograms).[27] Nevertheless, one distinguished historian of Latin America has uncovered information which indicates that in some years the total could have been substantially higher. In 1602 the Mexican authorities informed Madrid that annual shipments of silver from Acapulco to Manila usually amounted to five million pesos (143,750 kilograms) and that in 1597 the total had been twelve million (345,000 kilograms).[28]

Whether these enormous figures can be believed or not remains a matter for some debate. Although the extremely large vessels which made the Acapulco-Manila voyages could easily have carried that much silver, particularly since more than one galleon usually made the trip in each direction, it is by no means clear that sufficient quantities of bullion were being mined in the New World at that time to permit such huge amounts to be exported to the Philippines.[29] Nevertheless, it should also be noted that illegal mining, smuggling and official corruption were endemic in Spanish America, and in 1632, a time when Sino-Spanish trade is thought to have been long past its early seventeenth-century peak, the ecclesiastical council in Manila informed Philip IV that some 2,400,000 pesos (68,990 kilograms of silver) were being brought to that city annually from Acapulco.[30] Moreover, another well-placed observer of commercial activity in Manila during these years wrote that the "king of China could build a palace with the silver bars from Peru which have been carried to his country . . . without . . . having been registered and without the king [of Spain] having been paid his duties".[31]

A second route for New World silver going to China began with the famous *flotas de plata*, the Spanish treasure fleets which carried bullion every year from Mexico and the isthmus of Panama to Seville. From there substantial quantities of silver were transported illegally into Portugal where they were loaded, perhaps with other bullion which had been smuggled to Lisbon from Peru via Buenos Aires,[32]

[27] Ch'üan, *Chung-kuo ching-chi shih*, i, p. 444. In this article pesos have been converted to kilograms of silver at a rate of 0·02875 kilograms per peso.

[28] Borah, *Early Colonial Trade*, p. 123.

[29] See, for example, the estimates of Spanish-American bullion production in Brading and Cross, "Colonial Silver Mining", p. 579; Bakewell, "Registered Silver Production", pp. 92-7; Vilar, *History of Gold and Money*, p. 351.

[30] "Letter from the Ecclesiastical Cabildo to Felipe IV", in E. H. Blair and J. A. Robertson (eds.), *The Philippine Islands*, 55 vols. (Cleveland, Ohio, 1903-9), xxiv, pp. 254-5.

[31] Don Hieronimo de Banuelos y Carillo, "Relation of the Filipinas Islands", in *ibid.*, xxix, p. 71. Pierre Chaunu and O. H. K. Spate regard this statement with justifiable scepticism, but it is clear that illegal shipments of silver to Manila during this period were substantial. For the views of Chaunu and Spate, see Spate, *Spanish Lake*, p. 201.

[32] On this smuggling, see Cross, "South American Bullion Production and Export".

on to huge carracks and shipped around the Cape of Good Hope to Goa. In Goa these vessels unloaded European goods and took on additional silver, some of which had filtered through the Mediterranean and Middle Eastern trading areas into India. Proceeding first to Malacca and then to Macao, the Portuguese used this silver to purchase Chinese goods for the markets of Japan, India, the Middle East and western Europe.[33] As is true of the Acapulco-Manila trade at about this time, the illegal nature of much of this commerce makes its value and extent very difficult to estimate. Nevertheless, it seems likely that during the late sixteenth and early seventeenth centuries, Portuguese vessels carried between 6,000 and 30,000 kilograms of silver annually from Goa to Macao.[34]

The third and final route for Spanish-American silver going to China also began with the *flotas de plata*, but in this case some of the bullion which arrived in Seville was shipped, legally or illegally, to Amsterdam and London. During the early seventeenth century ships of the Dutch and English East India Companies began carrying some of this silver to South-East Asia, where it was used to purchase pepper, spices and, even more importantly for our purposes here, Chinese luxury goods such as silk and porcelain. Although the amounts of silver involved in this commerce were considerably smaller than those discussed above,[35] they were not unappreciated by the Chinese merchants who had long been a potent economic force in maritime South-East Asia. Indeed, the appearance of bullion-laden European ships in the waters of this region was greeted by an immediate expansion of Chinese commercial activity there. Before long the Chinese were exporting so many Spanish reals from Bantam that they "caused a shortage of these coins in the town, which indicates that the Dutch and English did not have enough barter commodities to obtain the Chinese goods, particularly the silk and porcelain".[36]

In fact, the Europeans probably did not have any "barter com-

[33] For a recent discussion of this trade, see C. R. Boxer, "Macao as a Religious and Commercial Entrepot in the Sixteenth and Seventeenth Centuries", *Acta Asiatica*, xxvi (1974), pp. 64-90.

[34] This estimate is based on Boxer, *Great Ship from Amacon*, pp. 62-4; C. R. Boxer, *Fidalgos in the Far East, 1550-1570* (The Hague, 1948), p. 6; Geoffrey Parker, "The Emergence of Modern Finance in Europe", in Carlo M. Cipolla (ed.), *The Fontana Economic History of Europe: The Sixteenth and Seventeenth Centuries* (Glasgow, 1974), p. 528; Vilar, *History of Gold and Money*, pp. 99-101.

[35] For figures on the amounts of silver shipped to Asia from Europe during these years, see Parker, "The Emergence of Modern Finance", p. 529; M. A. P. Meilink-Roelofz, *Asian Trade and European Influence in the Indonesian Archipelago between 1500 and About 1630* (The Hague, 1962), p. 378 n. 149; K. N. Chaudhuri, *The English East India Company* (London, 1965), p. 115; F. S. Gaastra, "The Exports of Precious Metals from Europe to Asia by the Dutch East India Company, 1602-1795" (paper presented at P.-M.W.M.H. Workshop, 1977).

[36] Meilink-Roelofz, *Asian Trade and European Influence*, p. 246. See also Leonard Blussé, "Chinese Trade to Batavia during the Days of the V. O. C.", *Archipel*, xviii (1979), pp. 195-213.

modities" in which the sophisticated traders from what was still one of the most technologically advanced nations on earth were really interested. As Antonio de Morga explained in his informative discussion of trade in Manila at this time, the Chinese accepted only silver for their products, "for they do not like gold, nor any other goods in exchange, nor do they carry any to China".[37] An English merchant put it more bluntly several decades later when he wrote that the Chinese "will as soon part with their blood as [silver], having once possession".[38]

DOMESTIC SUPPLIES OF SILVER IN MING CHINA

In order to appreciate why foreign silver was so prized by Chinese merchants during the late Ming period, it might be helpful to examine the figures in Table 1, compiled by Ch'üan Han-sheng. While these figures represent only official receipts from mining and not total production, Ch'üan, who has studied these problems for many years, believes that the Ming government took about 30 per cent of all the

TABLE 1

MING GOVERNMENT REVENUES FROM DOMESTIC SILVER MINING 1401-1520*

	Taels	Kilograms *(a)*	Annual average in kilograms
1401-10 *(b)*	1,299,167 (+)	48,719 (+)	5,413 (+)
1411-20	2,905,602	108,960	10,896
1421-30	1,993,591 (+)	74,760 (+)	7,476 (+)
1431-40 *(b)*	1,277,863 (+)	47,920 (+)	5,324 (+)
1441-50 *(b)*	289,752	10,866	1,811
1451-60 *(b)*	363,454	13,630	2,272
1461-70	614,680 (+)	23,051 (+)	2,305 (+)
1471-80	589,248 (+)	22,097 (+)	2,210 (+)
1481-90 *(c)*	802,396	30,090	3,009
1491-1500	530,552	19,896	1,990
1501-10	325,200	12,195	1,220
1511-20	329,200	12,345	1,235

* Notes and source: Ch'üan Han-sheng, "Ming-tai ti yin-k'o yü yin-ch'an-e" [Taxes on Silver Mining and Silver Production Quotas during the Ming Dynasty], *Hsin-ya shu-yüan hsüeh-shu nien-k'an*, ix (1966), pp. 246-54. The symbol (+) indicates that certain figures in the original source are imprecise and the totals are likely to have been slightly higher than those which appear in the Table.

(a) Taels are converted into kilograms at 0·0375 kg. per tael.

(b) No information is available for 1401, 1435, 1441-3 and 1450-4. Tael and kilogram totals for the decades in which those years fall are thus lower than they should be, and the annual averages have therefore been computed on the basis of those years for which information is available.

(c) From 1487 to 1520 government revenues from gold and silver mining were reported together. Since the amount of gold mined is thought to have been very small, however, the totals are listed as if they were only silver.

[37] Morga, *Philippine Islands*, p. 340.

[38] Cited in Boxer, *Great Ship from Amacon*, p. 1. See also Boxer, "*Plata es Sangre*", *passim*.

bullion mined in official and officially sanctioned mines.[39] Even if this admittedly speculative estimate is wildly inaccurate, and despite some problems with certain figures upon which the above Table is based,[40] it still appears reasonable to assume that under normal circumstances the Ming government and its agents would have taken as much bullion as possible without making the mines uneconomic to operate. Moreover, given the crucial role played by silver in Chinese economic life after 1400, a point which will be discussed below, it is unlikely that substantial amounts of bullion were mined without the government's knowledge or approval. For these reasons, it is the contention here that the figures in Table 1 can be taken as a very rough guide to the trends in Chinese silver production during the first half of the Ming period.

That guide indicates, and the research of Ch'üan, Liang Fang-chung and Momose Hiromu appears to confirm, that a very considerable boom in Chinese bullion mining during the early fifteenth century was followed by an erratic though substantial decline in silver production for the rest of the period up until 1520. Moreover, although reliable figures for the years after 1520 are unavailable, there does not seem to have been any significant increase in production above the low levels which are indicated in Table 1 for the late fifteenth and early sixteenth centuries. Indeed, it is likely that China's domestic output for the rest of the dynasty equalled but a fraction of the bullion imported into the country from Japan and the New World. In fact if Ch'üan's estimates are only approximately accurate, Chinese mines may never have produced as much silver annually as one Spanish galleon often carried across the Pacific from Acapulco to Manila

[39] Ch'üan Han-sheng, "Ming-tai ti yin-k'o yü yin-chan-e" [Taxes on Silver Mining and Silver Production Quotas during the Ming Dynasty], *Hsin-ya shu-yüan hsüeh-shu nien-k'an*, ix (1966), pp. 258-9. On Ming silver mining, see also the important studies by Liang Fang-chung, "Ming-tai yin-k'uang k'ao" [A Study of Ming Silver Mines], *Chung-kuo she-hui ching-chi shih chi-k'an*, vi (1939), pp. 65-112; Momose Hiromu, "Mindai no ginsan to gaikoku gin ni tsuite" [On Domestic Silver Production and Foreign Silver during the Ming Period], *Seikyū gakusō*, xix (1935), pp. 90-147.

[40] During the late fifteenth and early sixteenth centuries, official receipts from bullion mining in Ming China took on a perfunctory, bureaucratic quality with the same annual total sometimes being recorded for a number of years in succession. The reason for this would seem to be that official figures at that time no longer reflected fluctuations in bullion production but rather were tax receipts collected at fixed annual rates from mining districts throughout the empire. Although it is unlikely that actual production was lower than those figures, it is clearly risky to speculate about the former on the basis of the latter. Nevertheless, there is reason to suspect that considerably less silver was being mined in China during the 1490-1520 period than had been the case even a few years earlier. Not only do the raw figures decline in the early 1490s and drop sharply after 1498, facts which should not be dismissed out of hand, but the throne was informed on several occasions during the late fifteenth and early sixteenth centuries that mines in certain parts of the country had declined in fertility to the point that they were no longer economic to operate. See Wada Sei (ed.), *Minshi shokkashi yakuchū* [An Annotated Translation of the Economic Chapters of the Ming Dynastic History], 2 vols. (Tokyo, 1957), ii, pp. 793-801.

during the late sixteenth and early seventeenth centuries. And by 1636, when four Portuguese galliots transported more than 75,000 kilograms of silver from Nagasaki to Macao for use in the Canton market,[41] bullion production in China was apparently negligible.

In order to evaluate the significance of these facts for the Ming economy, it is necessary to relate them to certain features of the dynasty's monetary system. Following the collapse of Mongol rule in China during the mid-fourteenth century, the first Ming emperor T'ai-tsu (reigned 1368-98) attempted to establish a nationwide currency system based on paper notes and copper coins. Convinced by the experiences of earlier dynasties that such a system could succeed only if precious metals were removed from circulation, T'ai-tsu and his advisors imposed severe restrictions on bullion mining, prohibited the use of silver and gold in commercial transactions, and ordered that commercial taxes were to be paid in government-issued paper notes and copper coins.[42] These policies proved unworkable. Far too much paper currency was printed during the early years of the dynasty with the result that just over a decade after the first notes were issued in 1375, their value *vis-à-vis* silver had plummeted by 80 per cent and subsequently fell even lower.[43] At the same time, the government failed to produce adequate supplies of copper coins to serve as subsidiary currency for the world's largest and, in many ways, most sophisticated economy.[44] It is thus hardly surprising that silver continued to circulate in economically advanced regions of the country,[45] although the legal restrictions on mining and trading must have kept it in relatively short supply.

This situation changed dramatically during the reign of the third Ming emperor Ch'eng-tsu (reigned 1403-24). Not long after he usurped the throne during the summer of 1402, Ch'eng-tsu ordered sharp increases in mining quotas and opened or reopened bullion mines in seven provinces and in the Chinese-held portions of northern South-East Asia as well.[46] The impact of these policies on government finances was immediate and substantial. In 1403, for example, official receipts from silver mining rose to 80,135 taels (3,007 kilograms) from a previously recorded high of 29,830+ taels (1,119+ kilograms)

[41] Boxer, *Great Ship from Amacon*, pp. 145-58.

[42] P'eng Hsin-wei, *Chung-kuo huo-pi shih* [Chinese Monetary History] (Shanghai, 1965), pp. 632-9; Ray Huang, *Taxation and Governmental Finance in Sixteenth-Century Ming China* (Cambridge, 1974), pp. 69-70.

[43] P'eng, *Chung-kuo huo-pi shih*, p. 671; Huang, *Taxation and Governmental Finance*, pp. 69-74.

[44] On this point, see Ch'üan, *Chung-kuo ching-chi shih*, i, pp. 355-67.

[45] Wada, *Minshi shokkashi*, ii, p. 705. I am indebted to Dr. James R. Geiss for bringing this point to my attention.

[46] *Ibid*, ii, pp. 781-7. See also John S. Deyell, "The China Connection: Problems of Silver Supply in Medieval Bengal" (paper presented at P.-M.W.M.H. Workshop, 1977); J. K. Whitmore, "Vietnam and the Monetary Flow of Eastern Asia, Thirteenth to Eighteenth Centuries" (paper presented at P.-M.W.M.H.Workshop, 1977).

in 1390. By 1406 the annual total had reached 209,136 taels (7,843 kilograms), and, as the figures in Table 1 indicate, official receipts from mining remained high for several decades to come.[47]

If we accept Ch'üan Han-sheng's assumptions concerning the relationship between these receipts and total production, the above figures mean that substantial quantities of silver were being mined in early fifteenth-century China which were not flowing directly into government coffers. There is considerable evidence to suggest that much of this additional bullion eventually filtered into circulation, where it was traded by weight and contributed to the general prosperity which marked the first third of the fifteenth century. Indeed, it appears that continuing public distrust of paper currency, serious shortages of copper coins, population growth and an expanding economy led many people to rely more and more heavily on silver as a medium of exchange. Negative evidence for this can be found in the government's periodic attempts to ban the metal's use in commercial transactions.[48] These attempts failed and by the 1430s, if not earlier, China was firmly committed to the silver standard it was to maintain until very recent times.

However, it was also during the 1430s that domestic production of silver in China dropped sharply from the record levels of the previous three decades. This was due to a combination of factors which included radical changes in government economic policy, widespread corruption in and mismanagement of mining operations, and the declining fertility of many mines.[49] According to the figures which are currently available, between 1436 and 1520 the Ming government's annual receipts from silver mining exceeded 100,000 taels (3,750 kilograms) only four times and averaged less than 50,000 taels (1,875 kilograms) per year over the entire period.[50] The probable implications of this for total production need not be reiterated. What should be emphasized is the fact that by 1500 the population of the Ming empire probably exceeded 100,000,000 and the Ming economy was once again beginning to expand with some vigour. With domestic production of bullion clearly insufficient to meet that economy's monetary needs, it is hardly surprising that foreign traders found the Chinese demand for silver almost insatiable when they appeared off the country's south-eastern coast during the sixteenth century. Fortunately for the Chinese, of course, foreign demand for Chinese goods was equally intense and an extraordinary expansion in commercial activity ensued.

[47] For these figures and general discussions of bullion production in China at this time, see Ch'üan, "Ming-tai ti yin-k'o", pp. 246-7; Momose, "Mindai no ginsan", pp. 93-100.

[48] Wada, *Minshi shokkashi*, ii, pp. 703-19; Huang, *Taxation and Governmental Finance*, p. 71.

[49] Wada, *Minshi shokkashi*, ii, pp. 781-92.

[50] Ch'üan, "Ming-tai ti yin-k'o", pp. 249-54.

THE IMPACT OF JAPANESE AND SPANISH-AMERICAN SILVER ON THE LATE MING ECONOMY

Although statistical work on the pre-modern Chinese economy is still in its infancy, the figures in Table 2 are of more than passing interest for our purposes here. As Ch'üan Han-sheng and Li Lung-hua have noted in the article from which the figures are drawn, during the fifteenth and sixteenth centuries the T'ai-ts'ang vault (or treasury) became the chief receiving agency for central government revenues collected in silver. Despite accounting procedures at the vault which leave much to be desired,[51] Ch'üan and Li believe that the fluctuations in its annual receipts not only reveal a good deal about the state of government finances from year to year, but also reflect changes in

TABLE 2

SILVER RECEIPTS OF THE T'AI-TS'ANG VAULT 1528-1643*

	Taels	Kilograms		Taels	Kilograms
1528	1,300,000	48,750	1590†	3,740,500 (+)	140,269 (+)
1539†	2,000,000 (+)	75,000 (+)	1592†	4,512,200 (+)	169,200 (+)
1549	2,125,355	79,701	1593	4,723,000 (+)	177,113 (+)
1551	2,000,000 (+)	75,000 (+)	1604	4,582,000	171,825
1552	2,000,000 (+)	75,000 (+)	1605	3,549,000	133,088
1553	2,000,000 (+)	75,000 (+)	1606	4,000,000	150,000
1554	2,000,000 (+)	75,000 (+)	1609	4,000,000	150,000
1555	2,000,000 (+)	75,000 (+)	1612	4,000,000 (+)	150,000 (+)
1556	2,000,000 (+)	75,000 (+)	1613†	4,000,000	150,000
1557	2,000,000 (+)	75,000 (+)	1616	4,000,000	150,000
1563	2,200,000 (+)	82,500 (+)	1617†	3,890,000	145,875
1564	2,470,000 (+)	92,625 (+)	1618	6,000,031 (+)	225,001 (+)
1565	2,200,000 (+)	82,500 (+)	1620	5,830,246 (+)	218,634 (+)
1567	2,014,200 (+)	75,533 (+)	1621	7,552,745 (+)	283,228 (+)
1568	2,300,000 (+)	86,250 (+)	1622	4,968,795 (+)	186,330 (+)
1569	2,300,000 (+)	86,250 (+)	1623	7,893,137 (+)	295,993 (+)
1570	2,300,000 (+)	86,250 (+)	1625	3,030,725 (+)	113,652 (+)
1571	3,100,000 (+)	116,250 (+)	1626	3,986,241 (+)	149,484 (+)
1573	2,819,153 (+)	105,718 (+)	1628	7,064,200 (+)	264,905 (+)
1577	4,359,400 (+)	163,478 (+)	1630‡	9,136,357 (+)	342,613 (+)
1578	3,559,800 (+)	133,493 (+)	1631	12,249,195 (+)	459,345 (+)
1580	2,845,483 (+)	106,706 (+)	1634	12,812,000 (+)	480,450 (+)
1581†	3,704,281 (+)	138,911 (+)	1637	16,700,000	626,250
1583	3,720,000 (+)	139,500 (+)	1639	20,000,000	750,000
1585†	3,700,000 (+)	138,750 (+)	1641	21,451,736 (+)	804,440 (+)
1586†	3,890,000 (+)	145,875 (+)	1642	23,000,000 (+)	862,500 (+)
1589	3,270,000 (+)	122,625 (+)	1643	21,300,000 (+)	798,750 (+)

Notes and source: Ch'üan Han-sheng and Li Lung-hua, "Ming chung-yeh hou T'ai-ts'ang sui-ju yin-liang ti yen-chiu" [Studies on the Annual Silver Receipts of the T'ai-ts'ang Vault after the Middle of the Ming Period], *Chung-kuo wen-hua yen-chiu-so hsüeh-pao*, v no. 1 (1972), pp. 136-9. The symbol (+) indicates that the figure given for this year in the original source(s) is imprecise and is likely to have been slightly higher than that which appears in the Table.

† This date is given in the original source(s) as an approximate one.

‡ For the years 1630-43, see n. 72 below.

[51] See Huang, *Taxation and Governmental Finance*, pp. 269-77.

the role played by silver in the economy and in the monetary system.[52] For reasons which should become apparent below, I share this belief, although with some reservations and qualifications.

Unless one takes the sharp rise recorded in the T'ai-ts'ang vault's silver receipts between 1528 and 1539 as significant, and this would be risky considering the paucity of information, there seems to be little or no statistical evidence of the impact of foreign bullion on Ming central government finances before the 1560s. Even then, a perfunctory and bureacratic quality of reporting makes generalization difficult if not impossible. In 1571, however, something dramatic clearly happened as a huge increase is registered in the vault's silver revenues. This occurred, it should be noted, four years after the Ming government had relaxed its restrictions on maritime trade, one year after Sino-Japanese trade was boosted by the founding of Nagasaki, and the same year in which Sino-Spanish trade began in earnest with the establishment of Manila as the Spanish capital in the Philippines. By 1577 the amount of silver entering the vault was nearly double the highest total recorded for the 1560s, and until the very end of the dynasty silver receipts there probably never fell below 100,000 kilograms per year. While other factors were involved in these developments, including important changes in the system of taxation to be discussed below, it seems clear that the marked increase in government silver revenues during this period was directly related to an unprecedented increase in foreign trade and bullion imports.

Further support for this contention can be found in Table 3. During the late sixteenth century Yüeh-kang in Fukien's Hai-ch'eng

TABLE 3

SILVER COLLECTED AS FEES FOR LICENSES, POUNDAGE AND TONNAGE IN HAI-CH'ENG COUNTY

1567-1628*

	Taels	Kilograms		Taels	Kilograms
1567-72	3,000‡	113‡	1594	29,000 (+)	1,088 (+)
1573	6,000	225	1599†	27,000 (+)	1,013 (+)
1576	10,000	375	1615	23,400	878
1583	20,000 (+)	750 (+)	1628	23,400	878

* Notes and source: Ch'üan Han-sheng, "Ming-chi Chung-kuo yü Fei-lü-pin chien ti mao-i" [Trade between China and the Philippines during the Ming Period], in his *Chung-kuo ching-chi shih lun-ts'ung* [Studies on Chinese Economic History], 2 vols. (Hong Kong, 1972), i, p. 428. The symbol (+) indicates that the figure given for this year in the original source(s) is imprecise and is likely to have been slightly larger than that which appears in the Table.

† This date is given in the original source as an approximate one.

‡ This figure is given in the original source as an annual one for the years in question.

52 Ch'üan Han-sheng and Li Lung-hua, "Ming chung-yeh hou T'ai-ts'ang sui-ju yin-liang ti yen-chiu" [Studies on the Annual Silver Receipts of the T'ai-ts'ang Vault after the Middle of the Ming Period], *Chung-kuo wen-hua yen-chiu-so hsüeh-pao*, v no. 1 (1972), p. 124.

county became one of China's leading ports, with a trading network which stretched from Japan to South-East Asia. It is now generally agreed that the rapid rise in silver collections shown in Table 3 was due almost entirely to the expansion of Yüeh-kang's foreign trade during these years and is thus indicative of the increased rate at which Japanese and Spanish-American silver was entering the Fukienese and ultimately the Chinese economy.[53]

The impact of that silver is also reflected in the changes which occurred in China's gold-silver ratio during this period. These changes are given in Table 4 along with the ratios which existed in Japan and in the Spanish empire at approximately the same time. Despite the very large gaps in our current information, it can be seen from Table 4 that as increased silver production led to the widening of gold-silver ratios in Japan and the Spanish Empire, increased silver imports (and some gold exports) gradually had the same effect in China. Indeed, by the last few decades of the Ming dynasty the ratio there was approaching levels in other parts of the world. However, while this last fact may be of some significance in indicating the degree to which China was becoming integrated into an emerging "world

TABLE 4

BIMETALLIC RATIOS OF SILVER TO GOLD 1566-1644
(UNITS OF SILVER TO ONE UNIT OF GOLD)*

	China	Japan	Spanish Empire
1566	—	—	12·12
1568	6·00	—	12·12
1571	—	7·37	12·12
1572	8·00	—	12·12
1575	—	10·34	12·12
1581	—	8·92	12·12
1588	—	9·15	12·12
1589	—	11·06	12·12
1594	—	10·34	12·12
1596	7·50	—	12·12
1604	—	10·99	12·12
1609	—	12·19	13·13
1615	—	11·38	13·13
1620	8·00	13·05	13·13
1622	—	14·00	13·13
1627-44	10·00-13·00	—	13·13-15·45
1643	—	—	15·45

* Sources: P'eng Hsin-wei, *Chung-kuo huo-pi shih* [Chinese Monetary History], (Shanghai, 1965), p. 714; Tetsuo Kamiki and Kozo Yamamura, "Silver Mines and Sung Coins: A Monetary History of Medieval and Early Modern Japan in International Perspective" (paper presented at the Workshop in Pre-Modern World Monetary History, Univ. of Wisconsin-Madison, 28 Aug.-1 Sept. 1977), table 10; Harry E. Cross, "South American Bullion Production and Export, 1550-1750" (paper also presented at the above workshop), table 1.

[53] On these points, see Rawski, *Agricultural Change*, pp. 57-69; Kobata, *Kingin bōekishi*, pp. 243-83; Ch'üan, *Chung-kuo ching-chi shih*, i, pp. 427-9; Huang, *Taxation and Governmental Finance*, p. 235.

economy", it reveals less about the country's domestic situation than might be thought. This is because the very small amounts of gold mined in China were rarely used in commercial transactions. Instead, following the virtual collapse of the Ming dynasty's system of paper currency during the late fourteenth and early fifteenth centuries, the Chinese had developed a bimetallic monetary system which was based not upon gold and silver, but rather upon silver traded by weight[54] and copper coins issued by the government. Although one tael (approximately 0·0375 kilograms) of silver was theoretically equal in value to 1,000 copper coins, in fact the exchange rate varied widely over time.[55] This was due to a combination of factors which included the government's inability to control the amount of silver entering circulation, its failure to provide an adequate supply of copper coins for domestic use, and the export of some of those coins to Japan and South-East Asia.[56]

Despite these complications, it is clear that any assessment of the impact of foreign silver on the late Ming economy must take into account the changes which occurred in the silver-copper exchange rate. Unfortunately, the figures which are currently available on this subject are very sketchy. (See Table 5.) Until much more research has been done on all aspects of this problem, it is of course risky to speculate about the precise meaning of these figures. However, it should be noted that the substantial narrowing of the silver-copper exchange rate between 1368 and 1453 was probably related to the relatively high levels of domestic silver production during the early

[54] One of the great puzzles in world monetary history is the failure of pre-modern Chinese governments to coin precious metals except for essentially ceremonial purposes. Yang Lien-sheng suggests this may have been because "the demand for such a medium of exchange was not strong enough to induce the state to take action". While this explanation may have some validity for the period prior to 1400, it is less than satisfactory for the "Silver Age" which followed. It clearly was not the case that Chinese governments during the late imperial period were technologically or administratively incapable of producing gold and silver coins, but it may be that holding sway over a vast centralized empire far removed from those parts of the world where such coins were minted, they did not feel the same compulsion as governments elsewhere to advertise their prestige and legitimacy in this way. Moreover, they may have learned earlier than most of the difficulties involved in trying to persuade people to accept valuable coins at face value. As J. K. Galbraith has noted about late sixteenth-century Europe, "With numerous coins in circulation variously adulterated, clipped, filed, sweated, trimmed, and with the worst being offered first, coins became a problem. The path was now open for the next great reform, which was to go back to weighing. This decisive step was taken by the City of Amsterdam in 1609 . . .". This was a step the eminently practical Chinese did not have to take as they had never given up weighing. See J. K. Galbraith, *Money: Whence it Came, Where it Went* (Harmondsworth, 1976), p. 20. For Yang's views on gold and silver coinage, see his *Money and Credit in China: A Short History* (Cambridge, Mass., 1971), pp. 9, 27.

[55] For useful discussions of the relationship between unminted silver and copper coins during the late imperial period, see Ch'üan, *Chung-kuo ching-chi shih*, i, pp. 355-67; Huang, *Taxation and Governmental Finance*, pp. 74-81.

[56] On this last point, see Kamiki and Yamamura, "Silver Mines and Sung Coins", pp. 15-16, and table 2; Whitmore, "Vietnam and the Monetary Flow of Eastern Asia", pp. 15-17.

TABLE 5

BIMETALLIC RATIO OF COPPER TO SILVER 1368-1627*

	Ratio of copper to silver
1368	320 to 1
1453	266 to 1
1577	229 to 1
1597	152 to 1
1616	152 to 1
1621-7	112 to 1

* Source: P'eng, *Chung-kuo huo-pi shih*, p. 715.

fifteenth century. (See Table 1.) Similarly, the changes which occurred after 1577 almost certainly reflect the impact of silver imports on the economy and the monetary system.

If the statistical evidence for this contention is less than overwhelming, however, more impressionistic evidence is abundant.[57] For example, it was during the middle decades of the sixteenth century that the Ming government introduced the so-called "Single Whip Method" (*i-t'iao pienfa*) into its outmoded and appallingly complex system of taxation. Although this "method" was itself complicated and subject to regional and even local variation, it essentially meant that most land taxes, labour service obligations, and extra levies were commuted into silver payments. Given the depressed state of Chinese domestic bullion production at the time, it seems reasonable to assume that commutations on this scale would only have been possible with regular and substantial imports of foreign silver. Otherwise it will be difficult to explain why many of the most important early experiments with the "Single Whip Method" were conducted in Fukien and Chekiang, two of the provinces most directly involved in maritime commerce and bullion imports. Nor is it likely to have been merely a coincidence that probably the three most influential "Single Whip" reformers in sixteenth-century China, Hai Jui (1513-87), P'ang Shang-p'eng (1524?-81?) and Wang Tsung-mu (1523-91), were all natives of the south-eastern coast and thus were undoubtedly aware of the special conditions which existed there.[58]

In any event, there can be no question that the "Single Whip" reforms and the increased monetization which facilitated their implementation had a profound impact on the late Ming economy. And, as might be expected given the preceding discussion, nowhere was

[57] The following section draws heavily on my "Notes on Silver", pp. 5-8. See also Willard J. Peterson, *Bitter Gourd: Fang I-chih and the Impetus for Intellectual Change* (New Haven, Conn., 1979), pp. 64-80.

[58] On these points, see Liang Fang-chung, *The Single-Whip Method of Taxation in China*, trans. Wang Yü-ch'üan (Cambridge, Mass., 1956); Huang, *Taxation and Governmental Finance*, pp. 112-33; the biographies of P'ang, Hai and Wang which appear in L. Carrington Goodrich and Chaoying Fang (eds.), *Dictionary of Ming Biography*, 2 vols. (New York, 1976).

this more apparent than in the economically and technologically advanced south-east. There cities such as Nanking, Soochow, Sung-chiang, Ning-po, Chang-chou and Canton boomed as perhaps never before, smaller towns in the surrounding countrysides became thriving centres of specialized occupations, and the number of periodic markets in rural districts nearby rose dramatically.[59] At the same time, life in these areas became much more luxurious, at least for the favoured few. Merchants and entrepreneurs amassed and spent huge fortunes, sumptuary laws were tested and then ignored, and conspicuous consumption of all kinds increased significantly:

> Late Ming writers in Nanking and other cities were quite aware of changes in urban life-styles and social attitudes. They relate some of this change to the reform of the tax system . . . Some features of [this reform] . . . , particularly those involving commutation of labor services and special exactions are said . . . to have been conducive to ostentatious consumption in the cities. Urban commoners who had money were no longer under great pressure to conceal that fact; they could display their wealth in elegantly enlarged houses, gardens now free of restrictions on size . . .[60]

Of course, there is nothing in this discussion to suggest that the important economic and social changes which occurred in late Ming China can be attributed solely to the large amounts of foreign bullion which flowed into the country after 1530. Indeed, some students of Ming history might even argue that when compared to the size and scope of its domestic commerce, China's trade with Japan and the West during this period was of minor importance. To do so, however, would be to seriously underestimate the impact a substantial increase in the stock of precious metals could have on an economy which was imperfectly monetized and in which credit techniques were not highly developed.[61]

With a discredited system of paper currency, an inadequate supply of copper coins, and rudimentary banking and credit institutions, the economy of late Ming China would seem to be an excellent case in point. In fact, if Chinese domestic bullion production was as low at the time as now appears likely, the country's money supply may even have been in danger of contracting because of the wear and tear on

[59] Between 1491 and 1573, for example, the number of periodic markets in Fukien's Chang-chou prefecture increased by 245 per cent and continued to multiply right into the seventeenth century. Similar growth occurred in the countryside around Ning-po, although the rate of change there seems to have been somewhat slower, perhaps because the city had long been an important international trading centre and was thus under closer governmental supervision. This of course would have reduced the possibility or at least the profitability of illegal commercial activities. On periodic markets in the Chang-chou and Ning-po areas during this period, see Rawski, *Agricultural Change*, pp. 70-1; Shiba Yoshinobu, "Ning-po and Its Hinterland", in G. William Skinner (ed.), *The City in Late Imperial China* (Stanford, 1977), p. 399.

[60] F. W. Mote, "The Transformation of Nanking, 1350-1400", in Skinner (ed.), *The City in Late Imperial China*, p. 151.

[61] For a brief but informative discussion of this subject as it relates to sixteenth-century Europe, see Harry A. Miskimin, *The Economy of Later Renaissance Europe, 1460-1600* (Cambridge, 1977), pp. 156-7.

precious metals when they are used in commercial transactions.[62] That this did not happen and certain sectors of the Chinese economy grew with some vigour during the second half of the Ming period would seem to owe more than a little to the maritime trade which, after the mid-sixteenth century at least, the Ming authorities found themselves increasingly unable to control.

SILVER, FOREIGN TRADE AND THE COLLAPSE OF THE MING DYNASTY

Whatever the beneficial effects of that trade, and they were numerous,[63] the bullion which poured into China from abroad during the late Ming period could not solve all of the country's economic problems. Indeed, it even helped to create some new ones such as over-rapid urban growth, unbridled business speculation, and severe price inflation.[64] The economic and social instability which ensued was exacerbated by the fact that many merchants and producers at the time were induced by the inflationary spiral to rely too heavily on a continued expansion of the money economy. As has been suggested, however, that expansion was itself dependent to a very large extent on bullion imports to increase the money supply. Prior to the early decades of the seventeenth century that dependence did not pose a serious problem as Japanese and Spanish-American mines continued to produce prodigious amounts of silver, at least some of which was eagerly exchanged for Chinese goods in Manila, Macao, Canton, Nagasaki, Taiwan and South-East Asia. After that time, however, circumstances in various parts of the world began to change in ways which were to have profound effects not only upon the Ming economy but also on the course of Chinese history.

[62] On this subject, see C. C. Patterson, "Silver Stocks and Losses in Ancient and Medieval Times", *Econ. Hist. Rev.*, 2nd ser., xxv (1972), pp. 205-35. I am indebted to Professor John F. Richards of Duke University for bringing this article to my attention.

[63] It has long been acknowledged, for example, that the reforms of the chief grand secretary Chang Chü-cheng, who dominated Ming government from 1572 to 1582, put the dynasty on a sounder financial footing than it perhaps had ever known. Shortly before Chang's death in 1582 treasuries throughout the country were well stocked and the silver reserves of the T'ai-ts'ang vault alone are said to have been in excess of 6,000,000 taels (225,000 kilograms). Whether such reserves would have been possible without imports of foreign bullion must be regarded as doubtful, particularly since they were accumulated precisely when those imports began to soar during the 1570s and 1580s. For a recent discussion of Chang's reforms, see Huang, *Taxation and Governmental Finance*, pp. 294-301.

[64] James Geiss, "Peking Under the Ming, 1368-1644" (Princeton Univ. Ph.D. thesis, 1979), pp. 144 ff. See also Peterson, *Bitter Gourd*, pp. 70-3. It is instructive to compare Geiss's description of the impact of inflation on late Ming society and particularly on late Ming government with Ömer Lufti Barkan, "The Price Revolution of the Sixteenth Century: A Turning Point in the Economic History of the Near East", trans. Justin McCarthy, *Internat. Jl. Middle East Studies*, vi (1975), pp. 3-28. On p. 26 of this article Barkan explains how silver imports and inflation during the late sixteenth century led to "a period of severe financial and social agitation in the Ottoman Empire . . . By releasing a great passion for luxury, disturbing the public morality, and producing a sentiment of discontent and dislocation among many, the inflation constituted a source of great difficulties for the state".

Of particular importance here is the fact that Sino-Spanish trade in Manila experienced several periods of severe disruption between 1610 and the fall of the Ming dynasty in 1644. Some of this disruption may have been caused by declining silver production in the New World,[65] but, before 1630 at least, Dutch and English harassment of Chinese and Spanish shipping is likely to have been a much more important factor.[66] The situation in Manila improved markedly during the early 1630s, but this was followed by the imposition of stringent new trade restrictions by Philip IV in 1634-5, by a government crack-down on corruption in Acapulco in 1636, and by the Spanish massacre of more than 20,000 Chinese in the Philippines during the winter of 1639-40.[67] The massacre itself was partly the result of economic difficulties between the two communities and, as might be expected, it contributed to a sharp decline in trade between Manila and China during 1640-1.[68] There appears to have been some recovery in 1642-3, but this may have been offset by the fact that Portuguese merchants from Macao stopped trading in Manila in 1642 when they learned of Portugal's rebellion against the Spanish crown in Europe. Since the Manila-Macao trade is known to have been worth 1,500,000 pesos (43,125 kilograms of silver) in some years during the 1630s,[69] the economic consequences for both sides and for China were considerable.

Maritime trade in Asia was also affected during these years by Dutch blockades of Goa and Malacca, which virtually isolated Macao from the rest of the Portuguese empire. Then in 1639 the Portuguese lost their very lucrative role as middlemen in Sino-Japanese trade when the Tokugawa authorities refused to allow ships from Macao to trade at Nagasaki. Since those same authorities had prohibited Japanese nationals from trading overseas in 1635, this left only Dutch and Chinese merchants exporting silver from Japan during the early 1640s. And despite their strenuous efforts to maintain business at previous levels, by 1642-3 the total amount of silver leaving the country had apparently dropped to below 1,500,000 taels (56,250 kilograms) annually.[70] Whatever the exact figure, it was far below Ko-

[65] Levels of bullion production and economic activity in the New World during the seventeenth century are still subjects of considerable debate. See J. I. Israel, "Mexico and the 'General Crisis' of the Seventeenth Century", *Past and Present*, no. 63 (May 1974), pp. 33-57; John J. TePaske and Herbert S. Klein, "The Seventeenth-Century Crisis in New Spain: Myth or Reality?", *Past and Present*, no. 90 (Feb. 1981), pp. 116-35.

[66] See, for example, Kobata, *Kingin bôekishi*, pp. 259-64; Wills, "Maritime China", pp. 216-20.

[67] I have discussed these matters in more detail in my "Notes on Silver", pp. 10-13.

[68] "Relation of the Insurrection of the Chinese", in Blair and Robertson (eds.), *The Philippine Islands*, xxix, pp. 208-58; Chaunu, *Les Philippines*, pp. 157-9.

[69] Boxer, *Great Ship from Amacon*, p. 135 n. 284; Schurz, *Manila Galleon*, p. 132.

[70] Iwao, *Shuin-sen bôekishi*, p. 327; Oskar Nachod, *Die Beziehungen der Nederlandischen Ostindischen Kompagnie zu Japan im siebzehnten Jahrhundert* (Leipzig, 1897), appendix 63, p. 108.

bata's estimate of four to five million taels (150,000-187,500 kilograms) per year when Sino-Japanese trade was at its peak earlier in the century. Moreover, when this decline is considered in light of the depressed commercial situation in Manila, Macao and Malacca, it is clear that far less silver was even available for importation into China than had been the case just a few years earlier.

The implications of the above for the Ming economy would have been serious at the best of times. As it was, the Ming government had been forced to raise taxes seven times between 1618 and 1636 in response to military threats from the Manchus and from domestic rebels.[71] And although it is now clear that not all of the bullion recorded in Table 2 for the late 1630s and particularly the early 1640s was actually collected,[72] these tax increases undoubtedly drew silver away from those sectors of the Ming economy upon which much of the dynasty's earlier prosperity had been based.[73] Nor were those sectors aided by the fact that during the late 1630s and early 1640s, economic activity in many parts of the country was further reduced by widespread military operations and by a series of devastating natural disasters.[74] Under the circumstances, it seems reasonable to assume that rather than spend their money on non-essential items or invest it in increasingly risky enterprises, those in a position to do so hoarded even more bullion than usual and waited to see what would develop.[75]

Given the decline in bullion imports, the higher rates of taxation, and the probability of increased hoarding, then, there is reason to suspect that the amount of silver in circulation had been substantially reduced. That suspicion seems confirmed by the figures in Table 6.

[71] Ray Huang, "Fiscal Administration during the Ming Dynasty", in Charles O. Hucker (ed.), *Chinese Government in Ming Times: Seven Studies* (New York, 1969), pp. 117-19.

[72] By the late 1630s the Ming authorities were unable to collect many of the new taxes they had imposed, and the figures from the T'ai-ts'ang vault for the last decade or so of the dynasty's existence appear to have been projections of possible income rather than actual receipts. Professor Huang believes that tax collections during this period "were considered quite successful if any year's proceeds came close to 70 per cent of the projected quota. It is extremely doubtful that the actual collections in the 1640s ever met half of the assigned volumes": *ibid*., pp. 121-2.

[73] Under ordinary circumstances, of course, a substantial portion of the central government's silver revenues would have been spent on goods and services and would thus have re-entered the economy and the monetary system. During the late Ming period, however, large quantities of silver fell into the hands of powerful political and military figures, many of whom chose to hoard it rather than spend it on national defence. On this point, see Frederic Wakeman Jr., "The Shun Interregnum of 1644", in Spence and Wills (eds.), *From Ming to Ch'ing*, pp. 44-7, 77-8.

[74] Helen Dunstan, "The Late Ming Epidemics: A Preliminary Survey", *Ch'ing-shih wen-t'i*, iii no. 3 (1975), pp. 1-59.

[75] Whether the Chinese habitually hoarded a higher percentage of their precious metals than other pre-modern peoples seems questionable. Nevertheless, they were shrewd enough to know when silver was going up or down in value and to act on that information if they possibly could.

The abrupt reversal they illustrate in the long-term trend towards cheaper silver, the timing of which corresponds closely to what is currently known about the sharp decline in bullion imports discussed above,[76] had disastrous consequences for the late Ming economy. Without access to sufficient supplies of silver because of the rapidly deteriorating copper-silver exchange rate, many people in China were unable to pay their taxes or rents, repay loans, or in some cases even to buy food.[77] Finding it impossible under these circumstances to adequately pay or equip its military forces,[78] the Ming court gradually lost control of the northern portion of the empire, first to domestic rebels and then to the invading Manchus. The latter went on to defeat Ming loyalists in the south and to establish their own dynasty, the Ch'ing, which lasted until the revolution of 1911.

TABLE 6

VALUE OF ONE THOUSAND COPPER COINS 1638-1646 (IN TAELS OF SILVER)*

	Taels
1638	0·90 (+)
1640	0·50 (+)
1643	0·33
1646	0·17

* Note and sources: Yeh Shao-yüan, "Ch'i-chen chi-wen lu" [A Record of Things Remembered from the T'ien-ch'i and Ch'ung-chen Reigns], in *T'ung-shih* [A History of Sorrow] (Shanghai, 1911), *ts'e* 18, 2/6a; Chang Lü-hsiang, "T'ung-hsiang tsai-i chi" [A Record of Natural and Human Disasters in T'ung-hsiang], in Ch'en Heng-li (ed.), *Pu Nung-shu yen-chiu* [Studies on the Pu Nung-shu] (Peking, 1958), p. 325. The symbol (+) indicates that the figure given for this year in the original source is imprecise and is likely to have been slightly larger than that which appears in the Table.

CONCLUDING REMARKS

The preceding survey has attempted to demonstrate that during the sixteenth and early seventeenth centuries, domestic demand for imported silver and foreign demand for Chinese luxury goods in-

[76] It might be argued that since at least some foreign silver was still flowing into China during this period, bullion imports were essentially irrelevant to the trend revealed in Table 6. However, this would ignore the impact a sharp decline in those imports could have had on business and consumer confidence, particularly in the all-important south-eastern portion of the country.

[77] I have discussed these developments in more detail in my "Notes on Silver", pp. 17-21.

[78] In a recent study, J. A. Trevithick has written that "Governments, in an attempt to squeeze more and more out of a hard-pressed wartime economy, have often taken the easy way out and resorted to the printing presses to provide the requisite finance for the purchase of arms, the payment of soldiers, etc.". The Ming government was no exception, and in 1643-4 it made a desperate attempt to reintroduce paper currency into the monetary system. The failure of that attempt was explained by one high-ranking official in the following words: "The people may be stupid, but who is willing to exchange bullion for paper?". See J. A. Trevithick, *Inflation: A Guide to the Crisis in Economics* (Harmondsworth, 1977), p. 9; Chang T'ing-yü (ed.), *Ming-shih* [Ming Dynastic History], 28 vols. (Peking, 1974), xxi, pp. 6501-2. See also Peterson, *Bitter Gourd*, pp. 74-6.

volved China more directly in world history and particularly world economic history than ever before. The effects of that involvement were of fundamental importance to the subsequent development of Chinese civilization. Indeed, although the statistics currently available on this subject are often less than satisfactory, there is more than sufficient evidence to indicate that many of the economic advances, social changes and even cultural monuments of the late Ming period can be properly understood only in light of the vast amounts of silver which flowed into China from such faraway places as Iwami and the mountains of Upper Peru.

At the same time, however, the enthusiastic participation of many Chinese merchants and producers in international trade helped to cause economic and other problems which the Ming government was never able to satisfactorily resolve. The Ming dynasty did not fall in April 1644 simply because bullion imports had declined sharply from previous levels, but that decline certainly exacerbated its difficulties and helped to undermine its stability. The term "seventeenth-century crisis" may have been overworked in recent years, but there can be no question that China experienced one of major importance.[79] Moreover, although they did not understand the complexities of the situation, seventeenth-century Chinese writers were themselves aware that the country's economic and political problems were intimately connected with silver and maritime trade.[80]

Of course, neither that century nor its "crisis" lasted forever. Indeed, by the early years of the next century, if not earlier, the Chinese economy was once again expanding rapidly and the country was regaining its deserved reputation as the world's richest and most powerful nation. And, as might be expected, that expansion and reputation were again supported by large-scale imports of foreign silver, particularly from the New World.[81] Unfortunately, that intriguing and important story goes beyond the scope of the present article.

[79] On this point, see S. A. M. Adshead, "The Seventeenth-Century General Crisis in China", *Asian Profile*, i no. 2 (1973), pp. 271-80.

[80] See, for example, Ku Yen-wu, *Jih-chih lu chi-shih* [Collected Commentaries on the Record of Knowledge Gained Day by Day], ed. Huang Ju-ch'eng, 2 vols. (Taipei, 1971), i, p. 258.

[81] Perhaps the best discussion of this subject to date is in Ch'üan, *Chung-kuo ching-chi shih*, ii, pp. 475-508. See also Vilar, *History of Gold and Money*, pp. 253-62; K. N. Chaudhuri, *The Trading World of Asia and the English East India Company, 1660-1760* (Cambridge, 1978), pp. 385-410. It should be noted that improved domestic mining, a much more efficient government, a sounder monetary system and a rapidly growing population meant that silver imports had a far less dramatic effect on the Chinese economy after 1680 than had been the case a century or so earlier. Still, as Ch'üan has convincingly demonstrated, the importance of those imports to the economic expansion which occurred in eighteenth-century China should not be underestimated.

8
Silver and the Fall of the Ming: A Reassessment

Brian Moloughney and Xia Weizhong

Historians of pre-modern Europe have found concepts such as "Renaissance," "Crisis" and "Enlightenment" useful for portraying the dominant features of an age. Yet these concepts are not without problems and their value is often the focus of much debate. This has been the case with the so-called "general crisis" of the seventeenth century. The attempt to link the political instability of mid-seventeenth century Europe to social and economic factors continues, yet without any clear consensus emerging. Recently Peter Clark has described the crisis debate as being "rather like an elderly car with a defective battery and dirty points." While there are times when it seems the car is not capable of going anywhere, when it's usefulness is past, suddenly "with a splutter and a start it comes to life and some rather frantic progress is made before it subsides into taciturnity."[1] One way in which the debate has been given new life in recent years is the attempt to extend its focus out from Europe to include social, economic and political change in other regions of the world. This trend towards a more integrated "world economy" approach has led to the suggestion that China also experienced a general crisis in the seventeenth century.

While the years of political turmoil leading to the collapse of the Ming dynasty in 1644 and the associated widespread social upheaval, with an estimated population decline of around twenty per cent between 1600 and 1650, might suggest parallels with events in Europe, monetary factors have been to the fore in encouraging this transference of an analytical approach, the concept of a "general crisis," to the Chinese world.[2] The flow of gold and silver from Europe to Asia has long been known. Fernand Braudel, for instance, has written of how "precious metals once absorbed into Mediterranean life were fed into the stream that continually flowed eastwards," and that this remained unchanged for centuries.[3] Monetary metal, and in particular silver, was seen as the rope linking east to west. The extent of this flow of silver into Asia was revealed with greater clarity when Pierre Chaunu extended his study of the *Carrera de Indias* to include the famous Manila galleons. It has been estimated that the Spanish American mining regions of New Spain (Mexico) and Peru contributed as much as eighty per cent of the silver produced in the world between 1550 and 1880, and Chaunu suggests that "a little more than a third" of this silver was shipped either from Acapulco across the Pacific to Manila or via the Atlantic and the various trading routes to Asia, much of it then going to China.[4]

Silver was thus seen to link China into a Seville-centred trading system, and just as Spain suffered a restricted flow of silver from the Americas in the early years of the seventeenth century, so too, it was suggested, did China. In a short and provocative article S.A.M. Adshead proposed that the Ming "silver rush" of the late sixteenth and early seventeenth centuries, the dispatch of eunuch tax commissioners throughout the empire in search of new sources of the metal, was a direct consequence of a decline in imports of silver, a decline that was seen to coincide with the first major contraction of the Sevillean world economy.[5] Here we have Earl J. Hamilton's argument that the vicissitudes of international movements of bullion reflected directly upon the economic, social and political conditions in sixteenth and seventeenth century Europe, finding a parallel in seventeenth century China. The conception of a seventeenth century "general crisis" was now being used to describe events in late Ming China.

This perspective on events in late Ming China was developed by William S. Atwell who has revealed much of the complexity of silver flows in the East Asian world. Atwell has given greater emphasis than previous scholars to the importance of Japan as an alternative source of silver to China, yet still maintains that Chinese imports of the metal declined markedly in the latter years of the Ming dynasty, thereby contributing to its collapse.[6] Similarly, Frederic Wakeman, in writing of "China and the Seventeenth Century Crisis," reflects much of the above-mentioned scholarship when he states that "although at a considerable remove, the Chinese economy stood ready to be badly affected by the severe depression that struck the worldwide trading system centred on Seville between 1620 and 1660."[7] China's dependence upon foreign silver is seen as linking it into this wider "world economy" and thus exposing it to the "general crisis" of the seventeenth century.

Chinese scholars have also remarked on the importance of silver in Ming China and on the dramatic increase in imports of the metal from the mid-sixteenth century onwards. Liang Fangzhong was one of a group of young scholars who in the 1930s began to concentrate more upon social and economic questions in Chinese history. As a result of this work Liang published two seminal studies, one on silver mining and the other on international trade and its silver component.[8] The most significant contributions in this area, however, have come from the economic historian Quan Hansheng, who has published a great deal of work on Chinese monetary history, commodity prices and foreign trade. Quan, more than any other Chinese historian, has revealed the extent of China's dependence on silver produced in the mines of Spanish-America and Japan.[9] Recently, other Chinese scholars have begun publishing work on these questions, and Japanese scholars have long been aware of the importance of the trade in monetary metal in East Asia.[10] What these scholars have not done, however, is suggest, as their Western counterparts have, that this dependence upon foreign silver exposed China to the contraction of an emerging "world economy," to the "general crisis" of the seventeenth century. A dependence on foreign trade for monetary metal is not seen as a contributing factor to the collapse of the Ming dynasty.

That Chinese and Western historians working on these same questions should differ so markedly in their interpretations suggests that the historiographical perspectives within which these scholars have worked have been of major influence in determining their conclusions. With this in mind, we propose to re-examine this issue of silver and the fall of the Ming dynasty. We will look first at reasons for the increasing demand for silver during the Ming dynasty and at how domestic production was unable to satisfy that demand. We will then turn our attention to international movements of bullion, to centres of production and agencies of distribution, in order to gain an impression of the changing volume of silver imported into China throughout the late Ming years. New studies have given us a clearer picture of this trade in silver, yet it remains the case that no precise figures can be given for the total volume of silver imported at any particular time. The data does not allow such precision. What can be established, however, is an overall trend for the period, and it is on the basis of this that we will assess the claim that China's dependence upon foreign silver was a contributing factor to the collapse of the Ming dynasty.

* * *

> *The Chinese are very skilful in judging at Sight the finest Silver, and are scarcely ever deceived.*[11]
>
> Jean Baptiste Du Halde

Silver had been used both as a monetary medium and as a store of wealth in China at least since the beginning of the imperial epoch, so this remark of the Jesuit Father Du Halde is perhaps not surprising. The ability to assess quickly a particular quantity of silver and the extraordinary demand for the metal were noted by nearly all foreign commentators of the Chinese world in the sixteenth, seventeenth and eighteenth centuries. Manuel de Cômara de Noronha, a Captain-General of Macao in the 1630s, wrote that "when the Chinese smell silver, they will bring mountains of merchandise."[12] Such a demand has led one historian to describe China as a great suction pump, drawing in silver from the four corners of the world.[13] But silver had not always been as highly valued in China as it was at the turn of the seventeenth century. Prior to the fifteenth century silver use was restricted and at times the metal was considered to be worth much less than were other forms of monetary media. As well as silver, gold, grain, silk, hemp cloth, cowries, copper coinage and paper notes all had at times functioned as money.[14]

It was only during the Ming dynasty that silver emerged to be, along with copper cash, one of the two main currencies of the realm. Indeed, the evolution of a bimetallic currency system, a system which continued into the nineteenth century, was one of the most important features of the Ming period.

Initially the early Ming administration found that due to a paucity of monetary metal it was forced to adopt a paper currency system similar to that in use during the preceding Song and Yuan dynasties. A Supervisorate of Paper Money (*Baochao tijusi*) was established in 1374 and the next year began issuing notes in six denominations. Within a few years, however, inflation had become a serious problem and despite continued attempts to control the number of notes in circulation confidence in them could not be restored. By the middle of the fifteenth century the value of these notes had depreciated 1,000 fold and it was said that they piled up in stores without attracting the slightest attention of passers-by.[15] The government abandoned the printing of paper currency in 1450, allowing copper cash and silver bullion to take over as the currencies of the realm.

This transition meant the Ming currency system became a form of parallel bimetallism: i.e., a bimetallic system with floating exchange rates.[16] The two currencies, copper and silver, functioned distinctively within the economy, with cash providing the basis to the system and being used in most retail transactions, while silver gradually dominated in the wholesale sector and later came to be used in the payment of higher salaries and taxes. And as the value of copper was much less than that of silver the two currencies could not easily be substituted for one another. In contrast to the copper coinage, silver circulated as bullion and its value was determined by weight and fineness – the silver unit of account, the *tael*, being considered equal to the unit of weight, the *liang*. Silversmiths did mould the metal into ingots, or "shoes of sycee," to make its use as currency easier, but the metal was still generally weighed and assayed at the time of transaction. The Portuguese Father Gabriel de Magaillans, a visitor to China in the seventeenth century, wrote of how the ingots were "cut with steel scizors, which the people carry about them for that purpose, and divide them into pieces, bigger or lesser, according to the value of the purchas'd Commodity."[17] As well as scissors, merchants carried with them small sets of scales, upon which they could determine the amount of silver needed to satisfy each particular exchange. Father Du Halde noted that the Chinese

> own it would be more convenient to have Mony coined, and of a determinate value, but they are afraid it would be a Temptation to Clippers and Coiners .... It is easy to judge that there would be Debasers of Mony in China, if the Silver was coined as well as Copper, since their small Pieces of Copper are so often counterfeited ...[18]

Counterfeiting and the consequent devaluation of the copper coinage were indeed serious problems, particularly towards the end of the Ming dynasty. One of the attractions of silver was that it was a stable currency, a currency less susceptible to inflationary pressure.

Silver was also the preferred currency of travellers and traders. We know that credit facilities and the interregional remittance of funds did exist in late Ming China, but it is likely that they were not well developed. For instance, when

Matteo Ricci was offered a house in Nanjing by Liu Guannan, Father Cattaneo arranged for some form of bill of exchange (*huipiao*) to be sent from Macao to enable Ricci to fulfill his promise to complete the payment on the house within a year. This bill of exchange was possibly a *yinpiao*, or silver note, issued by one of the merchants in Macao who was involved in the moulding and assaying of silver and who had contacts in Nanjing. Yet when Ricci went to draw the money he required, he found it impossible to locate the merchant cited as assignee on the bill (back in Macao the Jesuit Fathers had no difficulty in recovering their money). As foreigners the Jesuits may have found greater obstacles to conducting such transactions than did native Chinese, yet this incident does suggest that interregional credit remittance was not very common at this time.[19] Thus, merchants and travellers generally carried their funds with them. Silver was preferred for this as it did not suffer the regional differences in quality and value which copper did, and, as a precious metal, it was simply less burdensome to transport than were large numbers of coins.

Changes in the state's fiscal policy were also instrumental in creating an increasing demand for silver in the late Ming years. The core to the tax structure instituted during the founding years of the dynasty had been a land tax imposed on all landholders and payable in grain. But this tax had fallen heaviest on the productive southeast of the empire and in time this was to lead to much evasion and increasing inequities within the fiscal structure. In 1393, for instance, the prefectures of Suzhou and Songjiang contributed 13.68 per cent of the total tax paid, yet they encompassed only 1.76 per cent of the cultivated land.[20] Further exactions were imposed on landholders in the southeast when, in 1421, the Ming capital was shifted from Nanjing to Beijing and surcharges were added to the basic tax quota so as to pay the costs involved in transporting grain to the new administrative centre of the empire. These surcharges were often as much as two or three times the amount of the original land tax, even though landholders were responsible for much of the transportation themselves. Inevitably, those who could afford to bribe local officials and those with influence inside the yamen office began to explore all avenues possible to avoid meeting their tax and corvée obligations. Gu Dingzhen described in a memorial submitted in 1527 how:

> The amount of tax loss can be counted by thousands and tens of thousands each year. The law-abiding households in counties and prefectures made up the losses year after year, and eventually fell into distress.[21]

Although such inequities were most pronounced in the southeastern region of the empire, they were sufficiently widespread by the early sixteenth century for there to have been a considerable impetus amongst officials to implement some form of restructuring to the state's fiscal machinery. This restructuring culminated in a body of policy which came to be known as the Single Whip Reform, or *yitiao bianfa*, a reform which was the most significant innovation in fiscal policy during the entire Ming and Qing periods.

In essence, the Single Whip Reform saw a shift in the unit of assessment from the household to the landholding and the integration of the land tax and corvée obligation into a single commuted exaction to be paid in silver.[22] Payments in silver had been introduced as part of the Ming fiscal structure as early as 1436, when about fifteen per cent of the land tax, known as the "gold-patterned-silver" (*jinhuayin*) and levied on landholders in six southern provinces and Nan Zhili, was commuted at a flat rate of 0.25 *taels* per picul of grain (in total around 1 million *taels* annually).[23] Then, in the late fifteenth century, merchants operating under the auspices of the government's salt monopoly were granted permission to make payments for their salt receipts in silver rather than grain or goods. It has been calculated that by the 1560s salt merchants were trading around 2.8 million *taels* of silver annually, about four times as much as at the beginning of the century.[24] Then came the Single Whip Reforms. Although the process of integrating corvée obligations with the land tax into a single silver levy was not completed until the Qing period, during the reign of the Yongzheng emperor (1723-1735), most provinces had implemented the reforms to some degree by the end of the Wanli reign in 1620. Ray Huang estimates that even by the end of the sixteenth century between 65 and 100 per cent of the land tax and labour services in the rich heartland of the southeast, Jiangnan, had been incorporated, with approximately 60 per cent integration in the south outside of the Jiangnan region, and between 40 and 50 per cent integration in north China.[25] The impetus for the Single Whip Reform was a general desire to make the Ming fiscal system more efficient and more equitable, and, to some extent, this was achieved. But the greatest consequence of the reform was the tremendous stimulus given to the circulation of silver within the late Ming economy. Taxes continued to be assessed on the basis of productivity, but the taxpayer now had to sell that produce in order to meet the exactions of the state. Thus, by the turn of the seventeenth century silver had become an integral part of both the monetary and fiscal machinery in Ming China.

This increasing importance which silver came to have for the Ming state and society was not reflected in an increased domestic production of the metal. What we know of Ming silver mining shows that the years of peak output were those of the early part of the fifteenth century. It was at this time that the Yongle emperor (r. 1403-1423) increased mining quotas and extended Chinese administration over two silver-rich areas to the south. The first of these came with the Ming occupation of Vietnam between 1406 and 1427. The mountainous Viet-bac region, north of the Red River delta, contained a number of gold and silver mines, which the Ming administration was able to control during its limited period of occupation.[26] The second silver-rich area was to be of more lasting benefit to the Chinese empire. Early in the fifteenth century Chinese control was extended south of Yunnan prefecture (the Kunming region) and beyond Yongzhang prefecture to an area previously administered by the Maoshan states of Upper Burma and known to them as "bawdwingyi," or "Great Silver Mines." Prior to Chinese occupation of this region much of its silver output had been carried by Dai traders southwest into Bengal. It has been estimated that the annual average output from the mines of the Bawdwin region during the 437 years of Chinese administration (1421-1858) was 3,300 kilograms of silver.[27] The mines of Yunnan province were the most productive in the Ming empire, generally producing more than fifty per cent of its total silver output.[28] Production at these mines was not continuous, however, as the administration frequently closed the mines through fear of rebellion by disaffected mining labour. Even with the so-called "silver rush" from 1596 until 1605, when new mines were brought into production and output from existing ones was stepped up, it seems that total production never again reached the level attained in the early years of the fifteenth century.

The work of Quan Hansheng on the revenue which the Ming administration derived from domestic silver mining confirms the trend outlined above. The period 1411-1420 saw an annual average income to the government of approximately 10,900 kilograms of silver. This was down to an annual average of approximately 5,300 kilograms for the period 1431-1440; 2,300 kilograms for the period 1451-1460; 2,200 kilograms for the period 1471-1480; 2,000 kilograms for the period 1491-1500 and 1,200 kilograms for the period 1511-1520. If the administration was collecting about 30 per cent of total production, as Quan suggests, then these figures would indicate a decline in overall production from approximately 36,300 kilograms annually in the period 1411-1420 to approximately 4,000 kilograms annually in the period 1511-1520.[29] Ray Huang notes that in 1548 the reported income from silver mining was 62,030 *taels* (approx. 2,345 kgs), which would indicate a total production for the year of around 7,800 kilograms. For the year 1601, Huang cites the reported income as being 110,210 *taels* (approx. 4,166 kgs), or a total production of around 13,880 kilograms.[30] With all the problems of under-reportage, corruption etc., these figures are highly speculative, but they do tend to indicate that domestic silver output remained limited throughout the entire Ming period. Indeed, as Quan Hansheng notes, the Spanish administration at the Potosi mines in Peru taxed production at a rate far less than did the Chinese, yet was able to collect more revenue in a week than the Chinese could in a year.[31]

Historians in the People's Republic have done much in recent years to reveal the great social and economic changes that Chinese society underwent from the mid-Ming period onwards.[32] Continued population growth through to the beginning of the seventeenth century, an expanding and diversifying economy and an increasingly monetised tax structure – all of this created a tremendous demand for silver. Yet domestic production was unable to meet that demand. Hao Jing, a Supervising Secretary with the Ministry of Revenue around the turn of the seventeenth century, questioned that while silver was becoming of increasing importance in the late Ming economy, it "did not fall like the rain from the sky, nor gush forth from the ground. [And as] it was not made by the people, how then would it not be exhausted?"[33] The answer to Hao Jing's question lay with foreign trade. Chinese merchants were obtaining more silver than ever before in their trade with foreigners and by the late sixteenth century had gained a reputation for wanting nothing but silver in exchange for their silks and ceramics. This reputation spread far and wide. So much so that a Florentine merchant, Filippo Sassetti, could write in a letter of 1586 that "the Chinese among all peoples of Asia, are wild about silver as everywhere else are men about gold."[34] And a factor with the English East India Company wrote in 1636 that the Chinese lust for silver was such that "they will as soon part with their blood as it, having once possession."[35] It is to this foreign trade that we now turn.

You shall understand that the Hollanders have here [in Japan] an Indies of money.[36]

Will Adams

An Indies of money it was! By the time the Englishman Will Adams wrote these words in 1611, Japanese mines were beginning to realise their true potential, producing perhaps as much as one quarter of the world's silver output in the early decades of the seventeenth century. During the 1530s a new process of extracting and refining silver, the *hafuki* (cupellation) process, was introduced to Japanese mines and the resultant increase in production leads Delmer Brown to conclude that by 1540 silver was Japan's principal export.[37] Then, around then turn of the seventeenth century, the Japanese learned of further sophistications to the smelting process from the Portuguese, enabling them to profit from the mining of much lower grade ores. The silver that flowed from these mines attracted not only Chinese but also others eager to profit from an involvement in the trade.

The most famous of these carriers were the Portuguese *naos*, or "great ships," which sailed annually between Goa, Macao and Nagasaki from the 1560s until 1618. Thereafter, Dutch piracy forced the Portuguese to shift to smaller and faster galliots.[38] Portuguese involvement in East Asian trade began in the 1540s, but it was not until they obtained the important base at Macao and then gained access to the biannual Guangzhou fairs from 1578 onwards that they became an important source of silver for the Chinese. Some of this silver was of Spanish-American origin, "pepper money" shipped annually from Lisbon to Goa, part of which was then sent on to Macao, or silver obtained through trade at the Spanish port of Manila between 1620 and 1644.[39] For the most part, however, the Portuguese trade between Macao and Nagasaki was a trade in Chinese silks, sugar and porcelain, and Japanese silver.

How much silver did the Portuguese bring to China? Ralph Fitch, an English merchant travelling in Asia between 1583 and 1591, reported that each year the Portuguese carried 600,000 *crusadoes* (approx. 15,336 kgs) of silver from Japan for their China trade. We also know from Father Valignano, a Jesuit resident in Nagasaki, that in 1593 the Portuguese shipped off around 20,000 kilograms of silver. And when the "great ship" disappeared on its return voyage from Nagasaki in 1598/9 the reported loss was 400,000 *crusadoes* (approx. 10,244kgs).[40] With the shift to the smaller galliots from 1618 it seems the trade also increased in volume, silver exports averaging approximately 30,000 kilograms per annum. Then with the introduction of the seclusion edicts by the Tokugawa administration from 1635 onwards, prohibiting Japanese ships from trading in foreign ports, Portuguese trade increased dramatically. In 1635 the Portuguese exported 1,500 chests of silver (approx. 56,700 kgs) from Nagasaki. The next year this rose to 2,350 chests (approx. 88,830 kgs), and in 1637 to 2,600 chests (approx. 98,280 kgs). Then came decline. Exports for 1638 were down to 1,600 chests (approx. 60,480 kgs) and in the following year Japan instituted the final phase of its seclusion policy by expelling all Portuguese from the country and putting an end to this trade.[41] On the basis of this information we estimate that in the second half of the sixteenth century the Portuguese imported around 500,000 kilograms of Japanese silver to China, in the period 1600-1609 200,000 kilograms, 1610-1619 200,000 kilograms, 1620-1629 300,000 kilograms and 1630-1639 450,000 kilograms. Altogether these figures suggest the Portuguese imported around 1.65 million kilograms of Japanese silver to China. If we add to this another 500,000 kilograms for silver brought from Goa or obtained through trade at Manila we get a total of a little over two million kilograms of silver imported to China by the Portuguese in the period up until 1639.[42]

This trade was obviously of considerable importance to China, particularly to the Canton delta region. Yet to claim that the exclusion of the Portuguese from Nagasaki in 1639 saw a sharp downturn in the volume of Japanese silver entering China, as Atwell and others do, is to underestimate the resilience of China's demand for silver. As Frank Perlin has noted with regard to South Asia, money is "a medium particularly subversive of conventional boundaries, both material (political and geographical) and intellectual ..."[43] Just as the Portuguese benefited when Japanese were prevented from partaking in overseas trade after the introduction of the seclusion edicts, so Chinese and Dutch merchants benefited from the exclusion of the Portuguese.

Despite the prohibition on trade with Japan, Chinese merchants conducted a flourishing trade with the Japanese throughout the late Ming period. There are references to Chinese junks sailing directly to Japan from the 1530s onwards,[44] and from the latter years of the sixteenth century through until 1635 Chinese and Japanese merchants enjoyed a vigorous entrepôt trade in the ports of Taiwan, Siam and Vietnam. John K. Whitmore has described the nature of this trade in the central Vietnamese port of Hoi-an (Faifo):

> The Chinese and Japanese each had their own enclaves, adjacent to each other. Coming with the northeast monsoon at the end of the (lunar) year, ships from Japan brought large amounts of silver and copper cash which went mainly for silk, sugar, aloeswood, deerskins, rayskin and ceramics.... The Chinese merchants gathered during this four month "fair" and traded their silk, copper cash, and tutenag for the Japanese silver and the goods of Southeast Asia, pepper in particular, sandalwood, camphor, and other aromatics.[45]

Determining the volume of Japanese silver that reached China via this entrepôt trade is extremely difficult, perhaps impossible. Kozo Yamamura and Tetsuo Kamiki have estimated around 11,250 kilograms of silver was carried from Japan each year in Chinese and Japanese ships in the forty year period to 1600, giving a total of some 450,000 kilograms of silver.[46] In other words, a little less than that carried by the Portuguese for the same period. Iwao Seiichi tells us that 350 ships sailed from Japan under license, the so-called *shuin-sen* or "red seal" ships, and that each year these ships carried around 30,000 to 40,000 kilograms of silver – altogether between 1 million and 1.4 million kilograms for the thirty-five year period to 1635. We favour the lower of these two figures. If we take Robert Innes' estimate that the average volume of silver exported by each *shuin-sen* was 500 *kanme* (1,880 kgs), then we have a total of 658,000 kilograms of silver for all the red-seal ships between circa. 1600 and 1635.[47] To this we must add the unknown quantity of silver exported in private Japanese ships prior to 1633, which would have been somewhat less than that exported in the licensed ships. On this basis, we feel that the volume of silver exported by Japanese ships between 1600 and 1635 might have reached 1 million kilograms, the lower end of Iwao's estimate.

More important for our discussion, however, were the Chinese junks sailing directly to Japanese ports during the seventeenth century. The number of Chinese resident in Nagasaki is said to have increased from about twenty in 1608 to several thousand by 1620, and the number of junk visits increased to between 30 and 60 per year by the 1620s.[48] Chinese junks also continued to trade at Hirado up until 1635, when all foreign trade was directed through Nagasaki. Of this trade at Nagasaki, we know that Chinese imports were valued at 2,600 *kanme* in 1623, 3,000 *kanme* in 1629 and 4,420 *kanme* in 1634. The Chinese dominated the silk trade at Nagasaki by 1634, importing 1,000 piculs of yarn to the port (and a further 1,000 piculs to Satsuma through trade in the Ryukyu Islands). In the same year the Portuguese imported only 200 piculs of silk yarn to Japan.[49]

The number of Chinese junks visiting Nagasaki increased from 36 in 1634 to 40 in 1635, 64 in 1637 and, with the expulsion of the Portuguese, to 93 in 1639. Even in the heyday of the Portuguese trade, only six galliots arrived at Nagasaki, and while each of these galliots carried away more silver than did a single Chinese junk, the huge numbers of junks visiting in these years shows the extent to which the Chinese dominated the trade. In 1640 the number of Chinese junks trading at Nagasaki dropped to 74, then rose to an all time high of 97 in 1641. The following two years saw the number drop back to 34, probably in response to the surfeit of Chinese goods in the preceding year. Then in 1644 the number rose again to 55, to 82 in 1645, down to 54 in 1646 and was still as high as 70 in 1650. The amount of silver traded with Chinese merchants increased from 9,500 *kanme* (35,720 kgs) in 1634 to 17,000 *kanme* (63,920 kgs) in 1646.[50] In the decade 1635 to 1645, the last decade of the Ming dynasty, these junks probably carried over half a million kilograms of silver back to China from Japan.

If we turn our attention now to the Dutch, the only other traders to ship Japanese silver to China on any significant scale, we find a similar pattern to that of Chinese trade at Nagasaki during the 1630s. In the very early years of their contact with Japan (1609-1620s), Dutch merchants exported only negligible amounts of silver. The level of this trade increased in the early 1630s and then rose substantially following the prohibition on Japanese trade and the expulsion of the Portuguese. In 1634 Dutch merchants had exported only around 8,500 kilograms of silver from Japan. In 1635 this rose to 14,031 kilograms; in 1636 to 30,124 kilograms; 1637 to 40,242 kilograms; 1638 to 47,538 kilograms; and in 1639 to a peak of 75,956 kilograms.[51] It seems the *Bakufu* administration were aware of the possibility of a falling off in trade in 1639 and had asked the Dutch to increase the volume of their imports to help make up for those of the banished Portuguese. The Dutch invested heavily for the next year's trade, only to find the Chinese had done the same. The Japanese administration also attempted to increase trade in Chinese products through Korea and the Ryukyu Islands, with some success. The Nagasaki market was flooded with merchandise and the Dutch were forced to accept much lower prices than they had expected. In relating this incident, Tashiro Kazui notes that "the exclusion of the Portuguese vessels did not immediately result in a reduction of the volume of the trade."[52]

The Dutch never had direct access to the Chinese mainland during these years, but from 1624 until 1661/2 conducted trade with the Chinese through their base at Casteel Zeelandia on Taiwan. It was through this trade that the Japanese silver they exported from Nagasaki entered China. Pieter Nuyts, governor of the base on Taiwan from 1627 to 1629, describes how the Dutch entrusted Chinese intermediaries with their money, requesting that they purchase products in demand in Japan. He notes that the Dutch had a contract with the Japanese to supply them with 740,000 *guilders* (approx. equal to 7,400 kgs silver) worth of produce and that there was no trouble in fulfilling this quota; in fact, "much more merchandise would doubtless be forthcoming if there was capital to purchase it."[53] The money was forthcoming and by 1640 the Dutch were spending more than 50,000 kilograms of Japanese silver in the purchase of Chinese goods through Taiwan.[54] By 1645 Dutch exports of silver had fallen back to be around 30,000 kilograms, and in 1647 dropped further still to less than 20,000 kilograms. The main reason for this seems to have been the dominance of the Chinese in the Nagasaki trade and the increasing turmoil and conflict for power along the China coast which followed the fall of the Ming.[55] The main maritime force with which the Dutch had to contend at this time was that of the Zheng family regime, particularly the junks and trading alliances controlled by Zheng Zhilong. It has been estimated that in 1643 80 per cent of the trade conducted by the Chinese at Nagasaki was controlled by the Zheng family and as is well known it was Zheng Chenggong who forced the Dutch to abandon Taiwan by 1662, bringing to an end their involvement in the profitable China trade.[56]

By collating this information we have regarding the traffic of Japanese silver to China it is possible to gain an idea of the changing volume of imports throughout the late Ming years. In 1610 the trade conducted by Portuguese, Japanese and Chinese ships saw around 80,000 kilograms of silver enter China. By 1620 this figure had risen to around 100,000 kilograms, and by 1630 it was up around 130,000 kilograms. With the prohibitions on overseas trade by Japanese ships in 1635 the volume of silver reaching China may have dropped back to around 110,000 kilograms, only to rise to an all time high of around 200,000 kilograms by 1637. The exclusion of the Portuguese from Japan saw the Chinese and Dutch increase the volume of their trade and in 1639 around 170,000 kilograms of silver was still being carried to China, only slightly less than the peak year of 1637. In 1641 the volume of this trade was still around 145,000 kilograms, falling off to around 70,000 kilograms in 1643 and 100,000 kilograms in 1645.[57] The trend shown by this information does not support the claim made by some historians that there was a significant long-term reduction in imports of Japanese silver to China in the years leading up to the fall of the Ming dynasty. Rather, the late Ming years saw this trade reach its zenith.

The king of China could build a palace with the silver bars from Peru which are carried to his country.[58]

Admiral Don Hieronimo de Banuelos y Carrillo

Such comments were heard often at the Spanish court in the early seventeenth century. Spain's merchants resented the trade which the Chinese enjoyed with the "New World," depriving them as it did of much profitable business, and frequently petitioned the throne for its prohibition. Juan Velazquez Modr, for instance, wrote in 1628 on this familiar theme, stating that:

> It is very pernicious to permit the importation of the silk of China and its trade, both in the Indias and in España. For although not more than two hundred and fifty thousand pesos de Tipuzque can be taken from Nueva España to the Filipinas annually, besides that sum another incalculable quantity of money is taken in reals of eight; for the said silk can be bought or traded for nothing else, nor will the Chinese give or exhange it for other merchandise. Consequently they manage to get hold of and carry away annually the greater part of the eight-real pieces which are made in the said Nueva España....[59]

Some even suggested that 10 or 12 million *pesos* were carried each year in the Manila galleons to the Philippines, and thence to China. But as Juan Grau y Monfalcon notes in his eloquent defence of the Philippine trade, the mines of Mexico did not produce this much metal, and even if they did how could two ships carry, let alone conceal, such a vast quantity of silver. The same merchants who made these claims were unwilling, and quite naturally so, to accept that one million *pesos* of unregistered silver could be hidden amongst the nine to ten million *pesos* carried by the *flota de plata*, the silver fleet of 30 to 40 ships which sailed annually from Portobello and Vera Cruz to Spain. "The fact is," he concludes, "that he who is looking for transgressions does not think he is accomplishing his purpose unless he increases them."[60] Exaggeration and misinformation aside, how much Spanish-American silver did reach China in these years?

Suggestions that the flow of this silver to China declined in the late Ming years were developed from earlier studies of the Spanish-American trade which had as their focus Seville. Earl J. Hamilton found that the records of the controlling body of this trade, the *Casa de Contratación*, showed the volume of precious metal entering Spain increased steadily throughout the sixteenth century, reaching a peak of 115 million *pesos* (approx. 2.9 million kgs) in the period 1591-1600. In the early years of the seventeenth century imports of silver declined gradually, until the 1630s, thereafter there being a "precipitous drop" in imports until 1660, when the compulsory registration of imported treasure was abolished and Hamilton's analysis ends. By the decade 1651-1660 less than 18 million *pesos* (approx. 46,000 kgs) of silver was imported legally from the Americas.[61] As the Spanish crown was believed to have exclusive control over the trade with the "New World," a decline in the volume of precious metal arriving in Spain was seen as indicative of a decline in output from the American mines. And as China was a destination for Spanish-American silver from the mid-sixteenth century onwards, any decline in the output of these mines would have affected it, just as it had Spain. Here we have the genesis of the proposition that China suffered from the "general crisis" of the seventeenth century world.

The problem with this perspective, however, is that it was based on the Spanish end of the trade network. More recent studies on the American end of this network, the records of the various *cajas*, or treasuries, of New Spain and Peru, reveal that a decline in imports in Seville does not necessarily imply a decline in the output of the American mines. Harry E. Cross has shown that production at the Peruvian mines did decline during the seventeenth century, falling from 3.5 million kilograms in the decade 1591-1600 to 2.5 million kilograms in the decade 1631-1640, and was down to around 1.5 million kilograms by the last decade of the seventeenth century. By contrast, output from the silver mines of New Spain increased during the century, although total production probably was never more than 1 million kilograms per decade.[62] This increase in production in New Spain was insufficient to offset the decline of the South American mines, yet the overall decline for the seventeenth century was much less dramatic than was the decline in the volume of silver arriving in Seville. There are several reasons for this discrepancy. Firstly, the proportion of gold and silver retained in the Spanish colonies for the purposes of administration and defence increased throughout the century. Secondly, Dutch, English, French and Portuguese merchants obtained an increasing share of the bullion trade, particularly from the 1630s onwards.[63] Thirdly, and most importantly for our discussion, evidence suggests that at the same time as Seville was suffering a reduced flow of silver from Spanish-America, the amount of Mexican silver shipped to the Philippines increased considerably. From a comparative study of the quantity of public revenue remitted from New Spain to Seville with that sent to the Philippines, John J. TePaske has found that between 1581 and 1800, eighty-three per cent went to Seville while only seventeen per cent (1,123,000 kgs) went to the Philippines. During the early decades of the seventeenth century, however, remissions to Seville declined, while those to the Philippines increased to around twenty-five per cent of the total. Then in the 1640s over forty per cent of the silver exported from New Spain went to the Philippines.[64] A Spanish crisis does not necessarily imply a Chinese crisis. In fact, what was Spain's loss was, at least in part, China's gain.

Chinese sources provide little quantitative data on the trade conducted with the Spanish in the Philippines. Following the lifting of the prohibitions on overseas trade in 1567 (except for trade with Japan), the income the administration derived from junks operating out of the Fujian port of Yuegang (Haicheng county, Zhangzhou prefecture) increased considerably; from 6,000 *taels* in 1573 to 20,000 *taels* in 1583, 29,000 *taels* in 1594, and levelled off to 23,000 *taels* in 1615 and 1628.[65] This revenue was obtained from junks trading throughout Southeast Asia, not just the Philippines, and includes income derived from the sale of licences, tonnage dues and a tax on all goods traded. For junks sailing to the Philippines the administration imposed an additional levy in recognition of the fact that they returned not with goods but with silver. While the data covers only a few years, it reflects an increase in the overall level of trade, much of which would have been trade in silver. But we must recognise that most of the silver entering China went untaxed. When Haicheng was made a county seat in 1565/6 and new taxes were introduced on trade through its port, many merchants shifted their bases to the island of Xiamen, thereby escaping government surveillance.[66] Islands and ports such as this, where administrative control was lax, were the focus of the silver trade along the China coast. Because of this Chinese sources give little idea of the changing volume of silver entering the empire in the late Ming years.

Records of the Spanish administration in Manila provide a much better indication of the nature of this trade in silver with the Chinese. From Pierre Chaunu's work on these records and that of John TePaske for the trade through Acapulco we have compiled Table One.[67]

In 1593 and 1595 the Spanish crown had issued decrees stipulating that the Manila galleons carry no more than 500,000 *pesos* worth of silver to the Philippines each year. Contemporary observers of the trade noted that this limit was usually exceeded by 1.5 million *pesos*. In 1602, for instance, Alonso Fernandes de Castro and Fray Martin Ignacio de Loyola both estimated that around 2 million *pesos* (approx. 51,122 kgs) was carried annually in the galleons, as did Juan Cerezo Salamanca in 1633.[68] From the official Spanish records cited in Table One we can see that even in the decade 1621-1630, when the accounts show the most silver being sent to the Philippines, the annual average was slightly less than 1 million *pesos* (approx. 25,000 kgs). As the official account failed to record all silver shipped to Manila it seems the actual figure would have been closer to that estimated by the contemporary observers. But even if around 50,000 kilograms of Spanish-American silver was shipped to Manila annually, and all of it passed into Chinese hands, this amounted to only 40-50 per cent of the volume of Japanese silver that was carried to China each year in the period from the 1620s until the fall of the Ming in 1644.

Within the overall trend in the flow of Spanish-American silver to China as portrayed by the data presented in Table One there were smaller fluctuations in the trade. The period 1618-1622 saw Dutch attempts to disrupt the trade, though apparently these met with little success. The Spanish began sending a fleet out to ensure Chinese junks could sail to and from Manila in safety.[69] The only significant short term decline in the trade came in the period 1636-1641. Complaints by Spanish merchants over the Philippine trade finally provoked a reaction from the court and in 1635 Don Pedro de Quiroga y Moya was sent to Acapulco to enforce the legal limits on the trade. Discovering that far more cargo arrived on the galleons than was declared, Quiroga placed an embargo on all goods and refused to allow them to be sold until taxes and fines were paid.[70] Trade was disrupted for the next few years and little silver reached the Philippines. This situation was not helped when one of the galleons sent out in 1639 was wrecked before reaching Manila. As a consequence, many Spanish merchants in Manila were unable to meet the debts they had accumulated in their trade with the Chinese and the resultant tension erupted in a violent conflict, with many Chinese killed.[71] In discussing this incident William Atwell states that "although the name Don Pedro de Quiroga y Moya is not a familiar one to students of late Ming history, it would be difficult to single out another individual who had a greater impact on the Chinese economy in the 1630s."[72] But this is overstating the importance to China of Quiroga and of the galleon trade in general. We must remember that these years of disruption to the Manila trade saw China's trade in Japanese silver reach its zenith. Chinese junks, restricted from trade in the Philippines, could easily have sailed to Japan instead. Indeed, the increase in the number of junks trading at Nagasaki in these years suggest this is what they did. Between 1637 and 1641 some 150,000 to 200,000 kilograms of Japanese silver was carried to China each year, three to four times as much silver as came from Acapulco in the best years of the Manila trade.

Some silver of Spanish-American origin may have reached China via Europe and its trading links with Asia; i.e. the long route, taking some 2-3 years at least. How much is difficult to determine. The northern and central land routes probably saw little trade in silver in the late Ming years. Silver did flow into Russia via the trade of the Baltic and East European markets, but little if any of this would have found its way to China, despite the increasing contacts between the two empires in the seventeenth century. The Russian demand for silver was almost as intense as was the Chinese and as late as 1719 the Board of Commerce issued its envoy to Beijing with the following directive:

> All potentates strive to bring silver, both in ingots and in the form of money, into their hands and to abstain from exporting it... and we fully recognise that this must be the main concern of all European nations.[73]

Sino-Russian trade in the seventeenth century was a trade in silks and furs, not silver. Similarly, the Central and Inner Asian trade routes had declined during the sixteenth century, principally, it seems, due to the rise in influence of the Safavid Reformation, turning Persia into a Shiite stronghold between the Sunni regions to the west and east. The English merchant Anthony Jenkinson noted that trade in Bukhara was poor in 1558 and no Chinese merchandise had been seen in the city for three years. Similarly, tribute trade reaching China from Central Asian powers had declined by the seventeenth century, with only one mission arriving from Samarkand between 1600 and 1644.[74] It is unlikely, therefore, that the Chinese obtained any silver from these east-west land routes.

Much silver continued to flow through the Mediterranean during the seventeenth century. In 1694 a Gujarati ship reportedly returned from Jeddha with 5 million *rupees* worth of silver (approx. 56,650 kgs), and Dutch records reveal that in the latter years of the seventeenth century Asian merchants alone returned from the Red Sea with around 6 million *rupees* worth of silver (approx. 67,980 kgs). This silver flowed into the Indian subcontinent, as did most of the £1,390,436 worth of coin and bullion which the English East India Company shipped to Asia between 1600 and 1640.[75] It is possible that some of this silver, once it had been minted into Indian currencies, was traded for Chinese merchandise. From Akbar's time (1556-1605) silver *rupees* were used to purchase Tibetan wool, salt and musk, while the Malla dynasty in Kathmandu and the administration in Cooch Bahar had both established mints by the 1560s and were using their own silver coinage in trade with Tibet. Similarly, the Assamese traded silver for Tibetan goods and even cast one of their coins with the Chinese characters for "Tibetan coin," *zangbao*. We also know from the journal of the Armenian merchant Hovhannes that in the 1680s merchants were carrying silver from India to Lhasa and on to Xining.[76] The geography of the region restricted the quantity of this trade, however, and it is difficult to imagine large amounts of silver reaching China in this way. Some silver also may have reached China via the trade routes linking Yunnan Province with Burma. Throughout the Ming period cowries continued to be used as currency in Yunnan, much to the frustration of the Ming administration, and these came to China from the Maldive Islands via the trade of the Bay of Bengal and Burma.[77] Silver could have been carried along these same routes, though most likely not in significant quantities.

It was only through Dutch shipments of silver from Amsterdam to Batavia that any appreciable quantity of Spanish-American silver which had followed the long route via Europe and the Indian Ocean finally passed into Chinese hands. Between 1602 and 1650 the Dutch East India Company exported 459,802 kilograms of silver from Europe to the Indies.[78] Some of this went to Persia, to Surat, Ceylon, the Coromandel and Bengal, and some was used to purchase Chinese merchandise. The English merchant John Jourdain describes how each year the Chinese brought lead coinage, which they used to gain control of pepper production in Java, "and carrie rialls of eight out of the country for them."[79] As well as the lead coinage, the *picis*, the Chinese brought porcelain, ironware and piece goods. Every year between 5 and 10 junks came to Java to trade and at most they probably carried off less than 10,000 kilograms of silver per annum.

Java was of secondary importance to the Chinese as a source of Spanish-American silver, the main trade being conducted through Manila, yet even when we combine the figures for trade in these two centres we can see that the volume of silver of Spanish-American origin obtained by the Chinese in the late Ming years was always much less than that of Japanese origin. Imports of Spanish-American silver probably averaged between 35,000 and 60,000 kilgrams per annum, while from the 1620s onwards imports of Japanese silver were always over 100,000 kilograms, and for the period 1637 to 1641 rose to somewhere between 150,000 and 200,000 kilograms per annum. Too great a concentration upon the Spanish trade network and its centre in Seville can lead to a false impression of the volume of silver entering China in the latter years of the Ming dynasty.

After 1645 the number of Chinese junks visiting Java declined and the Dutch merchants in Taiwan found it difficult to meet their contracts with the Japanese.[80] Chinese trade at Nagasaki and Manila also began to decline. By the late 1660s and early 1670s hardly any junks were sailing from the Chinese mainland to the Philippines or to Java, while the trade at Nagasaki was confined mostly to merchants from Southeast Asia and those of the Zheng maritime network based on Taiwan. The reasons for this downturn in trade lay with the fall of the Ming dynasty in 1644 and the ensuing conflict along the Chinese coast as the Qing administration attempted to establish its authority over the empire. In order to eradicate support for Zheng Chenggong and the Ming loyalists from 1661 onwards the Qing enforced strict maritime prohibitions, in some areas withdrawing the coastal population many miles inland and thus inhibiting trade. This coastal evacuation policy was particularly drastic in the provinces of Guangdong and Fujian, the very provinces which had been most active in the silver trade.[81] With Chinese junks no longer able to sail freely to foreign ports very little silver entered the Chinese empire. Thus, the nadir in the silver trade came after the fall of the Ming, not before it, and domestic factors rather than vicissitudes in the international movement of bullion were critical in determining the volume of this trade.

Conclusion

Those who argue that China suffered a restricted flow of silver in the late Ming years cite the rapidly deteriorating copper-silver exchange rates as evidence of a shortage of silver in circulation.[82] A more obvious reason for this, and one noted by contemporary commentators, was the increasingly poor quality of the copper coinage. During the final years of the reign of the Chongzhen emperor (mid 1630s-1644), the lead and sand content in both official and counterfeit coinage increased to such an extent that "a hundred coins piled up hardly measured an inch in thickness and when thrown to the ground broke into pieces." In writing of the great increase in the price of rice in the Shanghai district during these final years of the Ming, Ye Mengzhu states that the reason for this was the poor quality of the coinage, which from 1638 onwards "deteriorated daily."[83] He does not mention a scarcity of silver. Indeed, it is not until the 1670s that officials such as Mu Tianyan began complaining of the hardship caused by the dwindling amount of silver in circulation, requesting the administration relax its strict prohibitions on foreign trade.[84] Again we find that it was not until after 1661 and the introduction of the ban restricting merchants from trading in foreign ports, well after the fall of the Ming, that China began to suffer a significant reduction in the volume of silver in circulation.

Similarly, the revenue recorded by the Taicang silver vault, the main silver treasury of the administration, does not show a decline in the late Ming years, as might be expected if there had been less silver in circulation. The record of silver received by this treasury shows an increase from 2.3 million *taels* in 1570 to 4.4 million *taels* in 1577. For the next 40 years the recorded income of the treasury remained stable, around 3 million to 4 million *taels* per annum. In 1618 the record shows an increase to 6 million *taels*; by 1630 this was up to around 9 million *taels*; in 1631 to 12.2 million *taels*; 1639 to 20 million *taels*, and 1642 to 23 million *taels*.[85] This continual increase is in part a reflection of the shift from a grain tax to a silver tax with the gradual implementation of the Single Whip Reforms. From 1618 onwards the dramatic increase was a result of the addition of seven surcharges to the land tax to help defray the increasing costs of provisioning the armed forces. Ray Huang advises that these figures reflect expected rather than actual income, and that the administration probably only collected between 50 and 70 per cent of the recorded amount.[86] If we allow for this unreliability in the figures for the latter years of the dynasty, taking the actual amount of silver available to the administration in 1642 as being closer to 10 million *taels* than the recorded 23 million *taels*, this still represents a five-fold increase from 1570 in the state's silver income. The state certainly did not suffer a decline in its silver income in the late Ming years.

Another factor to be taken into account when discussing the importance of China's dependence upon foreign silver is the size of the Chinese economy. Even in late Ming times the Chinese economy was huge, and while silver use was expanding at this time, it is difficult to establish a direct causal relationship between the volume of monetary metal entering the empire at any particular time and the concomitant degree of political and social stability enjoyed.[87] We have already noted that it was the two decades following 1661 that saw the decline in China's trade in foreign silver, and yet it was during these years that the Qing completed their conquest of the Chinese empire. By contrast, the late Ming years saw an unprecedented volume of foreign silver flood into China, and these were years of dynastic decline. The fundamental causes of that decline lay not with an inability to maintain a high level of military expenditure, but rather with the essentially political problems that resulted in the continued need for such a high level of expenditure. The suicide of the Chongzhen emperor that marked the collapse of the Ming in 1644 was due more to the factionalism and nepotistic squabbling that infected the late Ming bureaucracy than to the vicissitudes of international movements of bullion.

Late Ming China was not an outpost of a Seville-centred world economy. In the first half of the seventeenth century Japanese silver was always more important to China than Spanish-American silver. Thus, the decline in imports of monetary metal to Seville did not automatically signify the decline of a world monetary system, or an ensuing "crisis" in China similar in nature to that of Spain. A common dependence on imported silver has encouraged Western historians working within a historiographical tradition that conceives of the seventeenth century as a time of "crisis" to find in China events that parallel the European experience. But the East Asian silver trade was substantially different to the silver trade of the Atlantic and Europe. Events in seventeenth century China were essentially unrelated to those in Europe at the time.

NOTES

1. P. Clark, ed., *The European Crisis of the 1590s: Essays in Comparative History* (London: George Allen and Unwin, 1985), p.3. While the crisis debate can be traced back to the work of F. Simiand (*Les fluctuations économiques à longue période et la crise mondiale*, Paris: Alcan, 1932) and Earl J. Hamilton (*American Treasure and the Price Revolution in Spain, 1501-1650*, Cambridge Mass.: Harvard University Press, 1934), it came to prominence through articles published in the two journals *Past and Present* and the *Annales E.S.C.* from the 1950s onwards. A selection of these articles can be found in Trevor Ashton, ed., *Crisis in Europe, 1560-1660* (London: Routledge and Kegan Paul, 1965) and

Geoffrey Parker and Lesley M. Smith, eds., *The General Crisis of the Seventeenth Century* (London: Routledge and Kegan Paul, 1978). See also Theodore K. Rabb, *The Struggle for Stability in Early Modern Europe* (N.Y: Oxford University Press, 1975). Randolph Stern gives an interesting historical perspective on the use of the term "crisis" in his article, "Historians and 'crisis'," *Past and Present* 52 (August 1971):3-22.

2. For studies of various aspects of the process of dynastic change in seventeenth century China see Jonathan D. Spence and John E. Wills Jnr., eds., *From Ming to Ch'ing: Conquest, Region and Continuity in Seventeenth-Century China* [hereafter *From Ming to Ch'ing*] (New Haven: Yale University Press, 1979). Paul K. C. Liu and Kuo-shu Hwang, "Population Change and Economic Development in Mainland China Since 1400" in *Modern Chinese Economic History*, ed. Hou Chi-ming and Yu Tzong-shian (Taipei: Academia Sinica, 1979), pp.81-82 and Table A.1, suggest China's population declined from 160 million in 1600 to 123 million by the 1650s.

3. Fernand Braudel, *The Mediterranean and the Mediterranean World in the Age of Philip II* (London: Collins, 1972), p.464. See also John Day, "The Great Bullion Famine of the Fifteenth Century," *Past and Present* 79 (May 1978):3-54.

4. Pierre Chaunu, *Les Philippines et la Pacifique des Ibériques XVIe, XVIIe, XVIIIe siècles* (Paris: SEVPEN, 1960), p.269. See also his article "Manille et Macao, face à la conjoncture des XVIe et XVIIe siècles," *Annales E.S.C.* 17, 3 (1962):555-580 and Louis Dermigny, *La Chine et l'occident, le commerce à canton au XVIIIe siècle, 1719-1833* (Paris: SEVPEN, 1964), vol. 1, pp.92-103, 118-119 and 124-136. The estimate of Spanish American bullion production is given by Harry E. Cross, "South American bullion production and export 1550-1750" in *Precious Metals in the Later Medieval and Early Modern Worlds* [hereafter *Precious Metals*], ed. J. F. Richards (Durham N. C.: Carolina Academic Press, 1983), p.397 and p.403, Table 2.

5. S.A.M. Adshead, "The Seventeenth Century General Crisis in China," *Asian Profile* 1, 2 (October 1973):276 and *China in World History* (N.Y.: St. Martins Press, 1988), pp.194-209 and 235-236.

6. William S. Atwell, "Notes on Silver, Foreign Trade, and the Late Ming Economy," *Ch'ing-shih Wen-t'i* 3, 8 (1977):1-33; "International Bullion Flows and the Chinese Economy," *Past and Present* 95 (May 1982):68-90 and "Some Observations on the 'Seventeenth-Century Crisis' in China and Japan," *Journal of Asian Studies* 45, 2 (February 1986):223-244. Atwell has reiterated these views in his chapter in *The Cambridge History of China: The Ming Dynasty, 1368-1644*, vol. 1 (Cambridge: Cambridge University Press, 1988), pp.585-640.

7. Frederic E. Wakeman Jnr., "China and the Seventeenth-Century Crisis," *Late Imperial China* 7, 1 (June 1986):3-4. Wakeman begins his study of the Manchu conquest of China with similar comments; see *The Great Enterprise: The Manchu Reconstruction of Imperial Order in Seventeenth-Century China* (Berkeley: University of California Press, 1985), vol. 1, pp.1-6.

8. Liang Fangzhong, "Mingdai yinkuang kao" [Silver mining during the Ming dynasty], *Zhongguo shehui jingji shi jikan* 6, 1 (June 1939):65-112 and "Mingdai guoji maoyi yu yin de shuchuru" [International trade and movements of silver during the Ming dynasty], *Zhongguo shehui jingji shi jikan* 6, 2 (December 1939):264-324.

9. For a brief biography of Quan and list of his publications see Wang Yejiang, "Quan Hansheng zai Zhongguo jingji yanjiu shang de zhongyao gongxian" [The important contributions of Quan Hansheng to the study of Chinese economic history], *Shihuo yuekan* 14, 11/12 (March 1985):444-458.

10. Sha Ding and Yang Dianqiu, "Zhongguo he Lading Meizhou de zaoqi maoyi quanyu" [Early trade relations between China and Latin America], *Lishi yanjiu* 4 (1984):107-127. Zhang Kai, "WanMing Zhongguo shichang yu shijie shichang" [The Chinese market and the world market in the late Ming period], *Zhongguo shi yanjiu* 3 (1988):3-15. Yano Jin'ichi, "Nagasaki bōeki ni okeru dō oyobi gin no Shina yushutsu ni tsuite" [Concerning Chinese exports of copper and silver through trade at Nagasaki], *Keizai ronsō* 26, 2 (1928):92-110. Kobata Atsushi, "Jūroku jūshichi seiki ni okeru kyokutō no gin no ryūtsū" [The circulation of silver in the Far East during the 16th and 17th centuries] in *Kobata Atsushi kyōju taiken kinen kokushi ronshū* [A collection of essays on Japanese history in honour of the retirement of Professor Kobata Atsushi] (Kyoto: Kobata Atsushi kyōju taikan kinen jigyōkai, 1970), pp.3-10 and also his *Kingin bōekishi no kenkyū* [Studies on the history of the bullion trade] (Kyoto: Hōsei daigaku shuppan kyoku, 1976).

11. Jean Baptiste Du Halde, *The General History of China* (London: J. Watts, 1736), vol. 2, pp.286-287.

12. For this quote from a letter of de Noronha see C. R. Boxer, *The Great Ship from Amacon* (Lisbon: Centro de estuados historicos ultramarinos, 1959), p.11, n. 6.

13. Vitorino de Magalhaes Godinho, *L'économie de l'empire portugais aux XVe et XVIIe siècles* (Paris: SEVPEN, 1969), p.531.

14. General discussions of Chinese monetary history can be found in L. S. Yang, *Money and Credit in China: A Short History* (Cambridge, Mass.: Harvard University Press, 1952); Ping Xinwei, *Zhongguo huobi shi* [A history of Chinese money] (Shanghai: Boxiu chubanshe, 1954) and Xiao Qing, *Zhongguo gudai huobi shi* [A history of ancient Chinese money] (Beijing: Renmin chubanshe, 1984).

15. Ray Huang, *Taxation and Government Finance in Sixteenth-Century China* (Cambridge: Cambridge University Press, 1974), p.73. For one of many references to the devaluation of paper currency see the comments of the censor-in-chief in 1403, Chen Ying, in *Ming shi* [Ming History], (Beijing: Zhonghua shuju ed., 1974), *juan* 81, p.1963.

16. Chen Chau-nan, "Flexible Bimetallic Exhange Rates in China, 1650-1850," *Journal of Money, Credit and Banking* 7, 3 (August 1975):359-367.

17. Gabriel de Magaillans, *A New History of China* (London: Thomas Newborough, 1688), p.137.

18. Du Halde, *General History*, pp.289 and 293.

19. Nicola Trigault, *China in the Sixteenth Century: the Journals of Matthew Ricci 1583-1610* (New York: Random House, 1953), pp.343-351. See also Huang Jianhui, "Qingchu shang yong huipiao yu shangpin jingji de fazhan" [The use of bills of exchange by merchants in the early Qing dynasty and the development of the commodity economy], *Wenxian* 1 (1987):3-16.

20. Mi Chu Wiens, "Changes in the Fiscal and Rural Control Systems in the Fourteenth and Fifteenth Centuries," *Ming Studies* 3 (Fall, 1976):56.

21. Ibid, p.64.

22. Liang Fang-chung, *The Single Whip Method of Taxation in China* (Cambridge, Mass.: Harvard University Press, 1956); Huang, *Taxation*, pp.112-113; and the entries for Zhang Juzheng (vol. 1, pp.53-61), Hai Rui (vol. 1, pp.474-479) and Pang Shangpeng (vol. 2, pp.1113-1116) in *Dictionary of Ming Biography, 1368-1644*, ed. L. Carrington Goodrich and Chaoyang Fang (New York: Columbia University Press, 1976).

23. Huang, *Taxation*, p.52. *Ming shi, juan* 78, pp.1895-1896. Zhao Yipeng, "Shilun Mingdai huobizhidu de yanbian ji qi lishi yingxiang" [An exploratory discussion of the development and historical significance of the monetary system in Ming times], *Dongbei shida xuebao* 4 (1985):41-46.

24. Wang Ch'ung-wu, "The Ming System of Merchant Colonization" in *Chinese Social History*, ed. and trans. E-tu Zen Sun and John de Francis (Washington D.C., American Council of Learned Societies, 1956), pp.299-308.

25. Huang, *Taxation*, pp.125-130.

26. John K. Whitmore, "Chiao-chih and Neo-Confucianism: The Ming Attempt to Transform Vietnam," *Ming Studies* 3 (Spring, 1977):71.

27. John S. Deyell, "The China Connection: Problems of Silver Supply in Medieval Bengal" in *Precious Metals*, pp.222-223.

28. Quan Hansheng, "Ming Qing shidai Yunnan de yinke yu yinchane" [Taxes on silver mining and silver production quotas of Yunnan province in Ming and Qing times], *Xinya xuebao* 11 (March 1976):5-6, Table 2.

29. Quan Hansheng, "Mingdai de yinke yu yinchane" [Taxes on silver mining and silver production quotas during the Ming dynasty], *Xinya shuyuan xueshu niankan* 9 (1966):246-255. We follow Atwell ("International Bullion Flows," p.76) in converting Quan's yearly data into decennial figures. On Ming silver production see also Liang, "Mingdai yinkuang kao," and Bai Shouyi, "Mingdai kuangye de fazhan" [The Development of Mining during the Ming Period], in his *Xuebu ji* (Beijing: Sanlian shudian, 1961), pp.74-114.

30. Huang, *Taxation*, p.243. One *tael* is considered as approximately equal to 0.0378 kgs of silver. See appendix for a full list of currencies and weights.

31. Quan, "Mingdai de yinke," pp.260-261.

32. A good recent contribution is the book of essays by Han Dacheng, *Mingdai shehui jingji chutan* [Preliminary investigations into the society and economy of the Ming period] (Beijing: Renmin chubanshe, 1986).

33. Sun Chengze, *Chunming mengyulu* [Fragmentary recollections of the Ming capital] (H.K.: Longmen Shudian reprint, 1965), *juan* 47, p.780.

34. Cited by Frank Spooner, *The International Economy and Monetary Movements in France, 1493-1725* (Cambridge, Mass.: Harvard University Press, 1972), p.77.

35. Boxer, *Great Ship*, p.1.

36. Cited in Thomas Rundall, ed., *Memorials of the Empire of Japan in the XVI and XVII Centuries* (London: Hakluyt Society, 1850), p.31.

37. Delmer M. Brown, *Money Economy in Medieval Japan: A Study in the Use of Coins* (New Haven: Yale University Press, 1951), pp.33 and 56. For the estimate of Japan's

contribution to world silver production in the early 17th century see Robert LeRoy Innes, "The Door Ajar: Japan's Foreign Trade in the Seventeenth Century" (PhD. dissertation, The University of Michigan, 1980), p.10. See also Kobata Atsushi, "The Production and Uses of Gold and Silver in Sixteenth and Seventeenth Century Japan," *The Economic History Review*, 2nd series 18, 2 (1965):245-266.

38. Boxer, *Great Ship*, passim. Michael J. Copper, "The Mechanics of the Macao-Nagasaki Silk Trade," *Monumenta Nipponica* 27, 4 (1972):423-433 and George Bryan Souza, *The Survival of Empire: Portuguese Trade and Society in China and the South China Sea, 1630-1754* (Cambridge: Cambridge University Press, 1986).

39. Godinho, *L'économie*, pp.330-334 suggests the Portuguese were shipping around 130,000 *crusadoes* (approx. 3,300 kgs) of silver annually from Lisbon to Goa between 1605 and 1626. For Portuguese trade at Manila see Souza, *Survival*, pp.74-75 and 78-86.

40. Ralph Fitch, "The Voyage of Ralph Fitch" in *The Principal Navigators*, ed. Richard Hakluyt (London: J. M. Dent and Sons Ltd., 1907), vol. 3, pp.309-310. Brown, *Money Economy*, p.64, and Boxer, *Great Ship*, p.61.

41. Innes, "Door Ajar," pp.382-383 and Boxer, *Great Ship*, pp.144-157. A chest contained approx. 1,000 *taels* of silver, see Boxer, *Great Ship*, p.64.

42. Our estimates concur with those of Souza, *Survival*, pp.56-57, who suggests the Portuguese exported between 36.6 million and 41.1 million *taels* (1.38 to 1.55 million kgs) of silver from Japan in the period from 1546 to 1638. Kozo Yamamura and Tetsuo Kamiki, "Silver Mines and Sung coins – A monetary history of Medieval and Modern Japan in international perspective" in *Precious Metals*, p.351, give an estimate of between 900,000 kgs and 1.5 million kgs for the 40 year period to 1600. We feel this is far too high. Innes, "Door Ajar," pp.376-377, also feels Japanese estimates to be too high. For other references to the volume of this trade see Quan Hansheng, "Mingdai zhongyehou Aomen de haiwai Maoyi" [The overseas trade of Macao in the latter half of the Ming period], *Zhongguo wenhua yanjiusuo xuebao* 5, 1 (December 1972):262-263.

43. Frank Perlin, "Precolonial South Asia and Western Penetration in the Seventeenth to Nineteenth Centuries," *Review* 2 (Fall, 1980):289. For the argument that the exclusion of the Portuguese from Japan saw a reduced flow of silver to China see the work of Atwell cited above in n.6, and also that of Wakeman cited in n.7.

44. Kimiya Yashuhiko, *Nikki bunka kōryū shi* (Tokyo: Fusan bō, 1955), pp.799-807, gives a list of references to the arrival of Chinese junks in Japan during the late Ming period. On this "illegal" trade see the discussion of three cases of smuggling by Wang Zaijin in *Yuejuan*, *juan* 21, cited in Xie Guozhen, *Mingdai shehui jingji shiliao xuanbian* [Selected historical documents on Ming society and economy] (Fuzhou: Fujian renmin chubanshe, 1981), vol. 2, pp.135-140 and Lin Renchuan, *Mingmo Qingchu de siren haishang maoyi* [Private maritime trade of the late Ming and early Qing periods] (Shanghai: Huadong shifan daxue chubanshe, 1987), pp.40-84.

45. John K. Whitmore, "Vietnam and the monetary flows of eastern Asia, thirteenth to eighteenth centuries" in *Precious Metals*, ed. Richards, p.380. See also Innes, "Door Ajar," pp.62-66.

46. Yamamura and Kamiki, "Silver mines," p.351.

47. Iwao Seiichi, "Japanese Foreign Trade in the 16th and 17th Centuries," *Acta Asiatica* 30 (1976):8-10. Innes, "Door Ajar," pp.120-121.

48. Iwao Seiichi, "Li Tan, Chief of the Chinese Residents at Hirado, Japan, in the Last Days of the Ming Dynasty," *Memoirs of the Research Department of the Toyo Bunko* 17 (1958):33. See also Iwao's detailed study, "Kinsei Nisshi bōeki ni kansuru sūryōteki kōsatsu" [A quantitative study of early-modern Sino-Japanese trade], *Shigaku zasshi* 60, 11 (November 1953):1-40 and Chen Ziqiang, "Lüelun Mingdai zhonghouqi Fujian de dui Ri jiaotong" [A brief discussion of Fujianese contact with Japan during the middle and latter part of the Ming period], *Haijiao shi yanjiu* 2 (1985):46-50.

49. Innes, "Door Ajar," pp.391-392. Nakada Yasunao, *Kinsei taigai kankeishi no kenkyū* [Japanese Foreign Relations in the Edo Period] (Tokyo: Yoshikawa Kōbunkan, 1984), pp.361-411 discusses the substantial increase of the trade in raw silk through Nagasaki, particularly by Chinese merchants, after the introduction of the seclusion edicts.

50. Iwao, "Japanese Foreign Trade," p.11. Innes, "Door Ajar," pp.634-642, Appendix 1. One *kanme* is considered approximately equal to 3.76 kgs of silver.

51. Kato Eiichi, "Unification and Adaptation, the early Shogunate and Dutch Trade Relations" in *Companies and Trade*, ed. L. Blussé and F. S. Gaastra (Leiden: Leiden University Press, 1981), p.225, Table 2. Kristof Glamann, *Dutch Asiatic Trade, 1620-1740* (The Hague: Martinus Nijhoff, 1942), p.58.

52. Tashiro Kazui, "Foreign Relations During the Edo Period: *Sakoku* Re-examined," *Journal of Japanese Studies* 8, 2 (Summer, 1982):293.

53. For Nuyts' "Short account of the Chinese trade," see W. Campbell, *Formosa Under the Dutch* (London: Kegan Paul, 1903), pp.51-60.

54. John E. Wills Jnr., *Pepper, Guns, and Parleys: The Dutch East India Company and China* (Cambridge, Mass.: Harvard University Press, 1974), p.10, n.22.

55. Innes, "Door Ajar," p.408, Table 21. John E. Wills Jnr., "Maritime China from Wang Chih to Shih Lang: Themes in Peripheral History" in *From Ming to Ch'ing*, pp.203-238.

56. Wei Nengtao, "MingQing shiqi ZhongRi changqi shangchuan maoyi" [Sino-Japanese maritime trade at Nagasaki during the Ming and Qing periods], *Zhongguo shi yanjiu* 2 (1986):52. Yamawaki Teijirō, *Nagasaki no Tōjin bōeki* [Chinese Trade at Nagasaki] (Tokyo: Yoshikawa Kōbunkan, 1972), pp.29-36.

57. As 36 Chinese junks exported around 35,700 kgs of silver in 1634 (approx. 990 kgs per junk) and 54 junks exported around 63,920 kgs in 1646 (approx. 1,180 kgs per junk), we take the average volume of silver exported per junk as being 1,000 kgs.

58. Admiral Don Hieronimo de Banuelos y Carillo, "Relation of the Filipinas Islands" in *The Philippine Islands*, ed. E. H. Blair and J. A. Robertson, [hereafter *PI*], 55 vols (Cleveland: A. H. Clark, 1903-1909) 29:71.

59. *PI* 22:79. For archaeological evidence of the use of Spanish coins along the China coast see Zhuang Weiji, "Fujian Nanan chutu waiguo yinbi de jige wenti" [Some problems on the foreign silver coinage unearthed at Nanan, Fujian], *Kaogu* 6 (1975):352-354 and Wei Hongtao, "Fujian Quanzhou diqu chutu de wupi waiguo yinbi" [The unearthing of five batches of foreign silver coinage in the Quanzhou district, Fujian], *Kaogu* 6 (1975):373-379.

60. *PI* 27:173-174. For the claim that in 1597 12 million *pesos* of silver was shipped to Manila see C. R. Boxer, "*Plata es Sangre*: Sidelights on the Drain of Spanish-American Silver in the Far East, 1550-1700," *Philippine Studies* 18 (July 1970):464.

61. Hamilton, *America Treasure*, pp.1-42. The trend portrayed by Hamilton was corroborated by the extensive and detailed work of the Chaunu's on the *Carrera de Indias*, the trans-Atlantic trade network of the Spanish: Pierre and Huguette Chaunu, *Séville et l'Atlantique 1504-1650* (Paris: SEVPEN, 1955-1959), 8 vols.

62. Harry E. Cross, "South American bullion production," p.409, Table 5. John J. TePaske and Herbert S. Klein, "The Seventeenth Century Crisis in New Spain: Myth or Reality?" *Past and Present* 90 (February 1981):124-127, Table One. See also the chapter "Mining in Colonial Spanish America" in *The Cambridge History of Latin America*, vol. 2 (Cambridge: Cambridge University Press, 1984), pp.105-151.

63. John Lynch, *Spain and the Habsburgs: Spain and America 1598-1700*, vol. 2 (Oxford: Basil Blackwell, 1969), pp.168 and 212-226. Michel Morineau, *Incroyables gazettes et fabuleux métaux: Les retours des trésors américans d'après les gazettes Hollandaises, XVIe-XVIIIe siècles* (Cambridge/Paris: Cambridge University Press, 1985), pp.238-250.

64. John J. TePaske, "New World Silver, Castile and the Philippines, 1590-1800" in *Precious Metals*, p.434.

65. *Xiamen zhi* [Xiamen gazetteer] (Taipei: Chengwen chubanshe reprint, 1967), *juan* 7 (2b), p.124. Zhang Xie, *Dongxiyang kao* [A study of the East and West Oceans] (Taipei: Zhengzhong Shuju reprint, 1962), *juan* 7 (2a-3b), pp.267-270. Quan Hansheng, "Mingji Zhongguo yu Feilübin jian de maoyi" [Trade between China and the Philippines during the Ming period], *Zhongguo wenhua yanjiusuo xuebao* 1 (September 1968):38-39. Chen Ziqiang, "Lun Mingdai Zhangzhou Yuegang de lishi diwei" [A discussion of the historical position of Yuegang, Zhangzhou prefecture, during the Ming period], *Haijiao shi yanjiu* 5 (1983):90-97.

66. Ng Chin-Keoing, *Trade and Society: the Amoy network on the China coast, 1683-1735* (Singapore: Singapore University Press, 1983), pp.46-48. Fu Yiling, "Mingdai Fujian haishang" [Fujianese Maritime Merchants in the Ming period] in his *MingQing shidai shangren ji shangyeziben* [Merchants and mercantile capital of the Ming and Qing periods] (Beijing: Renmin chubanshe, 1956), pp.107-160.

67. Chaunu, *Les Philippines*, pp.92 and 106. TePaske, "New World Silver," pp.444-445, Tables 4 and 5.

68. *PI* 12:46-47 and 59; *PI* 24:292.

69. *PI* 19:68-69; O. H. K. Spate, *The Spanish Lake* (Canberra: ANU Press, 1979), p.224.

70. *PI* 30:50-52.

71. *PI* 29:168-171, 194 and 208-258.

72. Atwell, "Notes on Silver," p.11.

73. M.L. Sladkovskii, *History of Economic Relations between Russia and China* (Jerusalem: Israel Program for Scientific Translations, 1966), p.29. For the trade in silver in the Baltic, Eastern and Southeastern Europe see Artur Attman, *The Bullion Flow between Europe and the East, 1000-1750* (Göteborg, Kungl.: Vetenskaps-och vitterhets - samhället, 1981).

74. Adshead, *China in World History*, pp.196-197 and 200-203. E. D. Morgan and C. H. Coote, eds., *Early Voyages and Travels to Russia and Persia by Anthony Jenkinson and other Englishman* (London: Hakluyt, 1886), vol. 1, pp.87-92. J. K. Fairbank and S. Y. Teng, *Ch'ing Administration: Three Studies* (Cambridge, Mass.: Harvard University Press, 1960), p.157.

75. Aziza Hasan, "Mughal Silver Currency – A Reply," *Indian Economic and Social History Review* 7, 1 (March 1970):155. Joseph J. Brennig, "Silver in Seventeenth Century Surat: Monetary circulation and the Price Revolution in Mughal India" in *Precious Metals*, p.481. K. N. Chaudhuri, "The East India Company and the Export of Treasure in the early Seventeenth Century," *Economic History Review*, 2nd series 16, 1 (August 1963):23-38. For a detailed discussion of the great influx of silver into the subcontinent during the seventeenth century and the effect this had on the Mughal economy, see Irfan Habib, "A System of Trimetallism in the Age of the 'Price Revolution': Effects of the Silver Influx on the Mughal Monetary System" in *The Imperial Monetary System of Mughal India*, ed. John F. Richards (Delhi: Oxford University Press, 1987), pp.137-170.

76. N. G. Rhodes, "The Development of Currency in Tibet" in *Tibetan Studies in Honour of Hugh Richardson*, ed. Michel Aris and Aung San Suu Kyi (Oxford: Aris and Phillips Ltd., 1979), pp.261-268. Lvon Khanchikian, "Le registre d'un marchand arménien en Perse, en Inde et au Tibet (1682-1693)," *Annales E.S.C.* 22, 2 (April-May 1967):241-244.

77. *Ming Xiaozong shilu* [The Veritable Records of Ming Xiaozong] (Taipei: Academia Sinica reprint, 1964), *juan* 197, pp.3644-3645. Xie Zhaozhe, *Dianlüe* [A brief sketch of Yunnan Province] in *Ming Qing shehui jingji shiliao*, vol. 2, p.177. Kuo Tsung-fei, "A Brief History of the Trade Routes Between Burma, Indochina and Yunnan," *T'ien Hsia Monthly* 12, 1 (August/Sept 1941):9-32.

78. F. S. Gaastra, "The Exports of Precious Metals from Europe to Asia by the Dutch East India Company, 1602-1795" in *Precious Metals*, p.475, Table 5.

79. William Foster, ed., *The Journal of John Jourdain 1608-1617* (Cambridge: Hakluyt 2nd series, 1905), p.316. Leonard Blussé, *Strange Company: Chinese settlers, Mestizo Women and the Dutch in VOC Batavia* (Dordrecht: Floris Publications, 1986), pp.35-48 and 111-121.

80. *Dagh-register gehouden int Casteel Batavia*, 1644-1645 (The Hague: Martinus Nijhoff, 1903), pp.142 and 161-169.

81. Chaunu, *Les Philippines*, pp.164-169. Xie Guozhen, "Qingchu dongan yanhai qianjie kao" [A study of the evacuation of the southeast coast in the early Qing], *Guoxue jikan* 2, 4 (December 1930):797-826. Mio Kishimoto-Nakayama, "The Kangxi Depression and Early Qing Local Markets," *Modern China* 10, 2 (April 1984):227-256.

82. Atwell, "International Bullion Flows," p.89.

83. *Mingshi, juan* 81, p.1969. Ye Mengzhu, *Yueshi bian* [A record of experience], *Ming Qing shehui jingji shiliao*, vol. 2, pp.178 and 184-186. Zhao Yifeng, "Shilun Mingmo caizheng weiji de lishi genyuan jiqi shidai tezheng" [An exploratory discussion of the historical origin and temporal characteristics of the financial crisis at the end of the Ming], *Zhongguo shi yanjiu* 4 (1986):55-68.

84. Mu submitted a memorial to the throne around 1681 in which he stated that while in the early years of the Qing dynasty (1649-1650) foreign silver was still found in markets

throughout China, after the implementation of the maritime prohobitions in the early 1660s it had soon disappeared from circulation. For this memorial, entitled "Qing kai haijin shu" see He Changling, *Huangchao jingshi wenbian* [Essays on Qing statecraft] (Taipei: Shijie shuju reprint, 1964), *juan* 26 (13a-15b), vol. 2, pp.964-967. The memorial is discussed in Quan Hansheng, "Meizhou baiyin yu shiba shiji Zhongguo wujia geming de guanxi" [American silver and the price revolution in eigtheenth century China], *Zhongguo yanjiuyuan lishi yuyan yanjiusuo jikan* 28 (1957):541 and Kishimoto-Nakayama, "The Kangxi Depression," p.233.

85. Quan Hansheng and Li Longhua, "Ming zhongyehou Taicang suiru yinliang de yanjiu" [A study of the annual silver receipts of the Taicang silver vault in the latter half of the Ming], *Zhongguo wenhua yanjiusuo xuebao* 5, 1 (1972):123-157. See also Sun Chengze, *Tianfu guangji* [An extensive record of imperial treasuries] (Beijing: Xinhua shuju reprint, 1985), pp.152-153.

86. Ray Huang, "Fiscal Administration During the Ming Dynasty" in *Chinese Government in Ming Times: Seven Studies*, ed. Charles O. Hucker (N.Y.: Columbia University Press, 1969), p.121.

87. In a recent article Jack Goldstone suggests the total volume of European trade with China represented less than one per cent of the volume of China's economic activity in late Ming times: see "East and West in the Seventeenth Century: Political Crises in Stuart England, Ottoman Turkey, and Ming China," *Comparative Studies of Society and History* 30, 1 (January 1988):115.

Table One

Indications of Sino-Spanish Trade in Silver via the Philippines, 1590-1650 (*pesos*)

Period	Annual average tax on silver shipped from Acapulco to Manila (a)	Silver Remitted to Manila		Revenue derived from Chinese trade at Manila (c)	
		Public	Private (b)	years	(*pesos*)
1591-1600	1,715	466,016	578,170		
1601-1610	7,467	1,174,782	3,516,513		
1611-1620	7,015	2,541,652	5,048,118	1611-1615	118,314
				1616-1620	120,850
1621-1630	8,245	3,620,573	5,423,822		
				1626-1630	120,850
1631-1640	5,723	3,672,874	3,509,871	1631-1640	151,200
				1641-1645	133,066
1641-1650	3,980	2,206,810	1,759,706	1646-1650	117,788

(a) This *almojarifazgo* tax was levied on legally registered silver only.

(b) TePaske notes that these figures, from the Acapulco accounts, probably represent only a portion of the private silver shipped to the Philippines. "New World Silver," p.437.

(c) This revenue was derived from licenses issued to Chinese merchants and from the *almojarifazgo* tax on the trade which they conducted.

Appendix

Currency and Weights

(approximate conversion rates)

Chinese	:	1 *tael* = 1 *liang*	= 0.0378 kgs of silver
Japanese	:	1 *kanme*	= 3.76 kgs of silver
Dutch	:	1 *florin* (guilder)	= 0.01 kgs of silver
Portuguese	:	*cruzado*, an unit of account generally taken to be equivalent to 1 *pesos de a ocho* in Asia 1 *picul* = 100 *catties* = 60.48 kgs	
Spanish	:	1 *pesos de a ocho* (272 maravedis) = 0.025561 kgs of pure silver	

Sources : Boxer, *The Great Ship*, pp.335-342.
Sumiya Mikio and Taira Koji, *An Outline of Japan's Economic History, 1603-1940* (Tokyo, 1979), pp.53-55.
TePaske, "New World Silver," p.440.

9

The Cowrie Currencies of West Africa

Marion Johnson

PART I

> What we call money being arbitrary, and its nature and value depending on a tacit convention betwixt men, these [cowrie] shells, in several parts of Asia and Africa, are accounted current money, with a value assigned to them. This is established by a reciprocal consent, and those who are pleased to show a contempt of them don't reflect that shells are as fit for a common standard of pecuniary value as either gold or silver.

So wrote an anonymous Dutchman in 1747.[1] From his time to the present, the serious study of cowrie currencies has been hampered by a feeling that a cowrie currency is in some way worthy of contempt, or of patronizing treatment as 'primitive money'. We hope to show in this paper that the West African cowrie currencies, though they have some unusual qualities, are in no sense a 'primitive' money, but a sophisticated form of currency capable of adaptation to the particular needs of West African trade. The currency is naturally best remembered in its latest and most debased stage, when it had practically lost its value as a result of massive inflation and the introduction of European currency; but it had behind it centuries of useful existence before its breakdown. A detailed study of cowries is necessary in order to interpret commodity values expressed in cowries; and this study also throws much light on the operations of West African trade.

The English word 'cowrie' (the singular can also be spelt cowry, but there is general agreement on the plural cowries) is derived from an Indian root. In West Africa there are nearly as many words for cowrie as there are peoples who used the shells—the European names were seldom adopted, the Arabic rarely.

THE SHELLS

The cowrie shells of West African commerce have been almost exclusively the shells of *Cypraea moneta* and *Cypraea annulus*, both Indian Ocean species, the former from the Maldive Islands, the latter from the East African coast and islands, especially Zanzibar and the neighbouring

[1] *A Voyage to the island of Ceylon by a Dutch Gentleman*, (London, 1754), quoted in *The Voyage of Francis Pyrard of Laval to the East Indies, the Moluccas and Brazil*, ed. A. Gray and A. C. P. Bell (2 vols., London, 1887, 1890). The author also writes in a less modern vein: 'The Dutch drive a considerable trade with the inhabitants of the Maldives for these little shells called cowries, where are prodigious quantities of them, and not only on the shore, but in the very ground, being probably deposited there at the time of the Flood, and left there when the Ocean receded from the land...'

islands. None of the species of cowries found in the Gulf of Guinea are at all similar to these, nor do any of these local species occur in sufficient numbers to serve as money in the way that cowries are used. Indeed, it could be argued that a cowrie currency would be unworkable if money could be picked up in quantity by every beachcomber. Small olive shells, somewhat similar to cowries, but lacking the characteristic lip, are found in large numbers on some West African beaches; it is possible that the shell currency which the Portuguese found at Benin[2] (some distance from the ocean beaches) at the beginning of the sixteenth century may have consisted of these shells. Imports to the Benin area from the Indian Ocean began at least as early as 1515;[3] the immediate acceptance of cowries by the population at Benin and in the Forcados River may be held to suggest that the currency which the Portuguese found in use there already consisted of cowries. If so, they would have come down the river Niger: cowries are known to have been imported into the Niger Bend countries by way of North Africa centuries before the Portuguese first brought them to the Benin coast. Cowries, however, were readily accepted in the Kongo kingdom, where the previous currency is known to have been an Atlantic mollusc. Moreover, a quite different type of shell currency was in use on the island of Fernando Po, where it survived until the later nineteenth century.[4] The question of the pre-Portuguese shell currency in Benin is one which can only be settled by archaeology.

Shells of the species *Cypraea moneta* have been in use as money in various parts of the world for at least 1,000 years; they continued in use, on a limited scale, in parts of West Africa and Bengal up to the present century, and are still in use, as a regular market currency for small purchases, in at least one West African area to this day. The main source of supply has always been the Maldive Islands, an extensive archipelago of islands and shoals which forms the perfect habitat for these molluscs, as will be evident when we consider the total quantities of shells exported from the islands.

Arab travellers to the islands in the ninth and tenth centuries A.D. noted the collection of these shells by the use of rafts of coconut leaves, to which the living cowries attached themselves; other writers describe fishing by islanders wading into the sea.[5] It appears to have been the usual practice to

[2] Pachecho Pereira, *Esmeraldo de situ orbis*, transl. and ed. G. T. Kimble (Hakluyt Society ed., 1936), 145: 'In the country of Beny... they use as money shells which they call Iguou, a little longer than these "zimbos" of Maniconguo. They use them to buy everything and he that has most of them is the richest.'

[3] A. F. C. Ryder, 'An early Portuguese trading voyage to the Forcados River', *J. Hist. Soc. Nigeria*, I, no. 4 (December 1959), 301 n. quoting Arquivo Histórico Ultramarino, Lisbon, São Tomé caixa 2.

[4] T. J. Hutchinson, *Impressions of West Africa* (London, 1858). The same shell money is described by Mary Kingsley, *Travels in West Africa* (London, 1891), 59, and O. Baumann, *Fernando Po und die Bubie* (Vienna, 1888), 83–4.

[5] Sulayman al Tajir and Al Masudi, quoted in M. Hiskett, 'Materials relating to the cowry currency of the Western Sudan, II,' *Bull. S.O.A.S.* XXIX, part 2 (1966). See also the sources quoted in the notes to the Hakluyt edition of Pyrard de Laval's voyage, op. cit. especially de Barros: 'Now the manner in which the islanders gather these shells is this—

bury the shells in the sand until the animals rotted away; but one early nineteenth-century writer states that they were shipped with the live mollusc still inside, 'so that such a cargo was said to be stinking and very unhealthy'.[6] It is to be hoped that this was only a temporary departure from earlier and better practice. European markets in the nineteenth century distinguished between 'Live Maldives' and 'Dead Maldives'; in the former the enamel was complete and glossy, in the latter (consisting of shells thrown up on the beaches, not fished from the sea) the enamel was partly worn and no longer glossy. Such shells were of course useless for ornamental purposes, and in some places were not acceptable as currency, or commanded a lower price.[7] There is no reason to think that 'Live Maldives' were necessarily shipped with the living shellfish inside.

Ibn Battuta, who spent some time in the Maldive Islands in the fourteenth century, wrote of cowries being sold in units of a *cotta* or basket of 12,000 and a *boustou* or sack of 100,000; these units were reported still in use in the nineteenth century. Ibn Battuta goes on to say:

> They sell them in exchange for rice to the people of Bengal, who also use them as money, as well as to the Yemenites, who use them instead of sand [as ballast] in their ships. These shells are used also by the negroes in their lands; I saw them being sold at Mali and Gawgaw at the rate of 1150 for a gold dinar.

In the Maldives, according to Ibn Battuta, a gold dinar would buy 400,000 and sometimes as many as 1,200,000 cowries. Even if this was the larger gold 'dinar' of India, the profits must have been considerable.[8]

This trade was presumably still going on when the Portuguese arrived in the Indian Ocean; it continued well into the nineteenth century with very little change. In 1515 the King of Portugal gave a licence to a Portuguese merchant to have 500 quintals of cowries brought from India each year for the São Tomé trade. The cowries were to come as ballast in His Majesty's ships, without charge for freight.[9] De Barros, in the middle of the sixteenth century, wrote:

> With these shells for ballast, many ships are laden for Bengal and Siam, where they are used for money just as we use small copper coin for buying things of little value. And even to this Kingdom of Portugal, in some years as much as two

they make large bushes of palm leaves tied together so as not to break, which they cast into the sea. To these the shell-fish attach themselves in quest of food; and then the bushes are all covered with them, they are hauled ashore and the creatures collected. All are then buried in the earth till the fish within have rotted away. The shells, *buzios* as we, and *Igoros* as the Negroes call them, are then washed in the sea, becoming quite white.' Capt W. F. W. Owen, R.N., 'On the Geography of the Maldiva Islands,' *J. Royal Geog. Soc.* II (1832), 82, gives an account which evidently owes much to de Barros.

[6] H. C. Monrad, *Gemalde der Kuste von Guinee* (København, 1822; Weimar, 1824), 162 n. Monrad also gives the only known instance of forgery of this virtually unforgeable currency: 'they are sometimes forged with a similar kind of smaller and yellower shell, which is found on the shore' (probably an olive, not a cowrie).

[7] E. Hieke, *Zur Geschichte des deutschen Handels mit Ostafrika*, I (Hamburg, 1939), 70.

[8] Ibn Battuta, quoted in notes to Pyrard de Laval, op. cit.; see also translation by H. A. R. Gibb (London, 1929), 243.

[9] Ryder, op. cit, 301 ff.

or three thousand quintals are brought by way of ballast; they are then exported to Guinea, and the kingdoms of Benin and Congo, where also they are used for money.[10]

At this period, $4\frac{1}{2}$ *cotta* were reckoned to the quintal, so that 2,000 quintals would be over 100 million cowries.[11]

Pyrard de Laval, a Frenchman shipwrecked in the Maldive Islands in 1602, at the end of the Portuguese period, found thirty or forty shiploads of cowries being exported from the islands each year. All went to Bengal, the only place which demanded a large quantity at a high price, because the people there used them as ordinary money. Merchants from other parts of India carried large quantities of cowries to Bengal; Pyrard saw a ship from the Portuguese town of Cochin in South India, which had come to buy shells for the Bengal market, paying for them in rice. The shells were sold in 'parcels of 12,000, in little baskets of coco leaves of open work, lined with cloth of the same coco tree, to prevent the shells falling out. These parcels or baskets of 12,000 are negotiated there as bags of silver are here, which between merchants are taken as counted, but not by others.'[12] These *cotta* sold for a *larin*, a Bengal unit which Pyrard equated with about eight French *sols* (20 *sols* = 1 *livre*). Evidently the price had gone up since Ibn Battuta's time.

After the Dutch took over Ceylon, cowries began to appear in the Dutch markets. They came in quantity from about the middle of the seventeenth century.[13] They were still coming packed in *cotta*; Barbot, writing of about 1680, says: 'They are commonly brought over from the East Indies, in packs or bundles well wrapped, and put into small barrels in England or Holland for the better conveniency of the Guinea trade.'[14] As late as the end of the eighteenth century they are described as coming in baskets or balls of 12,000 to Rotterdam.[15]

In the eighteenth century, cowries in the Dutch markets occurred in two qualities, sorted and unsorted;[13] in 1680, however, Barbot's cowries were evidently a very mixed lot: 'These cauris are of many different sizes, the smallest hardly larger than a common pea, and the largest as an ordinary walnut, longish like an olive; but of such great ones there is no considerable quantity in proportion to the inferior sizes, and all are intermixt, great and small.'[14] Barbot adds that 10,000 shells weighed 60 lbs. This, if correct, would give only about 166 to the lb., by contrast with the figure of around

[10] De Barros, in Pyrard de Laval, op. cit. II, 384, n.

[11] Nuñes, in Pyrard de Laval, op. cit. I, 236–9, n.

[12] Pyrard de Laval, op. cit. I, 236.

[13] K. Glamann, *Dutch-Asiatic Trade* 1620–1740 (*Copenhagen and the Hague*, 1958), 22.

[14] J. Barbot, *Description of the Coasts of North and South Guinea*, in Churchill, *Collection of Voyages*, v, (London, 1732), 338.

[15] J. E. Herz, 'Ueber Verwendung und Verbreitung der Kaurimuschel', *Mitteilung der Geographisches Gesellschaft zu Hamburg* (1881), 14 ff. See also E. v. Martens, 'On the uses of shells', *Zeitschrift für Ethnologie* (Berlin, 1872), 67, quoting Beckmann, *Waarenkunde* (Göttingen, 1794).

400 to the lb. which is usual for Maldive cowries, or the rather higher figure which Portuguese writers suggest. Barbot's account of cowries at Whydah and Ardra is confused and in places contradictory, but there seems no doubt that some of his cowries were larger than the usual Maldive shells. These may have belonged to larger Maldive types; but it is possible that the Portuguese, and the Omani Arabs after them, had been carrying East African cowries to the Bengal markets.

By the early eighteenth century, though some large cowries were still being taken to West Africa, they were recognized as of less value than the smaller Maldive cowries. Both Atkins[16] and des Marchais[17] give weights for the small cowries which correspond approximately to the standard figure of 400 to the lb. Cowries came to England by way of India, by the roundabout route through Chittagong and Calcutta.[18] French merchants, until the *Compagnie des Indes* began to import cowries from India, obtained their supplies from London, Holland and Altona (Hamburg).[19]

The quantities were very large. One source quotes the annual demand for Dutch supplies via Ceylon in the 1720s as about 400,000 lb. per annum; some actual figures of imports are: in 1689, 192,951 lb.; in 1721, 317,613 lb.; in 1780, 133,299 lb.[20]—figures which suggest that the estimate of 400,000 lb., though an exaggeration, was not a very large exaggeration. The English Royal African Company, in the 1680s, when its monopoly was fairly effective, was exporting well over 1,000 cwt. a year; its exports in later years were much lower, but may have been offset by those of private traders.[21] In 1717 the French merchants estimated their needs at 300,000 livres weight a year; at this date they were believed to be carrying a little less than one-third of the trade at Whydah.[22] The total export of cowries to Europe may therefore have been of the order of a million pounds weight a year for the years round 1720, when the Whydah slave-trade was at its height. These figures are in line with what is known of the number of slaves exported (some 20,000 per annum), the prices paid (180 lb. of cowries) and the proportion of that price actually paid in cowries (about one-third). This total export amounts to some 400 million shells a year, apart from the quantities that went to Bengal at the same period. It is hard to believe that any fishing grounds could have maintained such an export for very long.

The price of cowries in Europe in the late seventeenth century was subject to considerable variation; Dapper[23] and Barbot[24] mention the fluctuations in price. The books of the Royal African Company bear this

[16] J. Atkins, *Voyage to Guinea* (London, 1735), 112.
[17] J. B. Labat, *Voyage de Chevalier des Marchais*, II (Paris, 1730), 109 ff.
[18] Pyrard de Laval, op. cit. I, 236–9, n.
[19] S. Berbain, *Le Comptoir français de Juda au XVIIIe siècle* (Paris, 1942), 82.
[20] E. v. Martens, loc. cit., quoting 'Johnston' (?A. K. Johnston).
[21] K. G. Davies, *The Royal African Company* (London, 1957), appendix I, 357.
[22] S. Berbain, op. cit. 82.
[23] O. Dapper, *Description d'Afrique* (Paris, 1768), 305.
[24] J. Barbot, loc. cit. 338.

out: between 1674 and 1695 the average price for the year per cwt. varied between £3. 7*s.* and £5. 8*s.* In 1696 the price was £7 and in the following year £7. 10*s.*; in 1698 it had risen to over £14. The figures for the next two years are missing, but in 1701 it was over £9, and in 1702, £11. The price then dropped to £6. 15*s.* in 1703 and a little over £5 in 1704.[25] These are the last figures available from the Company; six figures taken from the accounts of trading voyages between 1714 and 1730 range between £5 and £8 per cwt. (it is evident from Atkins's figures that his 'hundredweight' in 1721 was 100 lb.).[26] The Whydah trade declined considerably after the 1720s, and it is probable that the cowrie trade declined as well. Cowries did, however, continue to arrive in some numbers, especially on the Whydah coast. 'Each ship brings 30–60 or even 80 thousand weight,' wrote Pruneau de Pommegorge, who had been governor of the French fort at Whydah in the latter half of the eighteenth century,[27] and his figures are borne out by the accounts of two French ships trading there.[28]

One of these ships was paying for cowries in Europe at the rate of 17 *sols* per livre weight (20 *sols* = 1 livre tournois = 10*d.* to 1*s.* at this date)—a value of £4. 15*s.* per cwt. of 112 lb. Adams, an English merchant whose experience of the Guinea trade was between 1786 and 1800 (though he published in 1823), gave the price in Europe as £40–£80 a ton, £2–£4 per cwt.[29] The 1789 Report on the Slave Trade gives higher figures, ranging from £4 to over £6.[30]

At the beginning of the nineteenth century the English were shipping some 100 tons of cowries a year to Africa.[31] With the abolition of the slave-trade in 1806 these shipments dropped to an average of about 6 tons a year between 1810 and 1815.[32] With the development of the palm-oil trade, however, the cowrie trade recovered; by the 1830s between 60 and 100 tons went annually to the Gold Coast alone, and from 1835 at least 50 tons a year to the coast east of the Volta.[33] By 1850 the Gold Coast was importing

[25] K. G. Davies, op. cit. appendix, 357.

[26] E. Donnan, *Documents Illustrative of the History of the Slave Trade to America* (Washington, 1930–1), II, 181, 246, 275, 324, 363.

[27] P. de Pommegorge, *Description de la Nigritie* (Paris, 1789), 204.

[28] S. Berbain, op. cit. 98.

[29] J. Adams, *Cape Palmas to the Rio Congo* (London, 1823), 235.

[30] *Report to the Commissioners for Trade and Plantations on the West African Trade* (London, 1789), appendix 1 to part IV (unpaged).

[31] Mungo Park, *Journal of a Mission to the Interior of Africa in* 1805 (London, 1815), appendix (by Rennell), lxxxvi.

[32] H. Murray, *Discoveries and Travels in Africa*, II (Edinburgh, 1818), vol. 2:

	cwt.	qtr.	lb.		cwt.	qtr.	lb.
1810	114	3	16	1813	—	—	—
1811	91	0	10	1814	18	3	0
1812	85	3	8	1815	291	1	11

These figures were doubtless affected by the war in Europe and the consequent shortage of shipping, both to West Africa, and to India.

[33] *Report from the Select Committee on the West Coast of Africa* (1842), appendix no. 36, 499.

at the rate of some 150 tons a year.[34] Most of these cowries were still coming from the Maldives and other East Indian islands (60 tons were reported to have come to Liverpool from Manila in 1848[35]), but at Whydah the first experimental shipments of Zanzibar cowries were received in 1845.

Laird and Oldfield, trading on the Niger in 1831, reckoned the value of cowries at about 1*s.* per 1,000 (about 45*s.* a cwt.), based on their value in Europe.[36] The Niger Expedition ten years later gave a careful costing of cowries at 6*d* a lb., 56*s.* a cwt.[37] By 1849 there are indications of a large price rise on the English market; cowries are said to have risen in price to £5 or even £6 per cwt. in that year.[38] Contemporary opinion was inclined to blame rises in the price of cowries on increased demand on the Niger—perhaps itself due in part to the new demand for cowries in Bornu. It is possible, however, that the rise in price is an indication that the Maldive fisheries, after at least a 1,000 years of productivity, were at last showing signs of exhaustion.

In 1844 a Hamburg shipper, Herz, influenced no doubt by rising prices and shortage of supplies of cowries, sent a small sailing-ship to the Maldives to try to buy cowries direct, and so cut the long journey by way of the Bay of Bengal.[39] Only a small quantity could be bought, and this partly for a wheeled cart on board the ship which caught the fancy of the king of the islands; the people were too much tied to the centuries-old exchange of cowries for their staple food, rice, to dare sell them for anything else. Possibly, also, cowries were less plentiful than they had once been.

On the return journey rather similar cowries were found at Zanzibar, off the East African coast, where they were burnt for lime. These were the shells of *Cypraea annulus*, a species very similar to *C. moneta*, but somewhat larger. Some of these shells were sent to West Africa, where they proved acceptable to the palm-oil merchants. In the first place they were sold for Spanish and South American gold doubloons, which had no currency inland of Whydah and Lagos. In a few years, when these gold coins became scarce, with the decline of the slave-trade, cowries were sold for palm oil at the rate of 3 gallons of oil to 1 silver dollar.

East African cowries had probably found their way to the West Coast on

[34] B. Cruikshank, *Eighteen Years on the Gold Coast* (London, 1853; went to press September 1852), II, 42; see also: D. Kimble, *A Political History of Ghana*, I, *The Rise of Gold Coast Nationalism* (London, 1960), 4 n. Kimble's figure is presumably based on an estimate by the governor in his annual report in 1850, and should not probably be taken as an exact figure. See also Swanzy in answer to qn. 4689 in the *Report from the Select Committee on the Western Coast of Africa* (London, 1865).

[35] E. v. Martens, loc. cit. quoting Woodward, '*Manual of Mell* (*sic*)', 121.

[36] M. Laird and R. A. K. Oldfield, *Narrative of an Expedition into the Interior of Africa by the River Niger 1832–4*, I, (London, 1837), 166.

[37] W. Allen and T. R. H. Thompson, *Expedition to the River Niger*, I (London, 1848), 459 (appendix). See also page 350: 'We had been supplied with a large quantity, bought in London by the ton, from the price of which we estimated the value of 1,000 to be about 15*d*. In the former expedition, Lander considered 1,000 to be worth 1*s*.'

[38] E. Hieke, op. cit. 72.

[39] J. E. Herz, loc. cit. 14, 19.

several previous occasions. We have seen that possibly some of Barbot's mixed cowries may have originated in East Africa. There were large cowries in the cargo of the French ship *Dahomet* in 1772, valued at four-fifths the price per shell of the small ones, but used mainly in purchases from Portuguese ships.[40] In 1777 the Frenchman Morice reported that the East African cowrie could only be used in Bengal: 'it is worthless on the Guinea Coast'.[41] There were some suitable cowries, he added, but in such small quantity as not to be worth collecting (possibly young specimens of *C. annulus*?). He reported that some small English ships had loaded cowries at Pate, presumably for the Bengal market.[42] It is thus likely that cowries purchased in Ceylon and Bengal included some East African cowries; it was in any case necessary to sort them at Amsterdam, 'sorted' cowries commanding the higher price.[43]

Herz was not the first to see the possibilities of importing East African cowries in quantity to West Africa. Fowell Buxton, quoting 'the captain of a merchant vessel who was long on the East Coast of Africa', wrote in 1836: 'A most profitable trade might also be carried on in cowries, which abound on the coast, where he has purchased them at 4*d*. a bushel; on the West Coast they are current coin, and are told out by the 100.'[44] During the period of the slave-trade, as we have seen, large cowries were accepted only in small quantities and at a considerable discount. In the palm-oil trade there was less resistance to them. Their cheapness recommended them to the palm-oil merchants, and between 1851 and 1869, when the trade was at its height, well over 35,000 tons of shells reached West Africa from Zanzibar and other parts of the East African coast and islands.[45] This represents more than 14,000 million shells in nineteen years, or nearly 750 million shells a year. Imports continued, on a diminishing scale which finally shrank to a trickle, until they were finally prohibited in 1904[46]—surely a classic case of closing the stable door after the horse has gone.

The effects of this enormous import of cowries on the value of the cowrie currency will be considered in the second part of this paper. By the 1860s the large cowries had displaced the small at Lagos and as far inland as Abeokuta.[47] They were carried northwards, probably in diminishing proportions as they went inland; in 1889 large cowries were reported at Walewale, in what is now Northern Ghana, selling at a discount of 10%.[48] By the 1890s most of the cowries in use in Togo and in Yoruba country were

[40] S. Berbain, op. cit. 81, and appendix.

[41] G. S. P. Freeman-Grenville, *The French at Kilwa Island* (London, 1965), 120.

[42] Ibid. 114.

[43] K. Glamann, op. cit. 22.

[44] T. Fowell Buxton, *The Slave Trade* (London, 1838), 228.

[45] E. Hicke, op. cit. 283.

[46] F. Lugard, *Political Memoranda* (Nigeria, 1905), 224 ff.; see also A. H. M. Kirk-Greene, 'The major currencies in Nigerian history', *J. Hist. Soc. Nigeria*, II, no. 1 (Dec. 1960), 132–150.

[47] J. E. Herz, loc. cit. 23, quoting Burton.

[48] L. Binger, 'Transactions, objets de commerce, monnaie des contrées entre le Niger et la Côte d'Or', quoted in R. Mauny, *Tableau géographique de l'ouest africain au Moyen Age* (Dakar, 1961).

of the Zanzibar species[49] though these were never accepted west of the lower Volta or in part of the Ibo country.[50]

The reason for the popularity of the Zanzibar species with the shippers is not hard to discover. When Herz made his first shipments in 1844, cowries in Zanzibar cost about $¾ per cwt., as against $8–9 in the Maldives for the small cowries. This price difference was only partly offset by the fact that the hundredweight of Zanzibar cowries contained rather less than half the number of shells contained in a hundredweight of Maldive cowries. Prices on the West Coast reflected the difference in number of shells: Zanzibar cowries fetched $8–9 per cwt., Maldive cowries 18. Table 1 sums up Herz's figures.

Table 1. *Prices of Maldive and Zanzibar cowries*, 1844, *after Herz*[51]

	Shells/cwt	Buying price/ cwt. ($)	Selling price/ cwt. ($)
Maldive cowries (*Cypraea moneta*)	45,000–48,000	8–9	18
Zanzibar cowries (*Cypraea annulus*)	18,000–20,000	¾	8–9

In its early years the trade in Zanzibar cowries was thus a very profitable one for the firms shipping cowries from Zanzibar. Much ingenuity and a notable lack of scruple were exercised by the Hamburg firm mainly concerned, O'Swald, in driving out of business its competitors, including the firm of Herz, which had pioneered the trade, and Diedrichsen, the Flensburg merchant, later naturalized as a citizen of Hamburg, who did much of the work of selling cowries on the West Coast. Considerable secrecy surrounded the trade. Cowries were referred to as 'coffee';[52] some French visitors were informed that the cowries they saw in Zanzibar were being shipped for the use of the porcelain industry at Dresden.[53]

This story reflects a very old linguistic usage; in medieval times cowries had been known to the Mediterranean world as 'porcelains' or some related word (meaning 'little pigs'). They were familiar to Marco Polo under this name, and it seems that Venice was one of the centres of the Mediter-

[49] O. Schneider, *Muschelgeldstudien*, ed. E. Ribbe (1905), 169–70, quoting Mischlich and Speith.

[50] A. G. Krause, *Fada Language on the Gabon River in Portuguese West Africa* (Berlin, 1895), 371 ff.; G. T. Basden, *Among the Ibos of Nigeria* (London, 1921), 198; O. Schneider, op. cit. 171.

[51] J. E. Herz, loc. cit. 15, 16. Herz's figures suggest that he has somewhat exaggerated the difference in number per cwt. of the two types of cowries: 1 cwt. of Maldive cowries would normally contain a little under 45,000 shells; a cwt. of Zanzibar cowries was variously estimated to contain 20,000 to 24,000 shells.

[52] E. Hieke, op. cit. 73 ff.

[53] B. Schnapper, *La Politique et le commerce français dans le golfe de Guinée de 1838 à 1876* (Paris, 1961), 190–1, quoting Loarer, *Rapport commercial sur la côte orientale d'Afrique*, 123, a MS dated 1851 in Arch CCM, série OK, explorations commerciales.

ranean cowrie trade. Cadamosto, who heard of their use inland from Arguin, recognized them as 'those small kinds which are brought to Venice from the Levant'.[54] A Portuguese pilot in about 1540 wrote of 'shells which in Italia are called *porcellette*—little white ones which we call *buzios* and which are used for money in Ethiopia [Africa]'.[55] They were still known as 'porcelains' in eighteenth-century Marseilles,[56] and the name survived in Germany into the nineteenth century.[57]

Though cowries had a distinctive name in the Mediterranean, there is no suggestion that they were ever used as currency in this area; they passed through Mediterranean markets on their way to Morocco and thence to the cities of the Niger Bend. Cairo is said to have been a major cowrie market in medieval times,[58] but Leo Africanus speaks of the Timbuctu shells as coming 'from the Kingdom of Persia'.[59] This may have meant no more than 'the East'; but it is not unlikely that some of the cowries came up the Persian Gulf rather than up the Red Sea—those used as ballast by the Yemenis[60] could well have reached the Mediterranean markets through Persia. There is no reason to think that these were cowries of a different kind from the usual *Cypraea moneta*. It has been shown, however, by Mauny and Jeffreys[61] that there was at one time, probably temporarily, a shell currency of *Marginella* shells in Timbuctu; local tradition has it that they were used when the normal supply failed to arrive, and were buried when the cowries arrived.

Hiskett has devoted many pages[62] to refuting the entirely improbable theory that cowries reached West Africa overland from East Africa. They are known to have come in medieval times to the Niger Bend through Morocco and Sijilmasa.[63] A similar route was in use throughout the eighteenth century for cowries which by this time came by way of European ports, though still originally from the Maldive Islands; the English are known to have been importing them into Morocco in about 1700,[64] the

[54] A. Cadamosto, ed. G. R. Crone, *Voyages* (Hakluyt Edition, 1937) 29.

[55] Portuguese pilot quoted in J. W. Blake, *Europeans in West Africa* (London, 1942), 157.

[56] S. Berbain, op. cit. 82.

[57] G. Rohlfs, 'Geld im Afrika', *Petermanns Mitteilungen*, 1889, 187 ff.

[58] E. W. Bovill, *The Golden Trade of the Moors* (London, 1958), 127; also editorial comment on Laing's letters in *Missions to the Niger* (Cambridge, 1964), 245: 'Cowries were imported into Africa from Asia, mostly through Cairo.'

[59] Leo Africanus, *The History of Africa*, trans. Pory, ed. Robert Brown (Hakluyt Society Edition; London, 1896), III, 825.

[60] Ibn Battuta, *Travels in Asia and Africa* (London, 1963), 243.

[61] R. Mauny, 'La monnaie marginelloïde de l'ouest african', *Bull. IFAN*, XIX (1957), 659–69; M. D. W. Jeffreys, 'The marginelle currency of Timbuctu', *Bull. IFAN*, XV (1953), 143–51.

[62] M. Hiskett, op. cit. This theory seems to have originated with W. Jackson, *Shells as Evidence of the Migration of Early Culture* (Manchester, 1916), and is based on a carefully constructed distribution map—a warning to archaeologists and others who draw conclusions from distribution maps.

[63] M. Hiskett, op. cit. 345 ff., gives the early Arabic texts.

[64] Ch.-A. Julien, *Histoire de l'Afrique du nord* (Paris, 1961), 234. The English were importing to Morocco in *ca.* 1700: 'Guinea cowries, which are shells serving as money in that country.'

Dutch at the end of the century,[65] and they are also known to have been passing through Meknes in the early eighteenth century.[66] Most of the cowries came through Morocco, though some may have come to Hausaland from Tripoli in the later eighteenth century.

Hiskett has suggested that the Hausa cowries may have come, not by way of Timbuctu, but from Nupe and ultimately from the Guinea Coast.[67] There is certainly evidence in the nineteenth century for the northward movement of considerable numbers of cowries to the Fulani states of Northern Nigeria in the form of tribute,[68] and Hiskett suggests that it was either as booty or in exchange for captured slaves that cowries 'came to Hausaland' in Sharifa's time 'because he was zealous and raiding' (Hiskett's amended translation of the relevant passage in the Kano Chronicle). In some respects, however, methods of counting cowries in Hausaland are more akin to those of the Niger Bend than to those of the Guinea Coast: except on the coast, Yoruba cowrie arithmetic has much in common with that of Hausaland; in the interior the strung cowrie of the Lower Niger and the coastal area was apparently unknown. Had cowries spread inland from the coast, it seems likely that the convenient method of counting them in strings would have spread with them. The question of the origin of the Hausaland cowries thus remains an open one. Hiskett's linguistic arguments do not appear very convincing, at least to the linguistic layman.

The cowries of the trans-Saharan trade must have been almost exclusively small ones of Maldive origin; heavier cowries would have been at a great disadvantage in the long and costly journey over the desert.

COST OF TRANSPORT

Cost of transport was always an important factor in the cowrie currency; in the absence of any external control, cowrie values would always have varied from place to place, owing to the cost of moving them from one place to another by land.

Transport of cowries by sea, if the shells were used as ballast, might cost

[65] J. Grey Jackson, *An account of the Empire of Marocco*, 3rd ed. (London, 1814), 24.

[66] Commodore Stewart quoted by Thomas Winterbottom, *An account of the native Africans in Sierra Leone* (London, 1803), 221: 'The goods they [the Moroccans] bring to Guinea are salt, cowries etc.' (quoted in O. Schneider, op. cit.).

[67] Hiskett, op. cit. 355. Lucas, from North Africa, at the end of the eighteenth century, believed that merchants at Katsina obtained cowries from the countries nearer the sea (*Association for Promoting the Discovery of the Interior Parts of Africa, Proceedings*, I, 105ff; R. Hallett, *Records of the African Association* 167, 186). See also E. Daumas, *Le Grand Désert: Itinéraire d'une caravane du Sahara au pays des nègres* (Paris, 1848), 241: 'Cowries, they tell me, come from the Bahar el Nil (Niger), which runs ten days' journey west of Katsina. The sultan has organized a system of customs posts which prevent individuals from bringing them from the interior without paying enormous duties. He has the monopoly.'

[68] E.g. W. Allen and T. R. H. Thompson, op. cit. II, 85: 'Gori pays an annual tribute of 360,000 to the Filatah king.'

little or nothing, and along inland waterways, if other freight was also being carried, cowries may also have been used as ballast in canoes; this may account for the relative constancy of value of cowries in towns along the middle Niger.

Some estimate of the actual cost of transport by canoe on the river Volta towards the end of the nineteenth century can be deduced from figures given by Bonnat[69] and 'Dobson'.[70] In 1876 Bonnat was able to buy a canoe, probably holding about a ton of merchandise, for 25 heads of cowries, which he valued at '10 dollars and one shilling'. (In the area in question, the head of cowries consisted of 50 strings of 35 cowries apiece, i.e. 1,750 to the head.) He also hired men at the rate of $3 a month, five or six men to a canoe. Additional men were taken on at each of the many rapids on the river, at 1*s.* a head, plus subsistence, 3*d.* a head. The normal journey took something under a month from the coast to Salaga (Bonnat himself took twice this time from Akwamu, but was engaged in much diplomatic activity *en route*); wages would thus work out at some $15–18 per canoe carrying 1 ton, plus perhaps as much again for additional labour at the various rapids. Bonnat reckoned that transport from Krachi to Salaga cost him $25 a ton, a figure he considered extortionate. Thus the cost of transporting a ton of cowries from Ada to Salaga would be in the region of $55–60, apart from the cost of the canoe (which could be partially attributed to capital expenditure). There was also a toll or present at Krachi, and sometimes at other places on the river.

The value of a ton of cowries at the coast at this date was about 450*s.* or about $100 for the smaller Maldive cowries; at Salaga they were worth nearly double this. This would leave a reasonable margin of profit to cover the risks of damage to canoes, loss of goods by overturning, and the hazards, in disturbed times, of loss by plunder. For the larger Zanzibar cowries, however, the value of a sack of cowries, 1 cwt, was by this time about 12*s.*, so that a ton was equivalent to 240*s.* at the coast, and double that at Salaga. It is evident that, at these rates, while it might be worth while to transport the smaller cowries inland, it could hardly pay to carry the larger ones. After 1876, when tolls at Krachi were abolished, and it became possible to use the upper reaches of the river between Krachi and Yeji, thus reducing overland transport costs, the transport of small cowries to Salaga was probably fairly profitable. It must be noted, however, that the costs quoted are for canoes and boatmen in the middle reaches of the river; both were dearer at either end of the journey. Moreover, the profit from carrying salt was almost certainly higher than the profit from cowries could have been. That cowries went north up the Volta in earlier times is almost

[69] M.-J. Bonnat, diary for 1875–6 in *L'Explorateur* (Paris), 1876–7.

[70] 'George Dobson', 'The River Volta, Gold Coast, West Africa', *J. Manchester Geog. Soc.*, VIII (1892), 21: 'a canoe 25ft. long by 3 ft. wide and 18 ins. deep requiring eight men to pole it up' (for the identity of 'George Dobson' see note in my 'M. Bonnat on the Volta', *Ghana Notes and Queries*, no. 10 (1968), 6).

certain—the exchange rates for 1820, quoted above, make sense only on the the assumption that cowries were coming into the system at Salaga.[71] Some Zanzibar cowries had reached as far north as Walewale by 1889, carried there apparently by Mossi traders from Salaga. These may well have come overland through Dahomey. Burton believed that cowries were going through Dahomey as far as Segu in the 1860s.[72]

On land routes the cost of transport was a considerable item, especially on the desert crossing. The load of a camel is variously given by early nineteenth-century writers as 4–5 cwt.; several writers in the later part of the century put the figure lower, at some three cwt.[73] Barth[74] says that a camel would carry 100,000 cowries, a very good camel 150,000, and with Maldive cowries at about 400 to the lb. this is equivalent to 250 and 375 lb. respectively, unless, as is possible, some of the large cheap Zanzibar cowries had already found their way so far north; the introduction of cowries into Bornu is known to have caused a shortage of cowries, which might have stimulated the movement northwards of these early imports, which had first reached the coast some five years earlier.

On the basis of figures given by Shabeeny[75] it is possible to make some estimates of the costs of cowries in relation to the desert crossing. Shabeeny reckoned that a camel carried 4–5 quintals, some 190,000–240,000 cowries. At the standard rate of exchange this would be some 64–80 *mithqal* or *mizan* of gold (probably the large 75-grain mithqal), so that a camel-load of cowries would be equivalent to some £40–50 in gold at Timbuctu. The cost of hiring a camel for the journey was £8–9, about one-fifth of the value of the load. Cowries were sold in southern Morocco, so Jackson tells us, at 20 Mexican dollars a quintal, i.e. about £5, so that a camel-load in southern Morocco would have cost some £20–25. Add to this the cost of hiring the camel, and the cost price in Timbuctu would be about £28–35, as against a gold equivalent of £40–50. But Jackson tells us that merchants expected at least 30% profit, and often 50–60% profit on the desert traffic to cover its very real dangers. Thus commercial importation would have been just possible, but not highly profitable, at the end of the eighteenth century. It is easy to see why it ceased when cowries began to decline in

[71] J. Dupuis, *Journal of a Residence in Ashantee* (London, 1924), CXIV.

[72] J. E. Herz, op. cit. 23, quoting R. Burton.

[73] Shabeeny, quoted in J. G. Jackson, *An Account of Timbuctoo and Hausa* (London, 1820), 2: 4–5 quintals; Denham and Clapperton, *Narrative of Travels and Discovery in North and Central Africa.* (1822–4): 4 cwt.; R. Caillié, *Journal d'un voyage à Temboutou et à Jenné* (Paris, 1830), 67: 500 lb.; C. H. Dickson, 'Account of Ghadames', *J. Royal Geog. Soc.* 1860: 4 cwt.; Rabbi Mordekkai, 'Reisen nach Timbuktu', *Petermanns Mitteilungen*, 1870: not over 150 kg.; Dr. Ollive, 'Schilderung von Tendouf', *Petermanns Mitteilungen*, 1880: 3 centner; Col. Mircher *ca.* 1865 (in C. W. Newbury, 'North Africa and Western Sudan trade in the nineteenth century, a re-evaluation', *J. Afr. Hist.* VII (1966), 233): 3 kantar = 150 kg.; J. L. Miège, *Le Maroc et l'Europe*, III, (Paris, 1961-2); 85 ff.: 3 quintals = 154–162 kg.; *ibid.* 258: 150 kg.; F. Lugard, in *Annual Report, Northern Nigeria* 1902: 3–4½ cwt.

[74] H. Barth, *Travels and Discoveries in North and Central Africa*, II (London and Gotha, 1857–8), 163.

[75] Shabeeny, op. cit. 2 ff.

value south of the desert. It is known, however, that the price of cowries rose whenever there was famine or food shortage in Timbuctu,[76] presumably because some of them would be sent out of the town to buy food further afield. If the exchange rate remained high for any considerable length of time it would be redressed by imports across the desert.

The other main means of land transport for cowries were by donkey and headload. A donkey's load in the Kano area was given by Monteil in the late nineteenth century as 50,000 cowries.[77] At the same period, Lugard was using donkey loads of 84 lb. each, two to the donkey, between Jebba and Nikki.[78] If these were the usual cowrie 'bags' they would contain 20,000 each, 40,000 for each donkey. Lugard's cowries are particularly interesting. They were originally packed in 53 donkey loads of 84 lb. each. In Borgu, he found it worth while, as the result of the loss of most of his donkeys, to have his remaining cowries sorted out; the small ones (presumably *Cypraea moneta*) he packed into sacks containing 16 heads (32,000 cowries), making a heavy headload of 76 lb. each. These work out somewhat below the standard weight of 400 cowries to the lb., perhaps because they did not include the heaviest of the 'small' cowries normally included; they may also have included small specimens of the Zanzibar cowries, which seem to have been more variable in size than good Maldive cowries (which had been sorted). These cowries, Lugard noted, were 'alone worth the transport since so many more go to the load'. The remaining cowries, presumably mainly of the Zanzibar series, he made up into donkey loads of 80 lb. or more, each containing only 7½ heads (15,000 cowries); these cowries thus weighed more than double the small ones.

A German official in Togo estimated 'a fairly heavy donkey load' at about 100 lb.[79] Transport donkeys were not usually hired like camels, but were bought outright. Where the trader was also a farmer, he might own the donkeys and breed them on the farm; but it would seem that these could only carry smaller loads than those bred by the Mossi people specially for the caravan trade.[80] The price of donkeys varied a good deal according to the demand; when kola was cheap and plentiful at Salaga, the price of

[76] H. M. Miner, *Primitive City of Timbuctu* (Princeton, 1953), 49 n.

[77] P. L. Monteil, *De Saint Louis à Tripoli* (Paris, 1895), 282. This value agrees approximately with Barth's estimate that a donkey carried 5,000–6,000 kola nuts (Barth, op. cit. v, 29).

[78] F. Lugard, *Diaries*, ed. M. Perham and M. Bull, IV (London, 1963), 120. See also page 117: 'It is these donkeys that throw everything out. Four more were dying this morning, and had to be left, and one yesterday = 10 loads.'

[79] Mischlich to German Government, Akte 3832, of 20 Mar. 1903, p. 44, in German Colonial Archives at Potsdam (I am indebted to Dr I. Sellnow for this reference). Mischlich describes 3,000 kola as a 'fairly heavy donkey load'. This is 1½ times the standard headload of 2,000 kola weighing about 65 lb.

[80] Polly Hill pointed out to me that merchants could also be farmers, in which case the donkeys were bred on the farm. If the Mossi donkeys used in the Mossi–Salaga–Hausa caravan trade were larger and stronger than the southern donkeys, or the farm-bred donkeys of Hausland, this would account for the higher prices, and the Hausa merchants' preference for Mossi donkeys.

donkeys rose.[81] In 1889 Binger believed that in 'normal times' the price of donkeys at Salaga would be 40,000–45,000 cowries[82] (the greater part of its load, by Kano standards). In 1850 the Salaga price had been reported at about 15,000 cowries, less than a third of its load;[83] at this date prices in Hausaland were reported to be only 5,000 cowries, though heavy tolls were payable on the road between. Binger put the cost in 'Hausa' at 25,000–30,000 cowries in 1889.

The price of donkeys in Nikki when Lugard was there in 1894 was 60,000 cowries; and this was conditional on the price being paid in cloth.[84] These donkeys would have originally carrying been 40,000 cowries each, before Lugard sorted out the shells; after they had been sorted, each donkey was carrying only 30,000 large cowries—half its own value. Transport by such means could only be economic if the animal had a good chance of surviving in a saleable condition at the end of the route; Lugard's donkeys nearly all died, though seven of them, in a dying condition, were sold for 63,000 cowries.

Table 2. *Change in cowrie value of* mithqal *with days' journeys*

Journey	Days	Change in value of *mithqal* (cowries)	change %	% per day
Salaga–Yendi	6	4,300–4,200	2·3	0.4
Yendi-Zogho	8	4,200–4,000	4·8	0·6
Zogho–Nikki	10	4,000–3,800	5·0	0·5
Nikki–Niger	13	3,800–3,500	7·9	0·6
		or 3,400	10·5	0·8

Similar conditions applied to transport by human carrier, when the carrier was a slave; but whereas a donkey could to some extent feed itself *en route*, the slave would have to be provided with food for the journey; either he would carry the food, or pay for it out of the cowries he carried; in either case his effective carrying capacity was reduced.

The best estimate of the cost of transport by slave carrier can be derived from the exchange rates between cowries and gold given by Dupuis, from information derived from Muslim merchants in Kumasi in 1820.[85] Comparison between these exchange rates and the number of days' journey between towns, as given in itineraries from the same source, yields Table 2. Apart from the last alternative figure, which perhaps refers to a town on the Hausa side of the Niger (and would thus involve ferry costs), the figures are about as consistent as is compatible with an exchange rate in round numbers of hundreds of cowries.

[81] H. Barth, op. cit. v, 29.
[82] L. C. Binger, *Du Niger au golfe de Guinée* (Paris, 1892), II, 103.
[83] H. Barth, op. cit. v, 29. [84] F. Lugard, *Diary*, IV, 192.
[85] J. Dupuis, op. cit. CXIV. These caravans were in fact mixed donkey and slave porter, in varying proportions; costs were probably similar for both.

If a man's load is taken as the sack of 20,000 cowries, the usual convention, then the cost of a day's journey ranges from 80 to 120 cowries. This is in line with Meredith's statement that on the coast a man could live well, in the plentiful season, on two strings of cowries (80) a day.[86] Park, a few years earlier in Bambara, where the value of cowries was higher, reckoned that he could feed himself and buy corn for his horse on 100 cowries a day.[87]

By the end of the nineteenth century cowries had fallen in value in parts of the coast to about one-tenth of their value at the beginning of the century. On this basis a carrier would thus need something like 1,000 cowries a day to subsist, and would consume his whole load in under three weeks, even without any payment. By this date, however, carriers were usually paid. To return to Lugard's cowries: his carriers were paid, nominally, 40 heads a month. Even with all his re-sorting of cowries, and the unusually heavy loads carried by the men, they could carry only 16 heads a man; so in less than a fortnight a man would have earned all he could possibly carry. In practice, of course, the men were usually paid off on their return, and travellers carried as much as possible of their resources in more portable trade-goods. To some extent this hampered the spread inland of the coastal cowrie inflation; it also meant that the cowrie system, as a long-distance currency, had largely broken down. (In fact, the '40 heads' earned by Lugard's men were a money of account, no longer bearing any relation to cowrie values or prices. Lugard found that a day's rations actually cost between 80 and 200 cowries on the road to Nikki.[88])

The final *reductio ad absurdum* is recounted by C. H. Robinson, who wished to sell a horse which was no longer serviceable: 'The trouble is that we cannot sell it, as its value in cowries would require 15 porters to carry, to whom we should have to pay all the money they carried, and a great deal more besides...'[89] The consequences of inflation with the cowrie currency are only a little more absurd than the consequences of run-away inflation under any other system.

THE COWRIE CURRENCY AREA IN WEST AFRICA

We are not concerned here with the use of cowries as ornaments or for ritual and magical purposes, which is far more widespread than their use as money. Neither are we concerned with other cowrie currency areas in Africa such as the Congo, East Africa, etc., none of which were continuous with the West African currency area.

The earliest currency area of the cowrie which we can trace in West Africa was on the upper and middle Niger in the medieval period.[90] The

[86] H. Meredith, *An Account of the Gold Coast of West Africa* (London, 1812), 183.

[87] Mungo Park, *Travels in the Interior Districts of Africa* (London, 1799), 199 n.

[88] F. Lugard, *Diaries*, passim. For the operation of this system, see below, p. 47.

[89] C. H. Robinson, *Hausaland* (London, 1896), 46.

[90] See Arabic sources quoted in Pyrard de Laval, op. cit.; in Mauny, 1961, op. cit.; and in Hiskett, op. cit. Hiskett gives several of the Arabic sources in Arabic as well as in translation.

cowries came in by way of Sijilmasa, but the area in which they were current was apparently cut off from any other cowrie area; it would therefore seem that they must have been introduced deliberately, as an act of state, as in nineteenth-century Bornu, or by a well-organized group of merchants. The area known to have been covered by the cowrie currency included, in the thirteenth century, the city of Ghana, and in the fourteenth century the Mali empire with its eastward extension to Gao; by the early sixteenth century it included Timbuctu, and there is no reason to doubt that cowries were in continuous use, at least in the major markets of the middle Niger area from at least the eleventh century (when El Bekri noted them as imports to Kougha). From the sixteenth century onwards, cowries were in use in Timbuctu and Jenne and probably also in the Bambara country, which occupied the western part of the old Mali empire. They never became established as currency in the area of modern Senegal, Gambia, Portuguese Guinea, Sierra Leone or Liberia, nor in any but the most northern part of modern Guinea, and only in north and north-east Ivory Coast.

Cowries reached Hausaland, if the Kano Chronicle is to be believed, in the first half of the eighteenth century, and Hiskett has argued that they reached there from the south rather than from the west.[91] They were known to be the currency of Katsina in 1790, when Beaufoy was reporting from North Africa, and there is no suggestion that they were of recent introduction at that date.[92] By 1822, when Clapperton was there, cowries had reached Katagum, but not Bornu;[93] we know that they were introduced into Bornu shortly before Barth's visit there in 1850, but had not in his time reached Adamawa.[94] By the 1890s cowries were in use in Adamawa also;[95] but they did not apparently extend as currency east of Lake Chad, except to a limited extent in the market towns of Bagirmi in the latter half of the nineteenth century.[96] Timbuctu seems to have been the northern limit of the use of cowries in the nineteenth century; they were at a discount there because they were not taken by the desert traders, and Caillié is quite definite that they did not circulate in El Arouan.[97] In Agades, when the gold trade was flourishing there, cowries were in use, and even in the nineteenth century prices seem to have been quoted in cowries, though few came to the market (Barth's evidence on this point is, for once, somewhat ambiguous[98]).

In Kaarta, round Nioro,[99] cowries were in use in the mid-nineteenth cen-

[91] M. Hiskett, loc. cit. 346 ff.

[92] Beaufoy quoted in Hallett, *Records of the African Association*. Mattra, in 1793, also stated that there was no money in circulation but gold dust and shells (ibid. 118).

[93] Denham and Clapperton, op. cit. 220.

[94] H. Barth, op. cit. II, 311.

[95] S. Passarge, *Adamawa* (Berlin, 1893–5), 214, 475 ff.

[96] G. Rohlfs, loc. cit.

[97] R. Caillié, *Travels through Central Africa to Timbuctu*, II (London, 1830), 94.

[98] H. Barth, op. cit., German ed. V, 22–3; see also R. Palmer, *Bornu, Sahara, and Sudan* (London, 1936), 67, for an eighteenth-century chronicle: 'the chief gave a horse and a million cowries to him'. Newbury, loc. cit. 233 ff., quoted Col. Mircher's information from a Ghadames merchant *ca.* 1862 who paid 10,000 cowries customs at Agades.

[99] A. Raffanel, *Nouveau Voyage au pays des nègres*, I (Paris, 1856), 233.

tury; it is not certain that they were in use there before the area came under the influence of El Hadj Omar. Park, who passed through Kaarta in 1799, makes no mention of the use of cowries there.[100] Lenz found cowries in use at Kala Sokolo, west of Timbuctu, in 1880,[101] but by the beginning of the twentieth century they had disappeared from the area north of Kayes–Nioro–Mopti–Dori–Sansanne Hausa, except in Timbuctu itself.[102]

Southwards from the Niger, cowries appear to have spread along well-defined routes. In the nineteenth century they were found along the Tengrela–Kong–Bonduku route, but apparently not more than a small distance to the south-west of that route.[103] They were also found along the routes leading southwards through Doré, the Mossi towns, Bobo Dioulasso, Walewale, Yendi and Sansanne Mangu, and south by way of Salaga and Kete Krachi and by way of Kintampo.[104] They were used on the west–east route through Bonduku, Salaga, Djougou, Nikki and Boussa[105] and its alternatives through Gurma or Bembereke. There were areas between the main routes, however, where cowries did not penetrate, at least not before the rubber boom of the 1890s—parts of the hill country of Togoland and along the present eastern border of Ghana, for example.[106] Barth believed that in his time the cowries of Doré in the Niger Bend were brought by Wangara merchants from Sierra Leone or Rio Nuñez.[107] There is no evidence for this; the normal route was certainly through Salaga, but the long-drawn-out troubles there may well have disrupted the normal supply route at this time, and it is known that cowries were sometimes carried to the cowrie currency area through countries in which they were not current. In the 1890s Monteil found that Doré cowries were being brought by Mossi traders; Binger had met these traders at Walewale.[108]

Meeting these northern cowries (which were never strung on strings or pieces of grass as was usual nearer the coast) were the shells which spread northwards from the Guinea Coast between Whydah and the Niger Delta. These cowries, brought in the early sixteenth century by the Portuguese, and in the first place by way of São Tomé,[109] were in use on the coast in the sixteenth century, in both the Benin and Forcados areas; by the seventeenth century they were in use at Whydah and Ardra,[110] and we have discussed above the possibility that cowries may have been already in use in these areas before the arrival of the Portuguese.[111] In the seventeenth

[100] Mungo Park, 1799, op. cit., passim.

[101] O. Lenz, *Timbouctou*, French ed. (Paris, 1887), 225.

[102] E. Baillaud, *Sur les routes du Soudan* (Toulouse, 1902), 70.

[103] R. Caillié, op. cit. French ed., II, 38, 72, See also L. C. Binger, op. cit. I, 207.

[104] L. C. Binger, op. cit. I, 498 ff. H. Klose, *Togo unter deutschem Flagge* (Berlin, 1899), 362 ff.

[105] J. Dupuis, op. cit. cxiv.

[106] H. Klose, op. cit. 362 ff. See also David Asante's diary of 1884 edited by Christaller, 'Journey to Salaga and Obooso', *Mitteilungen Geog. Gesellschaft zu Bern*, 1886.

[107] H. Barth, op. cit. III, 202.

[108] P. L. Monteil, op. cit. 123.

[109] A. F. C. Ryder, loc. cit. 301 ff.

[110] M. Polanyi, *Dahomey and the Slave Trade* (Seattle, 1966), 126 ff.

[111] Ibid. 181, and passim; see also Barbot, Atkins and Berbain, op. cit.

and eighteenth centuries Whydah was the main centre of cowrie imports, though cowries were in use as far west as Lay (a little west of modern Ada, west of the Volta) in the 1680s,[112] and in Christiansborg by the early eighteenth century, and probably in the later seventeenth century.[113] They reached Bereku and probably Winneba in the early nineteenth century,[114] and Anomabu, but never quite as far as Cape Coast, in the years of the palm-oil trade with the Fanti hinterland in the middle nineteenth century.[115] Eastwards, cowries do not seem to have established themselves in the eastern half of the Niger Delta, except to a limited extent in the late nineteenth century. Robertson, however, specifically stated in 1819 that cowries were formerly current in Calabar.[116] It is true that ships trading there sometimes kept accounts in cowries, but they seem to have been trading in other goods, and using cowries solely as a money of account;[117] apart from this, no other evidence has been found to support Robertson's assertion.

From the coast, cowries spread inland along the main arteries of trade. They were in use in the nineteenth century up to the lower Niger,[118] throughout Yoruba country,[119] and in northern Ibo country;[120] they may well have been in use long before this. They spread up the Benue only during the nineteenth century—in mid-century, they were current 'only if Hausa traders were met with'.[121] It is not known at what date they reached Nupe; if Hiskett is right in suggesting that they reached Hausaland from Nupe in the early eighteenth century,[122] they presumably were in use in Nupe before this.[123] They probably reached both Nupe and Yoruba country from Dahomey, where they were current inland by the eighteenth century, and probably earlier.[124] The inland country behind the Volta delta was using cowries in the nineteenth century, at least as far north as Nkonya;

112 J. Barbot, op. cit.

113 J. Rask, *Regse beskrivles Til og fra Guinea* (Tronhjen, 1754), 84.

114 H. Meredith, op. cit. 183. H. C. Monrad, op. cit. 262.

115 B. Cruikshank, op. cit. II, 58; see also G. A. Henty, *March to Coomassie* (London, 1874), 260.

116 G. A. Robertson, *Notes on Africa* (London, 1819), 297.

117 E. Donnan, op. cit. II, 529.

118 M. Laird and R. A. K. Oldfield, op. cit. 341: '[Cowrie] will purchase any article from Eboe to Boosa, and passes current in every part of the interior.'

119 E.g. Miss Tucker, *Abeokuta, or Sunrise within the Tropics* (London, 1853), 26, and P. B. Bouche, *Sept ans en Afrique occidentale* (Paris, 1885), 198 ff.

120 *Petermanns Mitteilungen* (1861), 77.

121 T. J. Hutchinson, *Narrative of the Niger, Tschadda and Binue Exploration* (London, 1855), 253.

122 M. Hiskett, loc. cit. 356 ff. It should be noted that Hiskett's 'first and...the only concrete evidence of their entry by a direct east to west route' (p. 357) is based on a misreading of Barth's reference. It is almost certain that Barth's reference to the Bahr al Ghazal was as the source of the natron, and not of the cowries as well; and, in any event, the reference is to the Bahr el Ghazal, which flows into Lake Chad from the north-east, which is north rather than east from Bagirmi.

123 E. M. Chilver, 'Nineteenth-century trade in the Bamenda grassfields, Southern Cameroons', *Afrika und Übersee*, XLV, no. 4 (1961), 233–58. I am grateful to Dr A. G. Hopkins for this reference.

124 A. Dalzel, *History of Dahomey* (London, 1783), 135.

there would seem to have been a general use of cowries in the outlying parts of Ashanti, but not in metropolitan Ashanti.[125] They are referred to as in use in Akwamu and Kwahu, and at the market centre in Salaga, but Bowdich seems also to refer to their use, probably for small market transactions, in Kumasi itself.[126] By mid-century Cruikshank stated categorically that cowries were not allowed in Ashanti,[127] and other writers shared his opinion.[128] It is possible that there was a change of policy in Ashanti, at the same time as Muslim traders were excluded from Kumasi, after Bowdich's time.

By the late eighteenth century, and very probably earlier, the northern cowrie zone and the zone of the coastal cowries had met. Adams, writing from experience gained on the coast in the last decades of the eighteenth century, but after Park's travels had revealed the use of cowries in the Senegal–Niger region, wrote that cowries were always in great demand at Whydah, Ardra and Lagos: 'At which places they are not only the medium of exchange, but from whence they are also sent to Dahomey, Hio, Hausa, Jaboo, and into the very heart of North Africa, where it is known they are the circulating medium.'[129] The figures given by Dupuis (discussed above) for the exchange rate between cowries and gold make sense only on the assumption that cowries were coming into the system at Salaga; and it is to be noted that his informant also gave the exchange rates between gold and cowries at Whydah and Benin.[130] While it seems impossible to obtain positive evidence, we shall suggest in a later article that there may have been a single gold–cowrie currency system from the Niger Bend to the Guinea Coast some three centuries or more before Dupuis's time. This would strengthen the suggestion that cowries were already in use on the Benin coast when the Portuguese arrived.

By the late nineteenth century, when Zanzibar cowries were coming in to the coast at Lagos and Whydah, these cowries were going at least as far north as Walewale, in what is now Northern Ghana, where they sold at a discount of some 10%.[131] Barth, in about 1850, was convinced that the demand for cowries consequent upon their introduction into Bornu had raised the price of cowries in Yoruba country.[132] In a few places, the Zanzibar cowries never penetrated. They were never in use on the Gold Coast west of the Volta, probably because the government refused to accept them; one trader recalled a shipment being thrown into the sea.[133] In Ibo country east of the Niger, apart from the town of Onitsha itself, the larger cowries never penetrated.[134] In both these areas the small cowries retained a

[125] M.-J. Bonnat, Diary, loc. cit.
[126] T. E. Bowdich, *Mission from Cape Coast Castle to Ashantee* (London, 1819), 86.
[127] B. Cruikshank, op. cit., 42 ff.
[128] J. E. Herz, loc. cit. 11–28.
[129] J. Adams, op. cit. 263.
[130] J. Dupuis, op. cit. cxiv.
[131] L. E. Binger, op. cit. II (1892), 51; also L. E. Binger, loc. cit. 1889.
[132] H. Barth, op. cit. II (German ed.), 161; see also J. E. Herz, loc. cit.
[133] O. Schneider, op. cit. 146; A. G. Krause, loc. cit. 371 ff.
[134] G. T. Basden, *Among the Ibos of Nigeria* (London, 1921), 198.

value double that of the larger cowries on the opposite bank of the river. Resistance to the introduction of the larger cowries is also reported from Adamawa.[135]

COWRIE ARITHMETIC

Corresponding to the two main routes by which cowries came to West Africa—across the Sahara to the Niger Bend, or by sea to the Guinea Coast between the Niger and Accra—there are two basic systems of counting cowries. In the north, from Bambara to Timbuctu and indeed into Hausaland, and from Timbuctu as far south as Salaga and Kong, and in much of Yoruba country, cowries used as currency were not strung, but counted individually, in groups of five. On the Guinea coast, on the lower Niger, in Dahomey, in the inland Ewe country, in Kwahu, cowries were generally counted in 'strings', the cowries being pierced and threaded on a string or grass-stalk, in 40s, 100s, or whatever number was the local convention. In Bornu, where cowries were introduced as an act of state in the middle of the nineteenth century, the counting system is unique, having some affinities with both the northern and the southern systems. Apparently it derived, in part, from a pre-existing system of counting copper coinage. Bornu cowries were counted in groups of four, and in so-called '*rotl*' (pounds) of 32 cowries. The Ibo, on the lower Niger, also had a unique system of counting cowries, with a basic unit of six cowries which is not found elsewhere.

It is not known whether cowries were strung in the Niger Delta at the time of the early Portuguese trade in cowries there; they were certainly counted in units of 40, corresponding to the commonest value for the later 'string', but named *galinhas* (hens), a name later used for the larger unit of five strings.[136] The Bubo of Fernando Po, a very backward Bantu-speaking people, used a quite different system of shell money, but counted in strings, right up to the later nineteenth century;[137] it seems not unlikely that the coastal traders may have been using some such system before the arrival of the Portuguese. Ryder remarks on the use in Portuguese trading accounts of a local word for cowries in monetary units; this would make more sense if these cowries were either strung or pre-counted in units corresponding to strings.

In the Niger Bend country, cowries seem to have been originally used as 'small change' for a gold currency. One of the disadvantages of gold as currency, as several writers have pointed out, is that it cannot be subdivided beyond a certain point; very small purchases have to be made in a different medium. The use of cowries in this way presents no particular difficulty—five or ten cowries for a handful of grain or for a ferry across a river are quickly counted. The use of cowries for larger transactions depends on a rapid system of counting them.

[135] O. Schneider, op. cit. 135, quoting Dr E. Zingtgraff.
[136] A. F. C. Ryder, loc. cit. 131 ff.
[137] T. J. Hutchinson, *Impressions of West Africa* (London, 1858).

For a period of two centuries and probably longer the 'normal' exchange rate betwen the gold *mithqal* and cowries at Timbuctu seems to have been 3,000 cowries to the *mithqal*; sums of more than 3,000 cowries would thus probably have been paid mainly in gold in normal retail trade. It is perhaps for this reason that the north never adopted the custom of having cowries ready-strung in larger units.

Cowries in the north were generally counted in groups of five, which could be withdrawn from a heap by the fingers of one hand. These groups were then put together in piles of 60, 80 or 100, according to local convention; to make matters more confusing, the pile was generally known by the word for hundred. In Bambara, the 'hundred' was 80, at least when cowries were being counted. There is no record of how old this system of counting may be; it was certainly in existence at the end of the eighteenth century—Park noted that in Bambara 'it is curious that in counting the cowries, they call 80 a hundred; whilst in all other things they calculate by the common hundred'.[138] René Caillié, some thirty years later, writing of the first Bambara village he reached (which was the first point at which he found cowries in use), says: 'The price of a fowl is 80 cowries, a number which is called *keme*. To describe our 100, they say 80 and 20, or *keme nimonga*.' [139] A little later Caillié refers to a transaction in which a Bambara man 'counted [him] down 100 cowries, adding to this 20 of his own, which were equivalent to 100 of our country'. Almost every French traveller has something to say about the Bambara counting system, which seems to have offended the logical French mind; the French administration, however, was forced to accept the system, as the Islamic state of Macina had done before them.

The fullest account of the Bambara counting system is given by the French officer Mage, who travelled nearly to Segu in 1866. Cowries, he writes, have a special system of numeration: 'They are counted by 10s and at first sight it seems that the system of numeration is decimal.' But they were counted according to the following table:

$8 \times 10 = 100,$
$10 \times 100 = 1000,$
$10 \times 1000 = 10{,}000,$
$8 \times 10{,}000 = 100{,}000,$

which meant that 100,000 was in reality only 64,000, that 10,000 was only 8,000, that 1,000 was only 800, and 100 only 80,

but with practice one learns to count fairly rapidly, even with this system. As to the people of the country, their mode of operation is quite simple. They count by five cowries at a time, which they pick up with dexterity and a speed only

[138] Mungo Park, 1815, op. cit. 145. Cf. E. Mage, *Voyage dans le Soudan occidental 1863–6* (Paris, 1868), who says that kola was counted like cowries in Segu. Binger, op. cit. 1, 142, further south, reports: 'In this part of the Soudan, when referring to kola, the first large unit is 100, whereas throughout Samori's state it is only 80.'

[139] R. Caillié, op. cit. 1 (English ed.), 373, 390.

acquired by long practice; when, taking them in this way, they have counted sixteen times five, they make a pile of them, that is 100. When they have five of these piles they put them together, make five more, join them all together, and that is 1000. Traders and women, to avoid mistakes, usually begin by making a lot of little piles of five cowries and put them together in eights to make a half-hundred...[140]

By Mage's time all payments were being made in cowries, the gold currency having completely gone out of use. A larger unit of '20,000' cowries, known here as a '*captif*' (i.e. a slave), had been added to the system. Prices of actual slaves ranged from 4,000 to 40,000 cowries; the conventional *captif* corresponded approximately to a man's load of cowries, and we shall meet a similar unit again in other areas under other names. Fractions of a *captif* were reckoned in cowries.

Lenz, who passed through the area in 1880, wrote: 'In Segou, Massina, and to the north as far as Timbouctou, the number of 16 ×5 shells forms a sort of unit which is called not 80 but 100.'[141] This seems to imply that the Bambara hundred was current in Timbuctu in Lenz's time, but Barth in the 1850s, and Caron in the 1880s, both stated that the true hundred was in use in Timbuctu.[142] Possibly it depended on which party was in power there, or even upon the quarter one stayed in and the origins of one's host. Lenz went on to remark on the speed with which sums of 40,000 or 50,000 cowries were counted, but he adds that such a method of payment would only be possible in countries where the value of time is unknown.

Two theories are held as to the origin of this Bambara system of counting. Monteil and Ganay,[143] pointing to the existence of a Mandingo hundred of 60, as well as the Bambara hundred of 80 and the Muslim hundred of 100, suggest that the entire counting system of these territories was originally to a base 6, 8 or 10, with a primary unit of 5 or 10, due to the use of the fingers in withdrawing cowries from the heap. Monteil believed that the Bambara system had been extending at the expense of the others, and that the *ba* had become 10 ×8 in place of an earlier 8 ×8.

Other writers imply that the reckoning of 80 as 100 is a form of discount designed to favour petty trade. Einzig holds this view explicitly,[144] basing his theory on Baillaud,[145] who is not quite so explicit. On this theory, if five bars of salt are sold in a single transaction for 'a hundred thousand' cowries, the buyer pays only 64,000; if the salt is re-sold retail in small quantities, the seller receives the full 100,000. Similarly, a petty trader can

[140] E. Mage, op. cit. 171.

[141] O. Lenz, op. cit. II (French ed.), 162.

[142] H. Barth, op. cit. IV, 289 n.; E. Caron, *De Saint Louis au port de Tombouctou*, 3rd ed. (Paris, 1893), 287: 'De Safai à Tombouctou on employe le système décimal.'

[143] Ch. Monteil, 'Le nombre et la numération chez les Mandes', *L'Anthropologie* (bound in *Soudan and Guinea*, collection of articles in the Balme Library, Legon), 498; S. de Ganay, 'Graphes Bambara des nombres', *Journal de la Société des Africanistes*, XX (1950), 299. See also M. P. Marti, *Les Dogon* (Paris, 1957), 30, recording a symbolic figure representing both Mande and Bambara systems.

[144] P. Einzig, *Primitive Money* (London, 1949), 191 ff.

[145] E. Baillaud, op. cit. 71.

buy goods for 80 cowries which he can re-sell for 100. Soleillet, who was in Bambara shortly after Mage, puts it that for 100 and above there is a reduction of 20%; 'for 99 cowries one pays 99, for 100 one pays only 80, and for 1000 only 800'.[146] If this theory is correct, on the basis of Mage's table there should be, in addition to petty traders buying in quantities of a few hundred cowries, a class of traders who buy in units of 'a hundred thousand' cowries, for which they pay 64,000 cowries. In Mage's time 100,000 cowries would buy perhaps two horses, five to ten cows and about five slaves, or five bars of salt. Livestock dealers, slave-traders and salt traders might therefore all be buying in units over 100,000 cowries and selling in units below this amount. (In Mage's time 100,000 cowries would be worth something of the order of £10.) A system such as is suggested would allow a profit for the dealer within a system in which prices are fixed. Such adjustment in the arithmetic of cowries and other currencies used by non-literate traders is not uncommon.

The existence of the 'Mandingo hundred' attested by Park and by later writers,[147] suggests that different basic systems of counting may originally have existed; we shall suggest later, in connexion with the value of cowries, a possible alternative explanation. It is interesting to note that in the early twentieth century, when the exchange rate had been established by the French administration at 1,000 cowries to the franc, fluctuations in the exchange were achieved by paying the cowries at the Mandingo rate (600), the Bambara rate (800) or the Muslim rate (1000), regardless of the area of the transaction or the origins of the merchants.[148] It is not impossible that a similar usage had grown up in the period in which the gold–cowrie exchange had been stabilized. Had the Bambara hundred been part of the basic arithmetical system of the country it would surely have been used in all counting; but Park stated that all other things than cowries were counted by the common hundred. In some areas, kola was also counted by the Bambara hundred, but in others it went by the 'kola hundred' of 100, which seems to bear out the contention that the Bambara hundred belonged properly to the cowrie system, and must be explained in terms of cowries. Whatever the explanation, it could be equally applicable to the kola trade in areas where kola was practically a currency.

Binger referred to the Bambara system as the 'Dioula' hundred, by

146 P. Soleillet, *Voyage à Segou en 1878–9* (Paris, 1887), 235.

147 Mungo Park, op. cit. 1815, 145: 'Sixty is called a Mandingo hundred.' L. C. Binger op. cit. I, 162, reports the Mandingo hundred at Benokhobougoula.

148 M. Delafosse, *Haut Sénégal–Niger*, III (Paris, 1912), 48: 'in most of the countries the usual exchange is the Banmana or Bambara exchange, i.e. 800 cowries for 1 franc; in some areas there is the Malinke exchange, i.e. 600 cowries for a franc; and in a certain number of towns inhabited by Dioula, the Muslim exchange, 1,000 cowries for a franc. In addition, it must be noted that, in a single country, the price of cowries may vary according to the abundance or shortage of this money, and change from the Malinke exchange to the Muslim exchange.' Delafosse adds that there are Africans who go in for cornering cowries to raise the exchange rate so as to dispose of their stocks at advantageous rates.

which name it was apparently known in Dafina—by contrast to the Mossi hundred of 100.[149] Dafina, to the west of the Mossi country, was a boundary area between the systems, where it was necessary to specify which system was to be used in any transaction. East of Timbuctu and Dafina, cowries were normally counted in true hundreds. At Kong, the system was based not on 10 and 100, but on 20 and 200, *Moukha* and *sira*; but multiples of ten were also used. The practice was to count cowries by fives into piles of 200. Binger, who gives a full account of Kong practice, adds that credit was granted from one great market to the next. The seller had to come and count and fetch the cowries from the buyer—'the seller, having the profit, must also have the trouble'.[150] The Yoruba system was similar to that of Kong: 20 were called *ogun*, 200 *igba*, 2,000 *egba*, and 20,000 *okekan* (one bag). But at Lagos, cowries were counted in strings of 40 (*ogoji*), heads of 2,000 (*egna*) and bags of 20,000 (*okekan*). The 200 unit was recognized, however, as a discount of two cowries was allowed on every five strings.[151]

In Hausaland cowries were always counted in tens or twenties and hundreds. Barth described a merchant at Tasawa, north of Katsina,

> who was just occupied in that most tedious of all commercial transactions in these countries, namely, the counting of shells; for in all these inland countries of Central Africa the cowries or kurdi (*Cypraea moneta*) are not, as is customary in some regions near the coast, fastened together in strings of 100 each, but are separate, and must be counted one by one. Even those 'takrufa' (or sacks made of rushes) containing 20,000 kurdi each, as the governors of the towns are in the habit of packing them up, no private individual will receive without counting them out. The general custom in so doing is to count them by fives, in which operation some are very expert, and then, according to the amount of the sum, to form heaps of two hundred (or ten Hawiyas [20, which Barth says 'seems originally to have been the highest sum reached by the indigenous arithmetic'] or a thousand each. Having at length succeeded, with the help of some five or six other people, in the really heroic work of counting 500,000 shells, our friend went with us...[152]

Baba of Karo, in the Zaria area, gives a particularly vivid account of counting cowries from the point of view of the women traders:

> To count cowrie shells they spread them out on the floor; they counted in groups of five. Ten groups of five were 50. They collected the groups of 50 into groups of 200. Ten compounds of 200 were 2000. When they were arranged in compounds you started to sew up matting and pour in the cowries. One mat, 20,000 cowries; that was a man's load, and a strong man would soon get tired of carrying it...Counting money was the work of a chief's retainers—you can imagine there were a lot of them.[153]

[149] L. C. Binger, op. cit. I, 498.

[150] Ibid. I, 309; M. Delafosse, op. cit. 48, gives a similar account of terms 'in Mandingo and countries under Mandingo influence' for packets of 10, 20, 100 and 200 cowries.

[151] A. Mann, 'Notes on the numeral system of the Yoruba Nation', *J. Anthrop. Inst.*, 1887, 59 ff.; J. E. Herz, loc. cit. 23, quoting *Lagos Almanac.*

[152] H. Barth, op. cit. II, 38 ff.

[153] M. F. Smith, *Baba of Karo* (London, 1954), 80.

The unit, 20,000 cowries or one bag, would have weighed 50 lb. while the Maldive cowries were in use exclusively—a reasonable man's load for long distances. As the heavier Zanzibar cowries came in, the bag became heavier and heavier; a bag filled with Zanzibar cowries would have weighed upwards of 100 lb. By the late nineteenth century bags were mixed, the northward movement of the heavy cowries being to some extent checked by the difficulty of carrying them. Indeed, as we have seen, cowries became so low in value by the 1890s that it was scarcely worth while to transport them at all.

Counting in unstrung cowries, generally in thousands, extended south to the traditional limits of the northern trade—to Salaga (where, at least after coastal traders had begun to penetrate the market, 1,000 came to be known as a 'head'); to Kete Krachi, to Atebubu and to Kintampo (all offshoots of Salaga, so far as the late nineteenth-century trading community was concerned); to Bonduku (where cowries, *mithqals* and Akan gold-weights were all in use together); and, judging by the figures for the cowrie–*mithqal* exchange quoted by Dupuis, all in round hundreds of cowries, to the Yendi–Nikki–Niger route. Between this and the area of strung cowries lies the 'watershed' or 'divide' between the northern and the southern trade.

In Bornu, where cowries were introduced in the middle of the nineteenth century along with the Maria Theresa dollar, a different system of counting was in use. The unit of count was the *rotl*, an Arabic word meaning a pound weight; this unit is believed to have belonged to the copper coinage which was minted in Bornu in the late eighteenth and early nineteenth centuries. The cowrie *rotl* consisted of 32 cowries, or four *gabaga* of eight cowries each (the *gabaga* was the local traditional currency unit, a strip of cloth, which had replaced the copper coinage in the nineteenth century). According to various writers, the 32-cowrie *rotl* had a nominal value of 33, the odd cowrie being set aside to help in the counting, as a sort of tally, or as a discount for the trouble of counting. Three *rotl* would thus make an approximate hundred. This looks very like an attempt to bring a system inherited from currency units which could be physically divided into halves and quarters into relation with the cowrie systems in use on the lower Niger, where strings of 66 and 100 were known. The relation between the *rotl* and the Maria Theresa dollar was never fixed; in Barth's time it was subject to manipulation by powerful speculators, and ranged between 45 and 100 *rotl* to the dollar. Nachtigal put it at 120–130 to the dollar, and Monteil in the 1890s at 135–160.[154] The actual counting of cowries in Bornu was done in groups of four, not in fives as elsewhere in northern West Africa; it is probable that this method of counting goes back to the small copper coins of Bornu.

Cowries on the Guinea Coast were usually counted in 'strings' and 'heads'. On the Forcados River by about 1520, cowries were being counted

[154] J. E. Herz, loc. cit., quoting Nachtigal, *Sahara und Sudan*. Basically similar accounts are given by Barth, op. cit.; Rohlfs, loc. cit.; and P. L. Monteil, op. cit.

in *galinhas* (hens) of 40, and *cabres* (goats), for which Ryder gives the value of 910 cowries. This latter unit is odd in not being a multiple of the smaller unit (as is usual, not only in cowrie units, but also in the conventional values of 'animal' units of exchange, a goat generally being counted as equivalent to a round number of hens). It is possible that a maximum price in *cabres* imposed by the Portuguese was evaded by enlarging the cowrie content of the unit; or possibly a retailer's profit had been built into the system.[155] The unit of 40 cowries long remained the basic unit of the system of reckoning used on the Guinea Coast, though the name *galinha* became attached to the next larger unit, sometimes called the 'bunch' of five strings of 40 cowries. Twenty *galinhas* made a cabesse, or grand cabess, ten galinhas a small cabess. The word 'cabess' is evidently derived from the Portuguese *cabeça*, 'head', and with the change from Portuguese to English as the language of trade on the Guinea Coast at the end of the eighteenth century the basic larger unit became known as a 'head'. The nineteenth-century 'head' was the small cabess of 2,000 cowries, conveniently equated with the silver dollar in the first half of the century.

The usual eighteenth-century cowrie table runs as follows:

40 cowries = 1 toque or toccy
5 toques = 1 galinha or gallina
20 galinhas = 1 grand cabess

The grand cabess is thus 4,000 cowries; in about the year 1726 it was worth £1 sterling, a quarter of an ounce of gold. The weight of a grand cabess of cowries was about 10 lb.[156] The Danes used the small cabess of 2,000 cowries in the eighteenth century, and it is this figure which was adopted for the nineteenth-century 'head', which was known in some parts as the cowrie-dollar. The coastal traders were always anxious to equate cowries with their own currency units. One early nineteenth-century writer calls the 50-string unit of 2,000 dollars a 'grand cabess', and introduced a new unit, the small cabess, of 40 strings.[157] This related to a period when gold was temporarily at a higher value, and this usage did not survive the gold-value to which it belongs. English writers in the mid-nineteenth century were valuing the string at 1*d*.,[158] and one of them goes so far as to write of a head of 48 strings, equivalent to 4*s*.—a valiant attempt to reconcile English currency and cowrie arithmetic.[159] The usual nineteenth-century table runs:

[155] A. F. C. Ryder, loc. cit. 301 ff.

[156] E.g. Bullfinch Lamb in Smith, *A Voyage to Guinea* (London, 1744), 178. The account in Barbot, op. cit., is unfortunately garbled and contradictory; Dapper, op. cit., gives the improbable value of 2 cowries for a *galinha*; if this is an error for 2 strings, this would make sense of one of Barbot's versions. Barbot's account was no doubt 'modernised' before publication half a century later, in this respect as in many others.

[157] H. C. Monrad, op. cit. 263.

[158] E.g. Hill to Grey, Cape Coast, 24 Apr. 1852 (despatch quoted in G. E. Metcalfe, *Great Britain and Ghana*, Accra, 1964).

[159] W. F. Daniell, 'Ethnography of Akkrah and Adampe', *Journal of the Ethnographical Society* (London), 1856.

40 cowries = 1 string
(5 strings = 1 bunch)
50 strings = 1 head[160]

The galinha or bunch was frequently omitted. In the later part of the century we begin to hear of the larger unit, 10 heads or 1 bag. This is the same as the Hausa bag or grass mat; but as it was a convenient unit for head-loading it may have arisen independently on the coast, though there is no reason to doubt the commerical connexion between the Dahomey and Lagos coast and the Hausa country. The bag may also have been connected with the hundredweight sack, containing 10–12 heads of cowries, in which the Zanzibar cowries were imported from the 1850s.[161]

On the lower Niger strings of different numbers were in use during the nineteenth century. Laird and Oldfield reported that the cowries were 'strung together in portions of one or two hundreds, each portion being on a separate string.'[162] According to the Rev. Samuel Johnson, in Yorubaland cowries were 'usually strung by 200 in five strings of 40 each, three of 66 or 2 of 100 each, with a discount of 1 per cent'.[163] In the 1880s, we know that at Lokoja on the Niger cowries were strung in bunches of 1,000, representing 1*s.* at that date. A woman was employed by the trading station at 10*s.* a month (in goods) to count cowries.[164]

An attempt was made by the unfortunate Niger Expedition of 1841 to reform the system:

The natives generally drill a hole, and string them in 100s, 10 of which are tied together, thus facilitating the operation of settling, which is a very tedious affair when detached, as both the buyer and seller insists on counting them; when they differ, it has to be done again. As ours were loose, it would have been an endless trouble to have counted them [the reference is to 160,000 cowries]. Captain Trotter therefore substituted measure for number, and it was easy to make the natives see, by giving a fair quantity, that it was to their profit...[165]

Unfortunately, like so much of the rest of the work of the expedition, this apparently sensible measure did not last. The use of measures instead of number and weight is also reported in eighteenth-century Whydah, and in the late nineteenth century on the Gold Coast. It is still remembered in parts of the Volta Delta, but for some reason it never became a generally established practice. Possibly it offered altogether too much scope for ingenious dishonesty.

[160] G. A. Robertson, op. cit. 274; cf. R. Burton, *Wanderings in West Africa* (London, 1863), 234 n.

[161] C. W. Newbury, *The Western Slave Coast and its Rulers* (Oxford, 1961), 58 ff.

[162] M. Laird and R. A. K. Oldfield, op. cit. 166.

[163] S. Johnson, *History of the Yorubas* (Lagos, 1921; reprinted London, 1966), 118–9. See also Herz, loc. cit. 23, quoting *Lagos Almanac.*

[164] M. Viard, *Au bas Niger* (3rd. ed., Paris, 1886), 231: 'The cowries are pierced and threaded in advance in bunches of 1000' (at Lokoja).

[165] W. Allen and T. R. H. Thompson, op. cit. 1, 350, 460.

In the Ibo country on the lower Niger a quite different system of cowrie counting was in use. Cowries do not seem to have been strung, and were counted in sixes, the group of six being confusingly called a 'head': 'In the counting of cowries the number six is the standard. The one counting squats on the ground before a heap of shells and starts by separating groups of six, and then sweeps ten such piles into one heap to form an *ukwu*.'[166] The basic unit of six is without parallel in West Africa, and can therefore hardly be used as the basis for any general theory of the provenance of cowries, as Jeffreys has tried to use it.[167] The unit of 60 does appear as the 'Mande Hundred', but in the Mande system the next unit is 600, while the next Ibo unit is 1,200.

The counting of cowries in strings and heads was subject to a number of adjustments. In eighteenth-century Dahomey, strung cowries were one cowrie short of the nominal 40, the reward to the stringer for the work of piercing and stringing the shells. Cowries strung at the king's palace by the women there, however, showed a deficiency of three to six cowries per string, so that the king, while nominally paying fixed market prices, obtained goods at a discount of up to 15%.[168] This 'royal' string may be compared to the royal gold-weights of the king of Ashanti, described by Bowdich, heavier than those used by ordinary people, and thus a form of taxation. Both were means of obtaining revenue without complex arithmetical processes, and within the framework of a fixed price system. In the later stages of the nineteenth-century cowrie inflation, part of the loss of value of cowries in Dahomey was taken up by increasing the number of cowries in a string, the number of strings in a head remaining 50. Thus, at Whydah, instead of 40 cowries to the string there were 50; at Alladah and Abomey, 46. The 'royal' string at this period was about 40.[169]

At a period when cowries were worth 1*s*. a head of 2,000 on the coast, and shillings were selling at 1,000 at Salaga, the Salaga 1,000 became known as a 'head'. An intermediate value, in Kwahu, Nkonya and the inland Ewe country, was achieved by using a string of 35 cowries instead of 40, giving a head of 1,750 instead of 2,000.[170] Another example of this type of adjustment is given by a German official who reported that 'on the road to Pessi' one head corresponded to 2,000 cowries; in Scogode (Sokode) and Fasugu to 800–1,000 cowries and in Salaga to 600–700 cowries.[171] Here the 'head'

[166] G. T. Basden, op. cit. 198; See also G. T. Basden, *Niger Ibos* (2nd ed., London, 1966) where the Ibo cowrie table is correctly given. For a very full account of Ibo counting systems, see Jeffreys, 'The diffusion of cowries and Egyptian culture in Africa', *American Anthropologist*, 1948, 45–53.

[167] M. D. W. Jeffreys, loc. cit.

[168] A. Dalzel, op. cit. 214.

[169] J. A. Sketcherly, *Dahomey As It Is* (London, 1874), 28.

[170] Peregaux, quoted Schneider, op. cit. 144–5. Bonnat, M.-J., 'Diary', Jan. 1876, *L'Explorateur*, 1876. While the 35-cowries string formed a convenient bridge towards the higher northern values, it may have originated in connexion with a series of gold weights which were seven-eighths of the weight of the standard troy weights.

[171] R. Buttner, 'Journey in Togoland', *Mitteilungen aus den Deutschen Schutzgebieten*, 1891, 416 ff.

had been equated to 1*s.* or 1 mark, in the way that earlier writers had equated it with a dollar.

The habit of equating round numbers of cowries to existing coins could give rise to very curious arithmetic. For example, in Accra in the 1870s, 12 strings were apparently equated with 6*d.*, giving a head of 2*s.* 1*d.*; but six heads were reckoned at 7*s.*, giving a head of only 1*s.* 2*d.*[172] Similarly, at Whydah in 1849, 120 cowries were equivalent to 3*d.*, but 25,000 cowries were reckoned as £2, not £2. 12*s.*[173] Similar relationships are shown in the following table by Burton, consul for the Bights of Benin and Biafra in the 1860s:

40 cowries = 1 string = $\frac{3}{4}$*d.* to 1*d.*
5 strings = 1 bunch = 3*d.* to 6*d.*
10 bunches = 1 head = 1*s.* 9$\frac{1}{2}$*d.* to 2*s.*
10 heads = 1 bag = 14*s.* to $4

Burton adds that 'the rates are extremely various',[174] and it is obviously very dangerous to base any conclusions about the value of cowries on the equivalences given for small sums. In Lagos by the 1870s this type of equation was complicated by the fact that copper coin was at a discount. Thus the bunch, by then, was worth 1$\frac{1}{2}$*d.* in silver, 2*d.* in copper; 40 strings were equivalent to 1*s.* in silver, whereas a shilling in copper brought in only 30 strings. By the end of the century, when exchange rates had dropped even lower, 1,000 cowries were valued at 3*d.* in silver; but a penny was worth 300 only.[175] By a natural process, cowrie units were applied to the counting of large numbers of other things; 2,000 men, or yams, would be spoken of as 'a head'. In Yoruba the process went a stage further, and the number could be expressed by its equivalent in English money, at this final and utterly debased rate of exchange. Thus 300 can be expressed as 1*d.*, and 1,000 as 3*d.* A man may speak of a yam farm with '3*d.* yams'; while losses in battle are remembered as 'two sacks of men'.[176]

THE COWRIE AS A UNIT OF ACCOUNT

Cowries were always cumbersome to carry, and increasingly so as their value declined in the nineteenth century. Their most important use may always have been as market currencies, to facilitate exchange within the market rather than to carry from one market to another; for this purpose, gold dust, of high value for its weight and bulk, other high-value merchan-

[172] Azettu, priestess of Dente, quoted in Blue Book, *Affairs of the Gold Coast*, no. 3, 1874, enclosure no. 6 in 156.

[173] F. E. Forbes, *Dahomey and the Dahoman* (London, 1851), 110; this might suggest that the string of 50 in place of 40 was already in use at this date for large transactions, as we know it was in the 1870s.

[174] R. F. Burton, *Wanderings in West Africa* (London, 1863), 274 n.

[175] O. Schneider, op. cit. 172.

[176] R. G. Armstrong, *Yoruba Numerals*, NISER, 1962.

dise such as kola and salt, and what Polly Hill has called 'ambulatory purchases'[177]—livestock and slaves—were always more profitable for the merchant. Even within the markets, literate merchants did not always handle actual cowries, but kept detailed accounts in terms of cowries. One such case is reported by Daumas from Hausaland.

You understand that we did not load up with these cowries [in which prices had just been quoted], which were actually only a representative value of the goods sold, of which each of us kept an exact record, where the names of the buyers were written, with their purchase and their debt...later we were re-imbursed by the exchange of objects equivalent to those which we had brought...[178]

By the late nineteenth century the Royal Niger Company had invented a similar system; all the company's accounts in Africa were kept in cowries, but these were given the value of 1*s.* 3*d.* per head—a fairly realistic value when the company started operations, but increasingly unreal as time went on; one traveller reports purchasing cowries at half this figure.[179] The same writer describes the operation of the system: barter, he says is simplified.

by using a head of cowries (value fixed as 1*s.* 3*d.*) as the base. The price to be paid for native produce is fixed for the district from time to time; thus, supposing that a native brings in 5*s.* worth of palm oil, the agent allows him to select Manchester goods, salt, or whatever he wants, to the value of 4 'heads' of cowries, the value of all the articles on sale in the store of course being fixed in English money.[180]

Real cowries, which could be obtained much more cheaply, were not allowed to come into the transaction at all.

The porters of Lugard's 1894 expedition were enlisted at 40 heads a month, nominally 50*s.*; but they were paid in goods at the Company's valuation, and Lugard reckoned that they actually got the equivalent of about £1 a month.[181] Thus the cowrie on the lower Niger had become simply a unit of account—a measure of value without being a unit of exchange—a fate that had long since overtaken the Sierra Leone 'bar' (originally a bar of iron), and the Trade Ounce (originally an ounce of gold).

THE LAST OF THE COWRIES

By the beginning of the twentieth century cowries were being replaced by coin, both in the south of what are now Ghana and Nigeria, and west of the Upper Niger. The process was speeded, both in Northern Nigeria and in the French territories, by the presence of considerable numbers of troops and police who were paid in coin. Lugard (some of whose earlier experiences with cowries have been referred to above) considered the stabilization

177 Polly Hill, 'Notes on the history of the northern Katsina tobacco trade', *J. Hist. Soc. Nigeria*, IV, December 1968, 480.

178 E. Daumas, op. cit., 239.

179 A. F. Mockler-Ferryman, *Up the Niger* (London, 1892), 306.

180 A. F. Mockler-Ferryman, *British Nigeria* (London, 1902), 88.

181 F. Lugard, *Diary*, 256.

of cowrie values of the greatest importance. In his *Political Memoranda* dated 1905,[182] a long note on cowries begins:

Cowries. The cowrie will always remain (as it does in India to this day) a medium of exchange for very small values, but it is most eminently desirable that it should be replaced for higher values by metal coins on account of (*a*) the bulk and weight and the great time taken in counting, (*b*) the wastage involved by breakage, leakage of bags, etc.; and (*c*) its fluctuating value.

In the meantime, Lugard intended to concentrate on the problem of stabilizing the value of cowries, the exchange rate for which in the southern provinces of Northern Nigeria was double the rate in the more remote northern areas. He first prohibited the further importation of cowries, and arranged that Lagos and Southern Nigeria should do the same. He then made provision for every Resident to hold a Reserve of cowries: 'The fluctuations of coinage are controlled and credit maintained in every country by the principle of a reserve, which enables a State to redeem its tokens (whether they consist of paper money, or of coins whose face value is less [*sic*] than their bullion value) at the nominal value which the State has placed upon them.' Residents were to give the standard rate (which Lugard hoped to stabilize at 1200 cowries for a shilling) at all times, whatever the market rate; they could buy if the market rate rose above the standard rate (taking care not to overstock themselves); while if the market rate remained low, cowries could be requisitioned from the Residents of the southern provinces. Cowries were to be transported as government stores, the government recovering the cost of transport by the profit of the exchange. 'This will be shown under Head of Revenue "Miscellaneous".' Lugard recognized, however, that if government could make a profit by playing the exchanges, so also could the traders; it would thus not be possible to reduce the rate in the southern provinces of Northern Nigeria at once. Lugard's memorandum is of interest as one of the very few attempts by a government officer to treat the cowrie as a genuine, if inconvenient, currency subject to the normal laws of economics. Things did not work out quite as expected—British coin was accepted sooner than he had anticipated, and the average rate at which cowries were exchanged remained, in Zaria at least, double the rate Lugard had hoped to establish.[183]

Lugard was nearly right, however, in thinking that cowries would 'always' remain in use for very small transactions. They are so used to this day in a number of West African markets—Wechau in north-east Ghana,[184] in many Bambara markets,[185] in the Dogon country,[186] and among the Senufo of Ivory Coast,[187] to quote only a few cases which happen to have

[182] F. Lugard, *Political Memoranda*, 1905, revised 1906 (Lagos, 1906), 224.
[183] M. Smith, *Government in Zazzau* (London, 1960), 342.
[184] Polly Hill, personal communication.
[185] V. Paques, *Les Bambara* (Paris, 1954), 42.
[186] M. P. Marti, *Les Dogon* (Paris, 1957), 29.
[187] B. Holas, *Les Senoufo* (Paris, 1957), 72.

been noted. In the Accra area, on the other hand, they had reached this position by the 1870s, and even the recent disappearance of the smallest coins (tenths of a penny were issued between the two world wars) does not seem to have brought the cowries back into the market-places, though they are still in use for ritual and magical purposes. In other areas, where cowries are used for bride-price payments and funeral offerings, this has tended to keep them in use also in the markets, at values well above those at the beginning of the century. It has even been suggested that there is an illicit trade in smuggling cowries across the northern frontier into Ghana.[188] During the economic crisis of the early 1930s, when cash receipts from the sale of cash crops dropped enormously, cowries were reported to have reappeared in some Nigerian markets;[189] it is not certain whether this could happen again now, in the event of another monetary crisis (let no one be misled by the adoption by Ghana of a monetary unit, the *cedi*, named after the Ashanti name for cowries).

A few cowries have thus survived to outlive the colonial régimes which tried so hard to oust them; but they are but a shadow of the currency which oiled the wheels of West African commerce for centuries; and most West Africans look on cowries as a shameful relic of a primitive past rather than as a memorial to their ancient commercial enterprise.[190]

[188] Personal communication from Ghanian informants.

[189] A. H. M. Kirk-Greene, loc. cit., 132–50.

[190] I should like to acknowledge the facilities made available to me at the Institute of African Studies, Legon, by courtesy of Mr Thomas Hodgkin; Professor Ivor Wilks has given me most generous help and encouragement with this study; and I must also thank the large number of people who have taken the trouble to give me information and references about cowries.

PART II

COWRIES AND GOLD

IBN Battuta reported, in the fourteenth century, that cowries were exchanged at 1,150 to the gold *dinar* at Mali and Gao.[1] If his *dinar* was the standard *mithqal* of 72 grains (which a learned Muslim would probably use), then the old 75-grain *mithqal* of the gold-traders would exchange at the convenient round number of 1,200 cowries; a single cowrie would be worth approximately half a pre-1914 farthing.

A very different value is given by Leo Africanus, in the second decade of the sixteenth century—400 shells to the 'ducat', which he elsewhere appears to define as $\frac{3}{20}$ oz., or a *mithqal*.[2] This gives cowries a value three times that of Ibn Battuta, and may be a famine value, such as is quoted in succeeding centuries. It is very unfortunate that this value is aberrant; Leo visited Timbuctu just at the period when the Portuguese were beginning to bring cowries to the Guinea coast by sea, and it would be useful to know the value in the Niger Bend before the effects of these imports could have penetrated so far. It would also be interesting to know if Ibn Battuta's value survived into Portuguese times.

After the Moroccan invasion at the end of the sixteenth century, the value of cowries settled at 3,000 to the gold *mithqal* at Timbuctu and Jenne.[3] Park found the same figure two centuries later at Sansanding, and Caillié seems to have found a similar value at Timbuctu and Jenne as late as 1828. Charles Monteil, studying Jenne in 1902, believed that this value persisted apart from famine values, until the time of Sheikh Ahmadu (*ca.* 1827–42) and then declined to 4,000 to the *mithqal;* by the end of the century the figure was 6,200.[4] A late eighteenth-century writer reported a figure of 2,500 cowries to the *mithqal* in Katsina;[5] if this was the small *mithqal*

[1] Ibn Battuta (Muhammad ibn abd Allah), *Voyages d'Ibn Batoutah*, IV (Paris, 1922), 122, 435.

[2] Leo Africanus, *Description d'Afrique* (Paris, 1956), 469. It is possible that Leo's 'ducat' may be a smaller coin than the gold *mithqal*. Caillié in the nineteenth century found a silver '*mithqal*' at El Arouan one-third of the value of the gold *mithqal*. If Leo's coin was of a similar value, not only would his exchange rate be the same as Ibn Battuta's; his salt prices would also be similar.

[3] Mahmud Kati, *Tarikh el-Fettach* (Paris, 1913), 319.

[4] Ch. Monteil, *Djenné* (Paris, 1932), 280; M. Park, *Journal of a Mission to the Interior of Africa in the year 1805* (London, 1815), 161; R. Caillié, *Journal d'un voyage à Tombouctou et à Jenné*, II (Paris, 1965), 220, 426.

[5] S. Lucas in *Proceedings of the Association for Promoting the Discovery of the Interior Parts of Africa*, I (London, 1810), 186.

of 60 grains, this would be equivalent to a value of 3,000 to the 75-grain *mithqal.*

On the Guinea coast, Polanyi's assertion that the rate of 32,000 cowries to a Troy Ounce of gold dates back to 1700 or earlier cannot be sustained; it is apparently based on cowrie 'values' given by Bosman (1,000 for 2*s.* 6*d.*) and Barbot (200 for 6*d.*).[6] These values are based on the prime cost of cowries, and correspond to those quoted by Atkins in 1721: 'Cowries... bought at 1/– and sold here at 2/6 per pound' (a pound of cowries was about 400). Atkins also refers to 'a grand quibess [4000 cowries] which answers to 25/–'.[7] A few years later, Smith was quoting the value of the grand cabess at Whydah at £1 sterling; this remained its nominal value for some years.[8]

It is only exceptionally that gold and cowries were found together at Whydah, or anywhere on the coast except at Christiansborg, where the gold and cowrie systems met. Rask, a Danish chaplain who was at Christiansborg in the early eighteenth century, gives the value of one cabess as two *rigsdaler*, or one-eighth of an ounce.[9] The Danish cabess is usually 2,000 cowries (as distinct from the *stor cabess* of 4,000 cowries). This exchange rate is thus 16,000 cowries to the gold ounce, similar to Smith's figure. Rømer, giving a list of current prices at Christiansborg in 1749, quotes 20 lb. of cowries at 8 *rigsdaler* or half an ounce;[10] 20 lb. of cowries would be 8,000 shells, so this again works out at 16,000 to the ounce, which is specifically stated to be an ounce of gold; the 'trade ounce' has not yet made its appearance.

From Rask we also learn that at Christiansborg, the 'string' of 40 cowries was known as a *damba*, and 80 cowries were a *taku*;[11] these are the names of two of the smaller Akan gold-weights; in the eighteenth century, 24 *damba* or 12 *taku* gold made one gold ackie, sixteen ackies made an ounce.

Cowries		Gold	
40 cowries	= 1 *damba*		
2 *damba*	= 1 *taku*	2 *damba*	= 1 *taku*
25 *damba*	= 1 *rigsdaler* (= 1 ackie)	24 *damba* or 12 *taku*	= 1 ackie
50 *damba* or 2 *rigsdaler*	= 1 cabess		
8 cabess	= 1 gold ounce	16 ackies	= 1 ounce.

Evidently the cowrie system has been rounded off, to simplify counting, into hundreds and half-hundreds instead of multiples of twelve. The small

[6] K. Polanyi, *Dahomey and the Slave Trade* (Seattle, 1966), 168.

[7] J. Atkins, *Voyage to Guinea* (London, 1735), 112. Atkins's figures are confirmed by Houston (1825), quoted in E. Donnan, *Documents Relating to the Slave Trade of America*, II (Washington 1930–1), 290: '...cowries which were always formerly current money at five ackies of gold for a grand cabess of bougies...'.

[8] W. Smith, *A Voyage to Guinea* (London, 1744), 178.

[9] J. Rask, *Rejsebeskrivlse til og fra Guinea* (Trondjen, 1754), 257.

[10] L. F. Rømer, *Nachrichten von der Küste von Guinea* (København und Leipzig, 1767), 277.

[11] Rask, *Rejsebeskrivlse*, 84.

discrepancy may be accounted for by the customary 4% brokerage charged on exchanging gold for cowries.

This relationship opens up an interesting speculation. In Coast Portuguese (the trade language of the Guinea coast up to the eighteenth century), a string of cowries was a *toque*, or *tokky;* could this have been originally *taku* (which Bowdich spelt 'tokoo')?[12] If one *toque* cowries was equivalent in the fifteenth or early sixteenth centuries to one *taku* gold, then the exchange-rate for one standard *mithqal* (three-twentieths of an ounce of 16 ackies) works out at 1,152 cowries. Ibn Battuta's figure, in the fourteenth century, was 1150. This seems almost too good to be true.

If it is anything more than a rather remarkable coincidence, we would have to assume not only the survival of the fourteenth century rate into the late fifteenth century, but also a long-distance trade connexion between the coast and the Niger Bend based on a constant exchange rate between cowries and gold. There is evidence for the trade connexion, but not, as yet, for the use of cowries alongside gold on the Guinea coast. The hypothesis of a constant gold/cowrie exchange in the years around 1500 might explain some of the oddities of cowrie arithmetic, and also the various gold units in use. The 75-grain *mithqal* would be twelve 'Muslim hundreds' or 1,200 cowries; the small *mithqal* or *gros*, twelve 'Bambara hundreds' or 960 cowries; the *soa* (basic weight of the Ashanti system, but probably originating north of the forest) would be twelve 'Mandingo hundreds', or 720 cowries; and the coastal ackie would be twelve strings or 480 cowries. An ackie of goods at the coast would sell for a *soa* at the first great market; for a *gros* at the second, and for a large *mithqal* on the Niger bend. At each market, goods could be exchanged for their equivalent in cowries, to provide small change for local purchases, and to pay tolls and brokerage.

This system implies a northward flow of goods in exchange for a southward flow of gold; so far at least as its extension to the coast is concerned, it would therefore probably date from the arrival of the Portuguese.

The large imports of cowries in the sixteenth century would tend to lead to a depreciation of cowries (as is known to have happened in the Congo); a system linked to gold would be much more stable, but only so long as gold continued to flow; in the event of an interruption to the gold trade, the whole system would, perforce, go off the gold standard. Something of the sort happened in the second half of the eighteenth century, when, through the medium of the Trade Ounce, the gold-value of cowries halved during a period of interruption of the normal gold trade. It would thus be interesting to see if there is evidence of such a breakdown in the sixteenth or early seventeenth centuries. It has been suggested that a punitive expedition was sent to the gold market at Begho in the sixteenth century 'occasioned by the

[12] T. E. Bowdich, *Mission from Cape Coast Castle to Ashantee* (London, 1819), 330.

cessation in the flow of gold from Begho to Mali', which may have been part of a more general breakdown of the gold trade.[13]

The exchange on the coast in the early eighteenth century, as we have seen, was 16,000 cowries to the ounce; in the Niger Bend, the exchange rate was 20,000 to the ounce (3,000 to the *mithqal*). It thus seems that, in spite of the much higher cost of importing cowries across the desert, they were actually of lower value in the Niger Bend than on the Guinea coast. This may, however, be an illusion. The whole Bambara country east of the inland delta of the Niger normally counted by Bambara 'hundreds' of 80, which would bring '20,000' to 16,000, the same figure as on the coast; this reckoning extended as far as Jenne, and may even have reached Timbuctu, before the Tuareg conquest.

Table 1. *Cowries to one ounce of gold*

	Niger Bend	Coast
16th century	8,000	—
c. 1500	(2,667)	?8,000
c. 1700–50	'20,000' ?16,000	16,000
c. 1780	20,000	32,000
c. 1800–16	20,000	25,600
c. 1820	20,000	32,000
c. 1850	25,000	38,400

In the mid-eighteenth century, on the Guinea Coast, there was a major interruption of the gold-trade; the price of gold rose, not only in terms of cowries, but also in terms of trade goods and of silver dollars. At the same time, there was probably a small decline in the value of cowries.[14] Two Danish valuations, for 1750 and 1780, in terms of the Danish (silver) dollar, may give the amount of the decline. In 1750, when the gold ounce was 16,000 cowries, the silver dollar was 1600; thirty years later, with the gold ounce at 32,000 cowries, the dollar was 2,000.[15] In the Niger Bend, a similar decline in cowrie values probably took place, represented by a change to 'Muslim hundreds' from 'Bambara hundreds'—it was one of the conveniences of this system of reckoning that it enabled fluctuations in the rate of exchange to be taken up without complicated arithmetic. There is no

[13] I. Wilks, 'A note on the chronology, and origins, of the Gonja Kings', *Ghana Notes and Queries*, 8 Jan. 1966, 28. If there had been an interruption in the flow of gold from the southern goldfields, both the Songhai bid to control the source of their salt supplies (normally obtained in return for gold) and the Moroccan invasion of 1591 would be understandable.

[14] M. Johnson, 'The ounce in eighteenth-century West African trade', *J. Afr. Hist.* VII, part 2 (1966), 202–3. The decline in real value of cowries parallels the small decline in the real value (as distinct from the gold value) of the trade ounce.

[15] G. Nørregard, in *Vore Gamle Tropenkolonier*, ed. J. Brønsted, II (Købenahaven, 1954), 563.

record of a change in local gold values comparable to that on the coast (this was a period of much disturbance in the trans-Saharan trade, owing to troubles in Morocco). On the Guinea Coast, the value of gold fell again, temporarily, in the first two decades of the nineteenth century; the value at Accra is given by several writers as 1,600 to the ackie (25,600 to the gold ounce).[16] By the 1820s, it was back to 32,000 to the ounce,[17] and by the 1840s it had reached 38,400 to the ounce on the coast.[18] This last figure seems to represent a real decline in the value of cowries, reflected also in dollar values, and extending also to the Niger bend, where by the 1850s the gold ounce had gone up to 25,000 cowries.[19]

Long before this, merchants, faced with a large and permanent difference between values of gold on the Guinea Coast and on the Niger Bend, had abandoned the attempt to maintain a single gold value for cowries; from about 1820, and perhaps earlier, cowries were on a dollar standard.

COWRIES AND DOLLARS

Silver dollars came into use in Hausaland in the second quarter of the eighteenth century, with or shortly after the introduction of cowries;[20] they were already in use at Timbuctu and in the inland delta of the Niger by the third quarter of that century,[21] and were in use on the Sierra Leone coast before the end of the century.[22] At Accra, they were introduced in quantity in the first quarter of the nineteenth century,[23] and during the first half of the century they came into regular use at Whydah and Lagos (though they may have been in use earlier at Whydah).[24] They were introduced into Bornu as a currency shortly before the middle of the nineteenth century. At Sansanding in 1805 they evidently still had a rarity value, judging by Park's value in cowries,[25] and Bowdich reported in 1816 that they had recently fetched higher prices in Ashanti.[26]

[16] Bowdich, *Mission*, 330; H. C. Monrad, *Gemälde der Küste von Guinea* (Kobenhaven, 1822; Weimar, 1824), 263; H. Meredith, *An Account of the Gold Coast of West Africa* (London, 1812), 183.

[17] G. A. Robertson, *Notes on Africa* (London, 1819), 235, 266, 274.

[18] E. Carstensen, *Journal* (Copenhagen, 1965), 124. W. F. Daniell, 'Ethnography of Accrah and Adampe', *J. Eth. Soc.* IV (London, 1856), 23–4; this paper was read in 1852.

[19] H. Barth, *Travels and Discoveries in North and Central Africa*, V (London and Gotha, 1857–8), 17.

[20] *Kano Chronicle* (amended translation of this passage in M. Hiskett, 'Materials relating to the cowry currency of the Western Sudan: II', *Bull S.O.A.S.*, XXIX, part 2 (1966), 355.)

[21] El Haj Shabeeny in J. G. Jackson, *An Account of Timbuctoo and Houssa* (London, 1820), 2. Recent work by Ivor Wilks suggests that Shabeeny's journey must be dated to the late 1750s and 1760s, and that his 'Houssa' is to be found in the interior delta of the Niger; see *African Perspectives*, ed. Allen and Johnson (Cambridge), forthcoming.

[22] T. Winterbottom, *An Account of the Native Africans in...Sierra Leone*, I (London, 1803), 175.

[23] Monrad, *Gemälde der Küste*, 263.

[24] G. W. Neville, 'West African currency', *J. Afr. Soc.* XVII (1917–18), 223: 'some years ago I had passed through my hands £1,000 of old Mexican and Portuguese dollars, sent by the king of Dahomey, many dating back to the seventeenth and eighteenth centuries'.

[25] Park, *Journal*, 161.

[26] Bowdich, *Mission*, 330.

Nominally, the silver dollar is the weight of a gold ounce in silver (the ounce is the small Spanish ounce or doubloon). On the Guinea coast, the silver dollar was regularly equated with one ackie, or one-sixteenth of a troy ounce, and reckoned at 15 to the doubloon; a similar reckoning seems to have been in use at Fez. In Hausaland and Timbuctu, the dollar followed the traditional Arab reckoning for the silver *dirhem*, at two-thirds of a *mithqal*, one-tenth of an ounce.

At the established exchange rate of 3,000 to the gold *mithqal*, the dollar was thus 2,000 cowries in the north; at 32,000 to the ounce on the coast, the

Table 2. *Cowries, gold and dollars, 1820–50*

	c. 1820		*c.* 1850	
	75 grains gold	Dollar	75 grains gold	Dollar
Coast	5,000	2,000	6,000	2,400–2,600
Salaga	4,300	—	—	—
Yendi	4,200	—	—	—
Zoughou	4,000	—	—	—
Nikky	3,800	—	—	—
Boussa	3,500–3,400	Oyo? 2,000	—	—
Katsina	3,000	2,000	4,000	2,500

dollar was likewise 2,000 cowries. In the 1820s, Lyon heard of this value at Katsina,[27] Clapperton found it at Kano,[28] and perhaps also at Oyo;[29] a Muslim corrected Park's very high figure at Sansanding to 2,000 'on the Niger'.[30] At Timbuctu we hear, not on very good authority, of a rate of 3,000 to the dollar, and Clapperton seems to imply such a rate at Kano on his second visit.[31]

A stable dollar–cowrie value would necessarily imply a varying gold–cowrie value between the coast and Hausaland. Dupuis's figures, which we have already quoted, and which relate neatly to the cost of transporting cowries, supply some of the intermediate stages.[32]

By the 1850s the dollar was 2,400 cowries on the Gold Coast;[33] at

27 G. F. Lyon, *Narratives of Travels in Northern Africa* (London 1819), 138.

28 D. Denham and H. Clapperton, *Narrative of Discoveries in North and Central Africa in 1822, 1823 and 1824* (London, 1826), 51 of Clapperton's narrative.

29 H. Clapperton and R. L. Lander, *Records of Captain Clapperton's Last Expedition to Africa* (London, 1830), 59.

30 W. Hutton, *A Voyage to Africa* (London, 1821), 12.

31 Tartar Wargee, *Royal Gazette and Sierra Leone Advertiser*, 15 Mar. 1823; Clapperton and Lander, *Records*, 173. Clapperton quotes 1,500 to the dollar as 'just one half', presumably of the rate he should have had; he also explains 250,000 cowries as less than 100 dollars. Lander, on the return journey, twice refers to 4,000 cowries as 'little more than a dollar', perhaps working from the same value. I am grateful to Mr Robin Law for drawing my attention to these values.

32 J. Dupuis, *Journal of a Residence in Ashantee* (London, 1824), CXIV.

33 B. Cruikshank, *Eighteen Years on the Gold Coast*, II (London, 1853), 42.

Wydah, though the nominal 2,000 was still quoted, silver dollars passed at 2,400 to 2,600 cowries,[34] and soon the position was 'regularised' by raising the number of cowries in a string from 40 to 50, giving a 'head' of 2,500 cowries. In Kano the figure was 2,500 cowries; in Bornu, where the exchange was never stabilized, it ranged from 1,440 to 3,200. In Timbuctu, however, the value was 3,000 to the silver dollar.[35]

THE GREAT INFLATION: GOLD, DOLLARS, SHILLINGS

By 1850, we are on the threshold of the great cowrie inflation. Values on the Gold Coast during the period of the great inflation are complicated by the existence of two systems of reckoning: one in use by the government, the other by the merchants. By Order in Council in 1843, the troy ounce of gold was valued at £3. 12*s.* 0*d.* in West Africa, by contrast with a mint value of nearly £4. The difference, nearly 10%, was officially supposed to cover freight, insurance and other transfer costs. The value of £3. 12*s.* 0*d.* was known as the sterling value, and was used by the Gold Coast government; the merchants who were actually buying gold by the grain, continued to use the (much more convenient) system of valuing gold at £4 an ounce, 2*d.* a troy grain, a value which was called 'currency'; they accounted their transfer costs separately.[36]

Under the same Order in Council, dollars were valued at 4*s.* 2*d.*; the merchants, for a time, continued to reckon the dollar at 5*s.* 'currency', but by the 1850s they usually reckoned it at 4*s.* 6*d.* 'currency'. According to the Gold Coast Blue Book for 1855: 'The dollar is reckoned at 5*s.* currency, but is paid by Order of Council at 4*s.* 2*d.* but in fact passes at 4*s.* 6*d.* sterling, so that in practice the denomination in currency exceeds that in sterling by one tenth.'[37] A visiting missionary wrote: 'the dollar is 4*s.* 6*d.* at Cape Coast, 5*s.* at Accra and elsewhere 4*s.* 2*d.*'[38]

After 1870, the value of silver began to decline rapidly in terms of gold; the silver shilling became a token coin based on the gold sovereign, but sometimes valued in West Africa only according to its silver content. The dollar continued to be payable at 4*s.* 2*d.* while its bullion value declined to about 3*s.* 6*d.* The consequences have been described by Dr Hopkins[39]—merchants had dollars made in England and shipped to West Africa, where the government was forced to accept them at 4*s.* 2*d.* and pay to send them back to England. The dollar was demonetized in 1880, and soon found its

[34] F. E. Forbes, *Dahomey and the Dahomans*, I (London, 1857), 36.

[35] Barth, *Travels*, II, 161; II, 395; V, 25.

[36] *Gold Coast Blue Book*, 1850: 'The current value of British coins is 10% above the sterling value, in consequence of gold dust which is the principal currency of the country being £4 per oz. currency, which on being transmitted to England and all expenses being deducted averages about £3. 12. 0 per oz. sterling'.

[37] *Gold Coast Blue Book*, 1855.

[38] Daniell, 'Ethnography of Accrah', 24.

[39] A. G. Hopkins, 'The currency revolution in South West Nigeria in the late nineteenth century', *J. Hist. Soc. Nigeria*, III, no. 3 (Dec. 1966), 479–80.

own value at 3*s.* 6*d.* The German mark and the French franc, like the English shilling, were based on a gold standard after 1870; one mark was equivalent to a shilling, one franc was worth a little over 9*d.*

THE GREAT INFLATION—ACCRA

By 1852, the value of cowries at Accra was 2,400 for the dollar or ackie of 5*s.*, and thus 1*d.* a string. Cruikshank wrote in 1852 that the 'sudden and extraordinary increase of late years' in the import of cowries (about 150 tons a year) had had 'no effect in deteriorating them'; he wondered how long the importation of 'a very serviceable currency might be carried, without affecting its circulating value'.[40] By the time his book was in print, the value of cowries had already begun to collapse.

When the Poll Tax was under discussion in 1852, the governor had quoted the value as 1*d.* a string.[41] The first year's tax, though less than had been hoped, had included nearly 20,000 heads of cowries (some 40 million shells). These were finally accounted at 2*s.* 9*d.* a head: the Governor wrote of an 'unexpected drop in the exchange rate' leading to a loss of some £900;[42] he had therefore been expecting a rate of some 3*s.* 8*d.* sterling per head, about 4*s.* currency.

At first sight, the collection of some 20,000 heads of cowries might be expected to have a deflationary effect; but in fact the effect was probably the opposite. Most of the cowries probably came from hoards (which might also help to explain the diminishing yield of the tax in subsequent years). Moreover, the government did not transport the cowries to the capital at Cape Coast, where they were not current, but exchanged them locally (hence the loss due to the fall in the exchange rate). Thus some 20,000 heads of cowries were thrown suddenly on the exchange market—sufficient to account for a decline in the value of a currency which had recently undergone a large increase in quantity. An export from the United Kingdom of some 300 tons of cowries to the Gold Coast is recorded in 1853.[43]

The loss of confidence in the cowrie currency, and the consequent loss of value of savings, must have been at least aggravating causes in the serious political troubles which followed upon the introduction of the Poll Tax in the Accra area.

The value of cowries did not recover. By 1859 the official value was 2*s.* 6*d.* sterling a head; this was still the official value in 1861. By 1868–9 the official value had fallen to 2*s.* 3*d.* but was subject to fluctuation and had been as low as 1*s.* 6*d.*[44] A further rapid loss of value followed, possibly

[40] Cruikshank, *Eighteen years*, II, 42.

[41] Governor Hill, dispatch dated Cape Coast 24 Apr. 1852, quoted in G. E. Metcalfe, *Great Britain and Ghana* (London, 1964), 237.

[42] D. Kimble, *Rise of Nationalism in the Gold Coast* (London, 1960), 176.

[43] *Annual Statement of the Trade and Navigation of the United Kingdom*, 1853, under 'British Possessions on the Gold Coast'.

[44] A. Adumakoh, quoting Governor's report, 1859, *Economic Bulletin* (Ghana), vol. VII, no 4; Ag. Governor Bird, despatch, 13 Jan. 1860, Admin. 1/456, 12 (Ghana National

connected with the war of 1869–74. By the 1870s the value was down to 1*s*. a head,[45] 8,000 to the dollar, and cowries went gradually out of use, still at this figure.

THE GREAT INFLATION—DAHOMEY

The story of the Accra inflation has been told first because, unlike the inflation further east, it was not complicated by the introduction of Zanzibar cowries. The inflation at Whydah began a little earlier.

The first Zanzibar cowries arrived at Whydah at the end of 1845. Though this consignment was used to finance a war, and thus to provide cargo for the slave-traders, most of the Whydah cowries were probably used for the purchase of palm oil. By 1847, cowries were reported to be unpopular, and would appear to have been at a considerable discount.[46] In 1850, however, silver dollars were passing at 2,400 to 2,600 cowries, a figure similar to that obtaining on the Gold Coast.[47] It was not until the 1850s, and especially after 1856, that cowries began to arrive in quantity on the Whydah coast, when the firm of Régis began importing them from Zanzibar. By 1857 the value had fallen to three heads to the silver dollar;[48] in 1862 the value was quoted as 2½ heads to the dollar, but these may have been heads of 2,500 shells.[49] By 1871 the value was four heads of 2,500 to the dollar.[50] The value is quoted as 2,000 for 6*d*. in 1889, and during the 1890s is variously quoted as 2,000 cowries to the franc and as half of that figure—the latter the lowest value in West Africa.[51]

At the inland capital of Abomey, the inflation had been a little less severe; in 1871, 6,900 cowries went to the dollar in Abomey, as compared with 10,000 on the coast. Even at Porto Novo, values were slightly higher than at Whydah in the 1890s, as the head was reckoned here at 1,900 instead of 2,000.[52]

On the Togo coast the story was similar; by the 1890s the value was some 4,000 to the mark, a little higher than the lowest Whydah figure.[53]

Archives); Executive Council Minutes, 16 July 1861 (I am indebted to Mlle F. Cournaert for this reference); Gold Coast Blue Books, 1868 and 1869.

[45] E.g. Th. Opoku, 'Eines Neger-Pastors Predigtreise durch die Länder am Voltastrom' (1877), *Evangelische Missionsmagazin* (Basel, 1885), 270–1; G. Dobson, 'The River Volta, Gold Coast, West Africa', *J. Manchester Geog. Soc.* VIII (1892), 21.

[46] Dos Santos letters, in P. Verger, 'Influence du Brésil au Golfe de Benin', in *Les Afro-Américains*, IFAN Mémoire, no. 27 (Dakar, 1953), 74. The values deduced by Newbury from the Dos Santos papers appear to be incorrect; these papers are very hard to interpret—it seems possible that the printed version may include misreadings or misprints.

[47] Forbes, *Dahomey*, I, 36.

[48] C. W. Newbury, *The Western Slave Coast and its Rulers* (Oxford, 1961), 42.

[49] R. F. Burton, *Mission to Gelele*, I (London, 1864), 143. Burton also quotes fourfold rises in the price of *kenke* balls (and in the remuneration of state prostitutes), 316.

[50] J. A. Skertchly, *Dahomey as It Is* (London, 1874), 28.

[51] A. L. Albeica, 'L'avenir du Dahomey', *Annales de Géographie* (1895), 185; ibid, *Établissements français au Golfe de Benin* (Paris, 1889), 149; E. Foa, *Le Dahomey* (Paris, 1895), 148; O. Schneider (ed. C. Ribbe), *Muschelgeldstudien* (1905), quoting d'Albeica.

[52] Skertchly, *Dahomey as It is*, 28; E. Foa, *Le Dahomey*, 148.

[53] Schneider, *Muschelgeldstudien*, 148–9.

THE GREAT INFLATION—LAGOS

The course of the inflation at Lagos was similar to that at Whydah, though a little later, and was similarly associated with the import of cowries to pay for palm oil. The head was reckoned at about 4*s.* in 1851–4; after 1856, when the first really large imports began, the fall was rapid, and value fluctuated wildly. As late as 1858, Hutchinson could report the traditional value of one dollar a head (thought he probably recognized that this was, at the time of writing, an overstatement).[54] By the following year cowries became for a time wholly unacceptable; they later recovered, but only to

Table 3. *Value of 2,000 cowries, reduced to shillings and pence, 1850–95: Accra, Whydah and Lagos*

	Accra	Whydah	Lagos
1850	4*s.* 2*d.*	3*s.* 8*d.* to 4*s.* 2*d.*	4*s.*
1853	2*s.* 9*d.* to 3*s.*	—	4*s.*
1855	3*s.* 4*d.* variable	—	2*s.* 6*d.*
1857	—	1*s.* 6*d.*	—
1859	2*s.* 6*d.*	—	1*s.* 6*d.* to 1*s.* 9½*d.*
1863	—	1*s.*4½*d.* to 1*s.* 9½*d.*	1*s.* 4*d.*
1870	2*s.* 3*d.*	1*s.* 3½*d.*	1*s.* 3*d.*
1874	1*s.* 3*d.*	—	—
1876	1*s.*	—	1*s.*
1881	1*s.*	—	9¾*d.* to 1*s.* 1*d.*
1882	1*s.*	—	6½*d.*
1889	—	6*d.*	—
1895	1*s.*	4½*d.* to 9*d.*	6*d.*

the value of 1*s.* 6*d.*–2*s.* a head.[55] Burton, in a table notable for the oddity of its arithmetic, quotes a value for 1862 of 1*s.* 9½*d.* to 2*s.* a head, but adds that the values are 'extremely various'.[56] In 1863 he gives the very precise figure of 3 heads and 20 strings (6,800 cowries) for the dollar, a value of 1*s.* 4*d.* a head;[57] official values were a few pence higher. After 1863 the value settled down at 1*s.* 3*d.* a head, falling by 1876 to 1*s.* a head, 8,000 to the dollar, the same figure as at Accra.[58] Dr Hopkins, who gives a more

[54] Newbury, *Western Slave Coast* (1961), 58; T. J. Hutchinson, *Impressions of Western Africa* (London, 1858), 7: 'cowries are the chief currency (at Lagos), varying in value as Government consols or Railway Shares do at home, but generally averaging 2,000—which is called a head—for a dollar'.

[55] Newbury, *Western Slave Coast* (1961), 58.

[56] R. F. Burton, *Wanderings in West Africa* (London, 1863), 234 n.

[57] R. F. Burton, *Mission to Gelele*, 1 (London, 1864), 143: 'Cowries, it must be remembered, are merchandise, and the price varies accordingly. At the present they are abundant, and therefore cheap. The dollar (4/6*d*) now buys 2½ heads at Whydah and Agbome, 3 heads and 20 strings at Lagos and Abeokuta.'

[58] Hopkins, 'The currency revolution', 476–8, 479; P. Bouche, *Côte des esclaves et Dahomey* (Paris, 1885), 198 ff.

detailed, but substantially similar, account of cowrie values from 1850 to 1880, has traced the story of the final decline after 1880 to half this value, which he attributes to monetary factors—the demonetization of the dollar in 1880 and the increasing use of English silver coin.[59] These were doubtless important causes at Lagos; however, a similar, and even greater, fall took place at Whydah (which used Zanzibar cowries, but did not undergo quite the same monetary changes), while no corresponding fall took place in Accra (with a similar monetary history, but no Zanzibar cowries). There were therefore probably also other factors at work at these places. Part of the explanation may lie in the fact that the cost price of Maldives cowries never fell as low as the final Lagos value of Zanzibar cowries. Imports ceased much earlier, and the inevitable losses by export and non-monetary use may have offset monetary factors at Accra.

THE GREAT INFLATION INLAND

Not many figures are available for the changing values of cowries inland as the great inflation spread. On the Lower Niger, until the arrival of European traders with steamers in the nineteenth century, cowries had circulated alone without any other currency, and the same was true of many areas between the coast and the east–west routes of the savanna country. There are therefore no meaningful exchange rates for such areas until the penetration of European currency, which began on any scale only in the 1870s. On the Lower Niger, rates were quoted for the 1831 and 1842 expeditions, which were based on the prime cost of cowries in Europe, and are thus much lower than the values on the coast at this date.[60] As late as 1857, Glover was estimating the value of cowries at 'Loom' on the Niger at a rate corresponding to 2,000 to the dollar—a rate already long out of date on the coast.[61]

In the 1870s we hear of a widespread rate of 1,000 cowries to the English shilling, half the coastal rate at that date. This rate is reported at Salaga and Kete Krachi in Ghana, and throughout much of what is now Northern Ghana; in parts of north Togo and on the Niger at Lokoja and Onitsha.[62] By about 1890, when rates at Lagos and Whydah had gone still lower, the rate at Lokoja on the Niger, Loko on the Benue, and at Djougou in

[59] Hopkins, 'The currency revolution', 480–3.

[60] M. Laird and R. A. K. Oldfield, *Expedition into the Interior of Africa* (London, 1837), 166; W. Allen and T. R. Thompson, *Expedition to the River Niger in 1841*, I (London, 1848), 350.

[61] A. G. G. Hastings, *Voyage of the Dayspring (1857–9)* (London, 1926), 188.

[62] Opoku, 'Eines Neger-Pastors'; F. Ramseyer, 'Eine Reise im Norden von Asante und im Osten von Volta', *Geog. Gesellschaft zu Jena, Mitteilungen*, IV (1886); G. E. Ferguson, 'Report of a Mission to the Interior', 9 Dec. 1892, enclosure in dispatch of Governor Griffith, 10 Jan. 1893, Secret, C.O. 96.230, no. 2199, P.R.O.; R. Buttner, 'Journey to Togo', *Mitteilungen aus den deutschen Schutzgebieten* (1891), 416 ff.; M. Viard, *Au bas Niger*, 3rd ed. (Paris, 1886), 231; A. Burdo, *Niger et Benue* (Paris, 1880), 137 n. M.-J. Bonnat, Diary, printed in *L'Explorateur*, IV (Paris, 1877), 36–7, claimed to have 'given' the value of 1*s*. for 1,000 cowries in Salaga in 1876.

Dahomey were 2,000 to the shilling and mark; at Ilorin the rate was about 3,000; the official rate of the Royal Niger Company was still 1*s.* 3*d.* per head of 2,000, about 1,600 to the shilling; the actual rates were often below this, though this was the market rate in Nupe.[63] By 1902 the Ilorin rate was 4,000 to the shilling, about the same as on the coast (a similar rate is remembered at Kiama, near the Dahomey frontier); at Lokoja the rate was 2,500, while at Bida it was 3,000. In Ibo country, where cowrie values were always peculiar, the rate was 1,500 for the shilling.[64]

Farther north, even fewer figures are available. At Zaria we have a value for the gold doubloon of 45,000 cowries in 1861;[65] this suggests a value of some 60,000 to the gold *mithqal*, and probably implies a dollar value of 3,000 cowries. By the 1890s the dollar had reached 5,000 cowries, a figure which seems to have been standard over a wide area of the north, from Yola to Kano and as far south as Zaria and Lokoja; the 5-franc piece, though smaller, had a similar value.[66] At Kano the mid-century value of 4,000 to the *mithqal* had given place by 1862 to a value of 5,000, and by the 1890s to a value of some 15,000;[67] the dollar, meanwhile, had gone from 2,500 to 5,000 cowries. By 1902, shillings were coming into use; the exchange rate was 2,000 to the shilling at Zaria, 1,200 at Kano, while at Illo on the Niger the rate was 1,000 to the shilling, which was still the rate at Salaga and Kete Krachi.[68] Cowries in Bornu were a law unto themselves; 32 cowries made 1 *rotl* (or pound weight, a unit derived from the former copper coinage); but the number of *rotl* to the dollar was never fixed, and could often be manipulated. Comparison with Kano figures suggests that the difference was not very great.

In Timbuctu the 1855 figure of 3,800 to the gold *mithqal* had given place by 1880 to a figure of 8,000 or 9,000; the dollar, however, had changed only from 3,000 to 4,500.[69] By 1891 the *mithqal* was 10,000–15,000 cowries; the dollar had disappeared, and in its place was the French 5-franc piece at 5,000 cowries. Timbuctu values for cowries are lower than elsewhere in the north in the later nineteenth century, because cowries were not acceptable in the desert trade, and Timbuctu eventually became an isolated outpost of the cowrie currency.[70]

[63] C. H. Robinson, *Hausaland* (London, 1896), 43, 85; Dr Wolf, 'Journey to Borgu', *Mitteilungen aus den deutschen Schützgebieten*, IV (Berlin, 1891), 1 ff.; A. F. Mockler-Ferryman, *Up the Niger* (London, 1892), 306.

[64] *Report on Northern Nigeria* (1902), Cd. 1768, 14; G. T. Basden, *Among the Ibos of Nigeria* (London, 1921), 198.

[65] W. B. Baikie, 'Journey from Bida to Kano', *Journal of the Royal Geog. Soc.* 1867, 92 ff.

[66] S. Passarge, *Adamawa* (Berlin, 1893), 551; Robinson, *Hausaland*, 85; P. L. Monteil, *De St Louis à Tripoli par le lac Chad* (Paris, 1895), 297.

[67] Col. Mircher in C. W. Newbury, 'North African and Sudan trade in the nineteenth century, a re-evaluation', *J. Afr. Hist.* VII, no. 2, 233 ff.; Monteil, *De St Louis*, 297.

[68] *Report on Northern Nigeria* (1902), Cd. 1768, 14; F. Lugard, *Political Memoranda* (Nigeria, 1905), 224.

[69] O. Lenz, *Timbuctu*, II (French ed., Paris, 1887), 162.

[70] G. Jaime, *De Koulikoro à Timbuctu* (Paris, 1894), 223 n.1; E. Caron, *De St Louis au port de Timbuctu*, 3rd ed. (Paris, 1893), 278.

In Jenne the value of 4,000 cowries to the *mithqal* from Sheikh Ahmadu's time had declined only to some 6,000–8,000 by the 1890s, though these probably represent a rather smaller *mithqal*.[71] In 1895 5-franc pieces had been imported (presumably for the use of French soldiers) in such numbers that the exchange was 2,500–3,000 to the 5-franc piece; by 1902 the value was 4,000–5,600 to the 5-franc piece.[72]

Table 4. *Cowries, gold* mithqals *and silver dollars in the north, 1845–1900*

	Hausaland		Bornu	Timbuctu		Jenne	
	Gold *m*	Silver $	Silver $	Gold *m*	Silver $	Gold *m*	Silver $
c. 1855	4,000	2,500	1,440–3,200	3,800	3,000	4,000	—
c. 1865	6,000 Zaria	4,000–5,000	3,840–4,800	—	—	—	—
c. 1880	—	—	—	8,000–9,000	4,500	—	—
c. 1890	15,000	5,000	4,320–5,120	10,000–15,000	5,000 (5 fr.)	6,000–8,000	—
c. 1895	—	—	—	—	—	—	2,500–3,000 (5 fr.)
c. 1900	—	—	—	—	5,000 (5 fr.)	—	4,000–5,600 (5 fr.)

COWRIES AND THE 5-FRANC PIECE

On the upper Niger and in the country to the west, the place of the dollar was taken by the silver 5-franc piece.[73] This coin was nearly 10 per cent smaller than the silver dollar, but until the late nineteenth century this was more than compensated by the greater premium on silver. A standard rate west of the Niger was one 5-franc piece to one *gros* ($\frac{1}{8}$ oz.) gold. At 20,000 cowries to the ounce, the value of the 5-franc piece should be 2,500 cowries; this was the rate at which Raffanel actually sold his coins in Kaarta in 1848.[74]

On the upper Niger, Mage quoted an approximate value of 3 francs for a Bambara thousand (800) in the early 1860s;[75] but it does not seem that 5-franc pieces were yet current there, and the basis of his calculation is uncertain. It is not until after 1870 that we have real exchange-rates in this area between cowries and French silver. By this time the coins no longer contained even approximately their face value of silver, so that it becomes very difficult to interpret the values; as the French advanced, cowrie values became increasingly related to the value of the (gold-based) French franc; and at the same time the high premium on silver declined as the French brought in more coin; meanwhile, the world value of silver was

[71] L. G. Binger, *Du Niger au Golfe de Guinée, 1887–9*, I (Paris, 1892), 375.

[72] F. Dubois, *Timbouctou la mysterieuse* (Paris, 1897), 179; R. Bluzet, 'La region de Timbouctou', *Bull. Soc. Geog. de Paris* (1895), 322 ff.

[73] Monteil, *De St Louis*, 268, 280.

[74] A. Raffanel, *Nouveau voyage au pays des nègres*, I (Paris, 1856), 233.

[75] E. Mage, *Voyage dans le Soudan occidental, 1863–5* (Paris, 1868), 191.

still declining. To add to the confusion, in the area west of Timbuctu all values are quoted in Bambara 'hundreds' of 80.

In 1878 the 5-franc piece was worth '3,500'–'4,000' (2,800–3,200 shells) at Guigné, north of Bamako, and at '5,000' (4,000 shells) near the river.[76] In 1880, in very disturbed political conditions, the value at Segou and in the adjacent area ranged from '2,500' to '5,000' (2,000–4,000 shells).[77] The lower figure, '2,500', was still current at Koulikoro, Yamina and Sandanding in about 1890,[78] though to the east of Timbuctu, at Bandiagara, the rate was 5,000 to the 5-franc piece. This was also the rate in Timbuctu,[79] and probably also Jenne.[80] After the French conquest this became the official French rate, and continued to be the ruling rate at Timbuctu, on the Upper Niger and in the inland Niger delta; however, the actual rate was 6,500 by 1902 (these were possibly Bambara 'hundreds' of 80). Farther east, however, in the area between Dori and Say, the rate was 3,500 to the 5-franc piece,[81] a rate similar to the 1*s.* a thousand quoted at Illo.

THE MIDDLE ZONE, ABOUT 1890

In what is now southern Guinée, 5-franc pieces were found by Binger in 1889, if not in circulation, at least with recognized rates of exchange for cowries. At Oulosebougou the rate was 1,600 cowries, farther east, at Tenetou, Tiong-i, Fourou and Benkhobougoula, 2,000 cowries to the 5-franc piece.[82] Farther east still, no silver was in circulation, though it was occasionally sold in the markets as a commodity bought by the jewellers; Binger had to express his prices in francs through the medium of gold-dust at the current (European) rate of exchange.[83]

Binger also gives a number of exchange rates for cowries and gold. These he gives in terms of the '*barifiri*', which he equates to four *mithqal;* in Salaga at least, the '*mithqal*' of this date must have been about nine-tenths of the Nikky *mithqal* of seventy years earlier;[84] it was now worth some 10,000 cowries, as against Dupuis's value of 4,300 cowries. This is the only

[76] P. Soleillet, *Voyage à Ségou en 1878–9*, ed. G. Gravier (Paris, 1887), 235.

[77] J. S. Gallieni, *Voyage au Soudan français, 1879–81* (Paris, 1884), 380.

[78] Jaime, *De Koulikoro*, 223 n.1; Caron, *Au port de Timbuctu*, 287.

[79] Caron, *Au port de Timbuctu*, 278.

[80] Monteil, *De St Louis*, 280. (Values in 1895 were anomalous, owing to the glut of French silver coin.)

[81] E. Baillaud, *Sur les routes du Soudan* (Toulouse, 1902), 70–1.

[82] Binger, *Du Niger*, I, 27; I, 54; I, 191; I, 206; I, 162. The apparently 'round' numbers of the exchange rate of Tenetou, Tiong-i, Fourou and Benkhobougoula are for the actual number of shells; these were locally expressed in 'Bambara' hundreds of 80 shells each, and in the case of the last, in 'Mandingo' hundreds of 60 each.

[83] Binger, *Du Niger*, I, 308: 'it is impossible to make comparison with silver, as there is none in circulation...when, in the course of my narrative, I speak of an object which costs 3, 4, 5 francs, that means that, with the number of cowries asked, I could have procured 3, 4, 5 francs in gold dust at the rate of 3 francs per gram'.

[84] Binger, *Du Niger*, I, 375; II, 103. For the weight of the *barifiri*, see also B. Menzel, *Goldgewichte aus Ghana* (Berlin, 1968), which gives the actual weights of specimens collected at Salaga within a few years of Binger's visit.

figure from Binger's series which is directly comparable with Dupuis's values. Binger's figures range from 24,000 to the *barifiri* at Kong and a little above at Bobo-Dioulassou, to 30,000 at Bonduku market (though gold could be obtained more cheaply in the gold-mining villages), 40,000 at Salaga, 55,000–60,000 at Kintampo, and 'nearly 80,000' at the 'posts on the coast'.[85] (This last would correspond to half a gold ounce at the Gold Coast rate of 2,000 cowries to a shilling; but this was a silver rate, and the rates for gold may have been different.) Kintampo, between Bonduku and Salaga, was at this date cut off from Ashanti trade, which would account for the high value of gold there.

Farther east, cowries were 2,000 to the silver mark at Djougou,[86] compared with 1,000 for the shilling at Salaga, thus reflecting the low value on the Dahomey coast. At Nikky, where the gold ounce had been about 25,000 cowries in 1820, Lugard gives a 'local value' in 1894 which would correspond to just double that figure.[87] This was apparently an island of higher cowrie values, as might be expected in an area where European penetration was very late.

COWRIES AND COMMODITIES

Values so far given for cowries have been in terms of gold or silver. Since in West Africa the precious metals are trade goods, they do not necessarily have constant value, either over time, or from place to place—indeed, to

Table 5. *Shillings and commodities in cowries, Salaga*[88]

	Shilling	Ox	Slave	2,000 kola
Pre 1874	—	9,000–15,000	36,000–70,000	12,000
1876	1,000	—	—	—
1877	1,000	—	60,000–80,000	6,000–20,000
1882	1,000	—	—	20,000
1884	—	—	—	30,000
1887	1,000	—	120,000	—
1888	1,000	30,000–40,000	140,000	—
1889		25,000–30,000	80,000–120,000	32,000–35,000
1892	1,000	40,000–60,000 (after rinderpest)	60,000–140,000	10,000–50,000

assume that gold in the Ashanti gold-fields has the same value as gold in London or Morocco is to make nonsense of the whole gold trade.

The danger of interpreting cowrie values in terms of their exchange rate

[85] Binger, *Du Niger*, I, 308, 375; II, 103, 142, 166.
[86] Wolf, *Journey to Borgu*, 1 ff.
[87] F. Lugard, *Diaries*, ed. M. Perham and M. Bull, IV (London, 1965), 192.
[88] Compiled from sources quoted in *The Salaga Papers*, compiled M. Johnson, Institute of African Studies, Legon, Ghana (1966).

for shillings can be illustrated from a series of cowrie prices from Salaga (in central Ghana) in the later nineteenth century. It seems that the depreciation of cowries, in terms of goods, at Salaga began after 1877, though the value of 1,000 cowries to the shilling remained constant; Bonnat claimed to have established this rate in 1876, when it must have represented a considerable premium on silver. As late as 1889, Binger still found a premium on silver *vis à vis* gold at Salaga.[89]

Table 6. *Cowries per bar of Taodeni salt, 1595–1902*[90]

	Timbuctu	Jenne	Sansanding	Segu
1595	20,000 (*c.* 1 oz. gold)	—	—	—
1759	(*c.* 1 *mithqal* gold) (3,000)	—	—	—
1805	—	—	8,000	—
1828	—	10,000, 15,000, 20,000	—	—
1855	3,000–6,000 (*c.* 1 *mithqal*)	—	(7,500) 2 *mithqal*	—
1865	—	—	—	(Yamina) 20,'000'*
c. 1870	—	—	20,000	—
1880	8,000–9,000 1 *mithqal*	—	—	—
1886–8	—	—	—	40,'000'–60,'000'*
1889	—	20,000–25,'000'* 3 *mithqal*	40,000 (uneconomic)	—
1894	—	—	50 francs (overland)	—
1895	30–40 fr. (15,00–18,000)	60 fr. (36,000)	70–80 fr. (42,000–48,000)	—
1901	16–20 fr. (16–20,000)	25–28 fr. (32,500–36,400)	— —	35 fr. (45,500)
1902	—	36,000	—	—

* These are Bambara 'thousands', i.e. units of 800.

Salt from the Sahara was imported for centuries in slabs of a more or less standard size suitable for headloading; it might be hoped therefore that the prices in cowries of a slab of salt would give some index to the 'real' value of cowries over this period. Up to the end of the sixteenth century, figures in cowries are rare, and the values in gold suggest either very large fluctua-

[89] Binger, *Du Niger*, II, 104.

[90] *Tarikh el Fattash* (1913), 319; Shabeeny, in Jackson, *Account of Timbuctoo*, 2; Park, *Mission*, 161; Caillié, *Journal*, I, 457; Barth, *Travels*, V, 23; Lenz, Timbuctu, II, 98; Gallieni, *Voyage*, 424; Binger, *Du Niger*, I, 374; Jaime, *De Koulikoro*, 219; Dubois, *Timbouctou*, 179; Capt. Lenfant, *Le Niger* (Paris, 1903); Monteil, *De St Louis*, 280.

tions or inaccurate information. After salt from Taodeni came into use in the later sixteenth century we have rather more information. The very high rates, in francs, for 1895 are explained by the importation of large numbers of silver coins across the desert, presumably for the use of the French forces in Timbuctu and Jenne; the glut of silver coin is also reflected in the cowrie exchange. It seems that the standard price in Timbuctu for over a century must have been a *mithqal* of gold. Prices at Jenne and Sansanding were much more vulnerable to political and military influences. From time to time communications between Timbuctu and Sansanding were cut, salt for Sansanding arriving overland, usually via Walata.

COASTAL PRICES AND THE SLAVE-TRADE

To discover the change in the 'real' value of cowries, as distinct from the value in terms of gold or silver, it would be necessary to find a series of prices for a commodity which is unlikely to have changed much in local demand over the period in question. On the Guinea Coast one such commodity is the domestic fowl, for which values are known from the early eighteenth century. It is probable that the currency unit known as the

Table 7. *Prices of hens and gold* taku *in cowries, ca. 1520–1850*[91]

	Galinha	Live hen	Gold *taku* (2½ troy grains)
Early sixteenth century	40	—	?40
Early seventeenth century	?80	—	—
Late seventeenth century			
Early eighteenth century	200	240–280 (Accra)	80
Late eighteenth century	200	266–333 (Whydah)	160
Mid-nineteenth century	—	200–280 (Allada)	200

'hen' or *galinha* represented the cowrie value of a fowl in the early sixteenth century, and that the transfer of the name to a larger unit by the seventeenth century represents a rise in the cowrie price of hens. Comparison with the gold *taku* suggests that, while the hen may have been worth a *taku* in the early sixteenth century and in the early seventeenth century, and again by the mid-nineteenth century, there was a period, at the height of the slave-trade, when either hens were very dear, or gold very cheap, in terms of cowries. Barbot, writing of the late seventeenth century, valued a hen on

[91] A. F. C. Ryder, 'An early Portuguese trading voyage to the Forcados River', *J. Hist. Soc. Nigeria* 1, no 4 (December 1949), 301; O. Dapper, *L'Afrique* (French edition, Amsterdam, 1686), 306; J. Barbot, 'Description of the coasts of North and South Guinea', in A. and J. Churchill, *Collection of Voyages*, v (London, 1737), 339; Rask, *Reksebeskrivese*, 257; S. Berbain, *Le Comptoir français de Juda au XVIII siècle*, IFAN Mémoire (Dakar, 1942); Forbes, I, 110. Barbot's references to the prices of fowls on the Gold Coast and at Wydah are at 217 and 330 of his *Description*.

the Gold Coast, where gold was the currency, at 'at least a shilling', while on the Slave Coast a hen cost only sixpence in goods, representing a prime cost of 3*d*. This might be taken to imply that gold was cheap at this date—perhaps the 'golden age' of cheap gold recalled by later traders.

It appears from the table that cowries depreciated very little, in terms of hens, between the early eighteenth century and the mid-nineteenth. This is remarkable, as cowries were being imported in large quantities during the eighteenth century, at a rate which can hardly be less than 150 tons a year over considerable periods, and may at times have been much greater. Some of the cowries were doubtless hoarded, some withdrawn from circulation for other uses, and yet others lost or broken; but the numbers were so large (nearly 900,000 to a ton, some 150 tons a year for most of a century) that some other explanation must be sought.

In the eighteenth century, cowries were imported in part-payment for slaves. Some of them may have been exchanged for the trade-goods which formed the remainder of the payment. Unlike shillings imported in the late nineteenth century, they were not accepted in payment for imported goods by the importers, and must have been retained within the West African economy. It is therefore certain that the cowrie-using economy must have been an expanding one during this period. The ramifications of the eighteenth-century slave-trade must have carried cowries into an enlarging area. The 'Malay' (Muslim) traders who reached the Whydah coast in the early eighteenth century are known to have carried cowries inland;[92] by the second quarter of the century cowries were coming into Hausaland—Hiskett may well be right in thinking that many of these cowries came from Nupe and ultimately from the coast, whether or not the original introduction of cowries came from this direction.

This must have been an expanding as well as simply an extending economy. The rise in slave prices in the eighteenth century—in terms of cowries, from about 100 lb weight of cowries at the end of the previous century to about four times this figure by the 1770s—must in part represent growing costs as slaves were brought from further away and by more expensive means; part of this increased cost must have been paid in cowries to sellers of local produce along the routes, and spent in other ways. Between one-third and one-quarter of the price of slaves was normally paid in cowries; this forms a rough measure of the proportion expended within the economy.

Since the nearer and more easily captured slaves would always be taken first, a slave-trading economy tends to suffer, like Marx's capitalist system, from 'inherent contradictions' which make it necessary to expand continuously (and often in an 'imperialist' manner). This expansion provided the continuing demand for cowries, and maintained their value. (It is interesting to note that Ashanti, which used a gold-dust currency, appears to have had a similar demand for increasing amounts of currency; although

[92] J. B. Labat, *Voyage du Chevalier des Marchais*, II (Paris, 1730) 281.

cowries, and not gold, were used as a market currency in markets outside metropolitan Ashanti, gold was actually being imported into Ashanti in the second half of the eighteenth century, presumably to meet the needs of its expanding economy).[93]

The expansion was eventually halted, not by internal strains reaching the breaking-point (though it is arguable that some of the cowrie-using states had reached this point by the late eighteenth century), but by the abolition of the slave-trade by the very nations that were importing most of the cowries. This, though it did not put an end to the slave-trade, did greatly reduce the imports of cowries. Not until the third decade of the nineteenth century did the import of cowries into West Africa again reach substantial proportions.

COWRIES AND THE INTERNAL EXCHANGE ECONOMY

The Gold Coast, which had not been a large importer of cowries during the slave-trade era (its main slave exports came from west of the cowrie area), imported increasing quantities of cowries to pay for growing exports of palm oil; by 1850 the Governor estimated the quantity at some 150 tons per annum.[94] The quantity of cowries imported depended on the export of palm oil; the quantity which the economy could absorb depended on the expansion of the local internal exchange economy. Up to 1853, cowries had lost little in value.

The economy of the cowrie-using area of the Gold Coast was expanding in a very literal way; the Krobo, who were the largest producers of palm oil, and who, for religious reasons, refused to use any currency but cowries, were engaged in buying land which they used for growing food and more palm oil; they were also establishing the first of their great produce markets, selling their surplus food production for cowries to buyers from neighbouring states. Other palm-oil producers were engaged in similar, if less dramatic, investment in agricultural expansion.[95]

In Nigeria the expansion took other forms. It is known that some of the cowries imported in the south went north as tribute from the towns on the Niger;[96] many more probably went north by way of trade. Barth believed that the value of cowries in Yorubaland had been maintained by the withdrawal of cowries northward as a result of their introduction as currency into Bornu.[97] The cowrie-using area was thus still extending, as in the eighteenth century, in the years before 1850. There must also have been a general expansion of trade arising from the increased sales of palm oil.

[93] M. Johnson, 'The ounce in eighteenth-century West African trade', *J. Afr. Hist.* VII, no. 2 (1966), 203.
[94] *Gold Coast Blue Book*, 1850.
[95] M. Johnson, 'Migrants' progress', *Bulletin of the Ghana Geographical Association*, IX, no. 2 (July, 1964), 21.
[96] E.g. Allen and Thompson, *Expedition*, II, 85.
[97] Barth, *Travels*, II, 161, 395.

By the early 1850s the period of expansion was over.[98] On the Gold Coast, cowrie imports were much reduced after 1854 and did not begin to recover until 1866. It is interesting to find that the total value of cowries imported after 1852 (the cumulative total valued at the current rate) remains nearly constant for the years 1854–66, within £1,000 or so of £33,000 sterling. This is a figure of the same order of magnitude as the local value of the palm oil exported in normal years, but must presumably represent only a part of the cowries in circulation, as cowries imported in earlier years must still have been in use.[99] There is thus evidence of a sizeable internal exchange economy, even during years of stagnation.

A similar picture, on a larger scale, emerges from the imports of Zanzibar cowries to West Africa. Omitting exports by French firms (largely to

Table 8. *Palm oil, sterling and cowries, 1851–65*[100]

	Gallons of palm oil per head of cowries			Sterling per head cowries	
	Lagos	Whydah	Gold Coast (producers)	At 3 gall. = \$ = 4*s*. 6*d*.	Gold Coast
1851	3	—	—	4*s*. 6*d*.	—
1855	1⅔	—	—	2*s*. 6*d*.	—
1857–8	—	1	2½	1*s*. 6*d*.	2*s*. 6*d*.
1859	1¼	—	—	1*s*. 10½*d*.	—
1861–2	1	—	2	1*s*. 6*d*.	2*s*. 6*d*.
1865	—	—	1½	—	2*s*. 3*d*.

Whydah), the cumulative total of exports by the three German firms, valued at the rate current at Lagos, retain a fairly constant total value from 1856, at about £250,000 sterling; increases much above this figure are quickly offset by a further decline in exchange value.[101] Imports from Britain, though on a smaller scale, show a similar picture. Here again there seems to be some sort of ceiling above which the economy could not absorb more currency.[102] Renewed expansion in the later 1860s raised this ceiling for Lagos and its hinterland to at least half a million sterling by 1869.[103]

[98] The rapid expansion of West African external trade in the thirty years before 1850, followed by a period of stagnation, has been noted by Dr. Newbury in a paper on prices and profitability in West African trade in the early nineteenth century, read at the Tenth International African Seminar of the International African Institute, Sierra Leone, 1969. There is no obvious explanation of this change, which extends far beyond the cowrie zone, and cannot be due to monetary causes connected with cowries, though these may well have been of importance in the Niger Delta and on the Gold and Slave Coasts.

[99] Calculated from figures in the *Annual Statement of the Trade and Navigation of the United Kingdom*, 1852–66, under 'British Possessions on the Gold Coast'.

[100] Calculated from data in Newbury (1961), *The Western Slave Coast*, and Burton (1862), *Mission to Gelele*.

[101] Calculated from table in E. Hieke, *Zur Geschichte des deutschen Handels mit Ostafrika, J. Oswald und Sohn* (Hamburg, 1939), 283.

[102] Calculated from *Annual Statement of the Trade and Navigation of the United Kingdom*, 1852–66, 'West Coast of Africa'.

[103] Calculated from table in Hopkins, 'The currency revolution', 405.

The rapid loss of value of cowries during the years of the great inflation made them unsatisfactory as a standard of value. The importers of Zanzibar cowries, faced with a shortage of coin by the early 1850s, had perforce to accept palm oil in payment at a standard rate of three gallons to the dollar, a rate which remained fairly constant over the inflation years.[104] The standard of value thus tended more and more to be a quantity of palm oil, each port having its own standard. The value of cowries can thus be expressed in terms of the quantity of oil which could be purchased by one head of cowries, rather than in terms of the number of shillings which could be purchased.

From 1866, cowrie imports began to increase again, with growing exports of palm oil and kernels. On the Gold Coast this was followed by a further decline in value, possibly associated with the 1869–74 wars, until by the 1870s the value was no higher than at Lagos, and imports had virtually ceased.

By this time, silver coin was coming into West Africa in some quantity. At first it was not acceptable in the internal markets; its increasing use in West Africa must therefore reflect an increasing tendency to spend the proceeds of export crops on imported goods rather than on local goods and services—a tendency which was naturally encouraged by those with imported goods to sell. At this period we find European firms increasingly trying to cut out the middleman by establishing inland depots. The Royal Niger Company and its predecessors, in substituting a cowrie of account for real physical cowries, was following the same line of development.[105] Despite theories to the contrary, the movement in West Africa was from an internal exchange economy monetized by the cowrie to something much more akin to a barter of exports for imports, whether mediated by a cowrie of account or by a silver coin which rapidly returned to the importer.[106]

[104] J. E. Herz, 'Ueber Verwendung und Verbreitung der Kaurimuschel', *Mitteilungen der Geog. Gesellschaft zu Hamburg* (1881), 14 ff.

[105] See part I of this paper, M. Johnson, 'The Cowrie currencies of West Africa', *J. Afr. Hist.* XI, no. 1 (1970), 'The Cowrie as a unit of account'.

[106] H. Myint, *The Economics of Developing Countries* (London, 1964), 75. Myint's model is as follows: 'In the first stage, the peasants produced the export crop only on a part-time basis while continuing to produce their subsistence requirements. Since they were self-sufficient in respect to locally-produced foodstuffs and other goods, they had no need to hold cash for local transactions. Since their sole reason for producing the export crop was to obtain purchasing power over imports, we may assume that they spent the whole of their cash earnings on imported goods. At this stage, the money which flowed into the country to pay for the peasants' exports would flow out again immediately to pay for their imports. No "localized" currency would be retained for long within the country. The function of money was merely to facilitate the barter between the exports and imports at agreed rates, and a separate local currency was hardly needed. As a matter of fact, in the earlier phases of trade in West Africa, British silver coins were extensively used for this purpose.' The circulation of cowries in the Lagos hinterland and elsewhere before the 1880s, and of other local currencies such as manilas, is the measure of how far Myint's model (which he evidently regarded as applicable to West Africa) fails to fit the West African facts. For the increasing use of British silver coin, see A. G. Hopkins, 'Creating a colonial monetary system: the origins of the West African currency Board', *African Historical Studies*, III (1970), 101–32.

The final decline of the cowrie was due to the decline, absolute as well as relative in the last quarter of the nineteenth century, of the internal exchange economy for which it had been the currency.

COWRIES IN THE TWENTIETH CENTURY

By the beginning of the twentieth century cowries had begun to go out of use in a number of areas where they had been established for centuries. English silver coin had long replaced cowries on the Gold Coast for all but the smallest market transactions, and in Nigeria cowries were being replaced by English silver for larger transactions at least along the coast. Francs were replacing cowries on the Upper Niger and to the west of the river; by 1914 prices nearly everywhere in French territory were being quoted in francs, though cowries survived in use in many markets.

Values in 1914 showed little change from those at the turn of the century —1,200 to the shilling in the far north of British territory, 1,000 to the shilling over wide areas of the middle north, rising to 2,000 to the shilling on the Gold Coast and 4,000 to the shilling or even more west of the river Volta and over much of southern Nigeria. The artificial French value of 1,000 to the franc was widely evaded in the western part of French territory by counting in Bambara 'hundreds' of 80,[107] while in Mossi a rate of 800 to the franc was similarly established after 1906.[108] Throughout French territory there was much seasonal fluctuation, due partly to the seasonal demand for coin for tax.

Few cowries remained in use after the First World War, though they reappeared in some markets during the depression years, when the decline in value of cash crop exports led to shortages of coin. In Bida and Abuja they returned at much the same rate (in terms of shillings) as at the beginning of the century.[109] The French franc, as a result of inflation in France, had declined to about one-fifth of its former sterling value. Many cowrie values for commodities remained much the same as before the war; at Jenne in 1902 the *ba* of 800 cowries was worth about one franc; in Mopti in 1928, the *ba* was equated with five francs. Local cloth and iron remained much the same in terms of cowries; cereals had gone up in price, imported cloth and salt had gone down.[110] Similar changes were found in other French territories; east of Bandiagara, 1,000 cowries were worth 4 francs in 1928, as against one franc or rather less in 1900;[111] in the Lobi country the pre-war rate of 800 to the franc was replaced by a rate of some 600 cowries

[107] M. Delafosse, *Haut-Sénégal-Niger* (Paris, 1912), III, 48.

[108] E. P. Skinner in *Markets in Africa*, ed. P. Bohannan and G. Dalton (Evanston, 1962), 260.

[109] A. H. M. Kirk-Greene, 'The major currencies in Nigerian history', *J. Hist. Soc. Nigeria*, II, no. 1 (December, 1960), 132–50.

[110] Monteil, *De St Louis*, 269, 280.

[111] W. B. Seabrook, *White Monk of Timbuctu* (London, 1934), 238.

(fluctuating seasonally to some 2,000 cowries) to the 5-franc note.[112] Despite the seasonal fluctuations, cowries had proved more stable than francs—owing, no doubt, to the fact that it was in cowries that daily needs were bought in the markets.

Cowrie values in Ibo country were always peculiar, owing partly to the unique system of counting in use. At the turn of the century cowries in Iboland were worth about 8*d.* a thousand, double for the small cowries of the eastern hinterland. Before 1914 the value had risen to about 1*s.* a thousand, and by 1924 had reached about 1*s.* 4*d.* a thousand. Values thereafter declined, until by 1935–7 they were fluctuating at about 6*d.* per thousand.[113]

The only area for which present-day cowrie values are available is north-western Ghana; cowries have become somewhat scarce, but are still in use in the markets, where they have a regular exchange value. The values in the 1960s varied between 10 and 20 to the penny—in some markets they could even rise to 5 to the penny—by contrast with the turn-of-the-century value of 100 to the penny.[114]

The wheel has come very nearly full circle. Ibn Battuta's thirteenth-century value was something like eight to the (gold-standard) penny.

[112] Basden, *Among the Ibos*, 198; M. M. Green, *Igbo Village Affairs* (London, 1964), 41 n. 2. [113] M. J. Herskovits, *Economic Anthropology* (New York, 1952), 247–50.

[114] J. L. Dougah, *Wa and Its People* (Institute of African Studies, Legon, 1966): 20 to 1*d.*; other Ghanaian informants said: 1950, 140–150 to 1*s.*; 1965, 120 to 1*s.* at Wa; at Lawra, 5–10 to 1*d.*; at Wechau, Mrs M. E. Humphreys found cowries at 120 to 1*s.* in 1965. I must thank these informants, and many others who have taken the trouble to bring or send me information about cowries during the past five years.

10

A History of Money in Asian Perspective

Frank Perlin

A History of Gold and Money 1450–1920 by Pierre Vilar. Translated by Judith White. London: New Left Books, 1976. Pp 360.

Vilar's important book arose from a lecture course at the Sorbonne but, far from portraying the usual excesses of such exercises, has all the powers of concentrated wisdom, organisation and careful thinking typical of his best work. His reputation is best served by his fine study of Catalonia, which in fact is one of the half-dozen famous theses published during the great period of 'Annalist' production, and 'a marvellous work' [*Vilar, 1962; Carr, 1966: 712*]. It is an extraordinary combination of empirical research, analysis and synthesis, one of a small number of works which has served to redefine the way in which history should be written—even what should compose its content. Its importance, like *Beauvais et le Beauvaisis* (Goubert, 1960), or *An Economic Theory of the Feudal System* (Kula, 1976, on Poland, but already discussed in depth in India), extends far beyond Spain, or even Europe. But even in such a small and popularly-oriented book as his short *Spain. A Brief History* (1967) his power of coupling detailed information and generalisation—of imposing a meaningful and original order on several centuries of a nation's history in a tiny compass, is admirable. Vilar has also published numerous articles on historical demography (thus his interesting idea concerning the relationship between low prices and demographic growth), peasant movements, and on general features of economic and social history during the early modern period (his much-lauded, but in my view overrated 'Temps de Quichotte', a rather unfortunate exercise on marxism and history, and his much more interesting 'Problems on the Formation of Capitalism' [*1956 a & b, 1972*], and numerous others). It is time that those concerned with peasantry became more familiar with his work.

This brief summary of his accomplishments precedes a review which sadly ends in criticism. It is essential, therefore, to emphasise that this reviewer approaches the text not as an expert on monetary history (far from it—but nor is the 'expert' the most appropriate audience) but as a historian of third world peasantries, seemingly peripheral to the issues raised by Vilar's text, but in fact provocative of questions that made this book one hoped-for source of enlightenment in a field of almost complete obscurity. Readers who wish a sympathetic discussion of the main arguments of the text and its importance for economic thought should rather look to Kolm's interesting review [*1975*].

The aim of *A History of Gold and Money* is modestly stated as an attempt 'to provide some clarification of the problems which historically money has raised'(7). More specifically, he gives this as a wish to study the history of the present distinction between 'circulating money' and 'internationally valid currency'(15). His chosen audience seems to be the economist concerned with the present, rather than the historian seeking synthesis: it is a lesson in history given to those accustomed to the confining and deceptive circlings of logic on the necessity of historical knowledge and understanding in the construction of explanations for current monetary difficulties; and thus it is a lesson in the making of theory, very much representative of current concerns to historicise the social sciences. Putting aside any question of the veracity of specific views and solutions advanced by Vilar in the course of the text, it is a model of how this general problem should initially be approached. Indeed, the extent to which its value transcends the detail of its particular arguments puts it, like Kula's book, into a category of history-writing characterised by the stimulating seriousness of its reflection on the task in hand, its concern for evidence, its careful progress through the difficulties involved in particular arguments (or set up by the historiography), and, finally, its honesty in describing obstacles and impasses, instead of smoothing them over or hiding them from the reader's sight. Vilar is so aware of what history is about, thus of his professional responsibilities in approaching it, and consequently of the minimum conditions necessary for successfully advancing the state of knowledge, that, despite his intentions, the book must be recommended as valuable reading for historians too—all historians, not just those concerned with money or the 16th century.

A History of Gold and Money is precisely that. Vilar sets out to explain the role of precious metals by describing the history of their use as a monetary medium from 'primitive' and 'Ancient' times up to the early twentieth century.[1] More precisely, he is concerned with the role of money in the formation and workings of capitalism, thus with a Europe increasingly dominated by merchant capital. It is from about the later fifteenth century that monetary institutions begin to play an ever more fundamental role in the organisation of society and polity, and this is in the context of the extraordinary changes affecting Europe during early modern times (sixteenth to eighteenth centuries)—its expansion overseas, its internal growth and transformations, the connections between the two.

Vilar's framework is chronological. He moves step by step through classical times to the twentieth century, a good half of the book, however, being concentrated on the problems surrounding the overseas discoveries, the rapid acceleration of imported precious metals during the sixteenth century, and the sixteenth century price rise. There is justifiable reason for such emphasis, because it is at this time that the history of prices takes on a dramatic new shape, with important social and political consequences. It is also the time when overseas discoveries and early overseas trading ventures, and an equally dramatic increase in specie imports, went hand in hand, giving rise to E. J. Hamilton's classic thesis of the relationship between imports and the European price rise. From this start a vigorous and voluminous historiography has developed and its sophistication makes the exploration of this early period, at the impetuous beginnings of the growth of a capitalist system, peculiarly worthwhile. A wide range of explanations can be assessed and an archaeology of the complex relationship between money and society be reconstructed. The bulk of the rest of the book is concerned with subsequent developments in the seventeenth and eighteenth centuries, especially with attempts to maintain the flow of precious metals, the rise of new institutions for the management of money (banking, etc) and the actual history of circulation in Europe itself. At the same time that Vilar treats of social and economic history, he analyses the ways in which contemporaries coped with the new problems presented by money and prices, for a part of this whole history is the spate of works attempting to explain monetary problems and to dictate governmental policy from the sixteenth century on. The book with its abundant data, maps, graphs and narrative, is always problematical and theoretical; thus the concluding chapter 'In Search of a Global but not One-Sided Expanation' appropriately culminates a book-long argument.

Vilar is admirable on the complex, changing relationship between the African and American discoveries and the needs of the European economy, and on the structural mutations necessary for American gold to have become serviceable in the latter (costs of production less than its exchange value in the markets of import). His salutary emphasis on the detail of the production and marketing processes is important; that on forms of labour used in the production of silver (mita, etc) is a specific example of the close links which developed in different ways and with more or less haste between the histories of third world societies (traditionally treated as self-organising and independent) and Europe's own economic history. The approach, as we would expect, is broad and interdisciplinarian, the arguments clear, but with sensitive awareness of the complexity of the historical process: 'The historian should not then give money some special significance, but should observe it closely, and this requires research'.(19) So far as explanation is concerned, history comes before logic: in response to a hypothesis put forward by Cipolla, he replies, 'This in no way proves that the monetary factor "came first". Only history can verify that'. (18) This is a lesson apparently not learned by many social scientists in their approach to historical phenomena. This is not empiricism but that same concern for 'thorough historical research' evinced by Marx in his own theoretical attempt to understand money(7).

This is all I shall say on the content of the book. It is a book for historians, and this is why I looked to it for help on problems of money in the distant and remote eighteenth century Deccan (India). So let us elucidate.

The fact is that *A History of Gold and Money* is flawed. It is not so much a history of gold and money as a history of gold and money in European society and economy. This Eurocentric bias is not just a question of missing facts and details, or of the absence of a useful chapter, one that would do justice to current interest (in fact very tardy) in the non-European world. Rather, in our view, the failure to assess the history of gold and money in other parts of the world (so far as it is known) may ultimately lead to inadequate descriptions and explanations for European phenomena. These are severe remarks coming as they do from the edge of the crowd, but speaking generally it is time that the natural conservatism of our disciplinary idioms be subjected to some intense attack.

Originally published in 1969 in Spanish, republished in French in 1974, and now in English in 1976 (allowing adequate time for revision between editions, one would think) it seems to us astonishing that, excepting for the treatment of certain important centres of production of precious metals, or sources of them by trade, the third world—and especially Asia—figures only in a handful of brief, fragmentary references. A partial exception occurs between pages 93–101, where Vilar cursorily describes the currency systems of Asia, establishes (but without explanation) their need for imports of precious metals, and discusses the main characteristics of European trade in precious metals to the East. He remarks that 'Metaphors about chasms or cyclones have been used to express the attraction exerted by China or India for gold and silver, and to explain the influence of such distant economies as those of the Far East on the economies and currencies of Europe. It is essential to be clear about what this means. What establishes the different pressures is the disequilibrium between the relative values of commodities on the one hand and precious metals on the other, depending on whether one is in America, Europe or Asia'. (98) He then refers to three general subdivisions of the world: the dynamic and productive zones of Europe and monsoon Asia; the backward zones containing exploitable mineral resources, Africa and America; and Portugal which takes advantage of the price differentials between the first two categories.(98) And this is all. It hardly disposes of the problems set up by these structural differences—especially their relative stability over time; nor does it dispose of the question and character of interdependent relationships subsequently set up. The metaphors remain because the problem is not successfully set aside by simple categorisations. Indeed, Vilar's keen sense of problems, his concern to locate cause and explanation, to take nothing for granted, seems to lapse at this point. Thus, 'João de Barros, the great Portuguese expert, pointed out that the Malabar coast was marked by an absence of gold, and was therefore hungry for it. But this was true of the whole of the Deccan and of Hindustan. They were all part of the attraction area. The next point to examine . . .'.(98) But why? Why in this sense were they attraction areas? And what of the internal problems indicated by our need for metaphors?

In Chapter 21 some further brief references to Asia occur, but these are mainly concerned with problems of production, except for some much too abbreviated and purely schematic remarks on the relationship of overseas trade to European events. Finally, the importance of eighteenth century silver exports to Asia are briefly mentioned on page 285.

This is in spite of Vilar's statement that 'This book is . . . concerned with gold in the world as a whole . . .'(92). Indeed his remarks at this point lead the reader to believe that his intention was precisely to 'establish the vast scale of the non-European aspects of the problem'. (92) But subsequent treatment belies these expectations. As points of attraction for precious metals produced via the European system, Asia has a passive presence; Magalhães-Godinho's description of India as a low pressure area is as true historiographically as it is historically (Vilar, 98; quoting Magalhães-Godinho, 1969: 285).[2]

Yet it is precisely in the past 15 or so years that the whole question of the role of Asia in the growth of Europe during the Early Modern Period has begun to raise its awkward and ambiguous head—at least to raise itself as a very large question mark over the boundaries of convenience which have for so long confined our disciplines. Firstly, the role of gold and silver exports to Asia (India and China especially) in Europe's own economic growth; its central role in facilitating the development of an incipient world market structure; Europe's precocious search for overseas sources of precious metals; the many-sided response by different facets of European life to overseas expansion . . . all require a history of money to be more concerned with the role of the East as an essential (and formative) part of the subject matter. We repeat that this is even so with respect to treatment of Europe's own history. Precisely as a zone of attraction, Asia may, ironically, have had a far from passive role in Europe's growth.

Specie moved from America in two directions into Asia from the sixteenth century on: directly across the Pacific via Manila, and indirectly eastwards, through Europe, over land and by sea. Atwell [*1977*] sees the Manilla route as having been vitally important in the general development of China during late Ming times. In the seventeenth and eighteenth centuries Asia provided high prices for precious metals, mainly silver, and also for copper, and their import into India and China has occasioned an increasing, if never adequate discussion. With respect to India, indeed, Habib, and later Hasan, early put forward a translation of the Hamilton thesis into Indian terms [*Habib, 1963; Hasan, 1969*].[3] Such views are important because however questionable their particular theories, they serve implicitly (and in a pioneering manner) to equate the post-sixteenth century histories of Asia and Europe in a degree of relative inter-dependence that has yet to be properly phrased. Indeed a complex and portentous realm of questioning arises therefrom, the price being that from the perspective of the given frameworks of discussion one seems to pass beyond the bounds of credulity, even in merely asking such questions.

It can hardly be disputed that trade in precious metals was a fundamental lever of European trading ventures in Asia, and that the rapid growth in the organisation of overseas trade thereby depended on the zones of reception provided by Asia's economies for coinable metals. Given the world-wide trading dependencies that gradually developed during these centuries [*e.g. Mauro, 1961*] European activities overseas require to be understood in the facilitating context of the readiness of Asia's merchants to pay well for silver and copper and to exchange Asian manufactures and agricultural products for them. It is thus reasonable to suggest that European history—given the increasing importance of overseas trade to that history—must come to be considered more in the light of such inter-dependencies with the rest of the world.[4]

What of this receptive medium? The whole question of the contribution of this trade to price movements in India is disputed, even the fact of inflation itself. Thus Raychaudhuri argues that company trade 'probably accounted for a fraction of the total economic activity of Bengal in the 17th century', and this is surely true of India generally. He also suggests that indigenous expansion of output could well have accounted for price phenomena in Bengal [*Raychaudhuri, 1977: 530*]. Indeed, we could go further in arguing that there seem to be excellent grounds for postulating India and China as providing independent sources of growth capable of absorbing, and even coming, more or less gradually, to require heavy imports of specie from outside. In the fifteenth and sixteenth centuries, as in Europe, indigenous sources of precious metals were more intensively worked and became comparatively exhausted. We can thus suggest the probability of a two-stage development: in the seventeenth century imports into India were of especial significance for European trade and thus for Europe. In the eighteenth century a rather complex series of dependencies developed as imports accumulated and accelerated.[5] Atwell's hypothesis is surely excessive, and yet at a secondary level of gradually-instituted dependencies between independently-growing regions of the world, subsequently coupled in trade, a dramatic reduction in the flow of metals may indeed have resulted in the crisis phenomena which he documents for Ming China in the 1630s. A table worked out recently by Michel Cartier shows a degree of common development in different regions of East Asia between the sixteenth and eighteenth centuries, that is astonishing in the light of the highly compartmentalised historiography (Cartier, 1978). The evidence for India also fits into this conspectus of long term growth (despite unevennesses and reverses, regional crises, etc). India indeed provides us with a useful exemplar for a discussion of indigenous sources of growth.[6] Between the late fifteenth and eighteenth centuries there is sufficient evidence to argue the case for extensive expansion of peasant orders throughout the subcontinent,—expanding, that is, out of nuclear foci and belts of settlement concentrated along coasts and fertile river valleys. This was also a period of impetuous state development, and of equally impetuous experimentation in modes of taxing the countryside and of progressive extension of monetisation. According to Perkins [*1969*], Atwell [*1977*], Elvin [*1973*] and others, such processes may be discernable in China as well.

So far as peasantries are concerned, the development of a realm of questioning, now beginning to dissolve the old certainties about pre-colonial Asia, has also come to affect old judgments about the role of cash economy and peasant autonomy. Peasants of the seventeenth and eighteenth century, in most of India—under Mughals or Marathas—were caught up in a web of contextual conditions dominated by the prior existence of money, by an extensive and pervasive financial organisation, and by monetary accounting systems running right through the whole hierarchy of society from merchant and noble to village and peasant. Chronic local and seasonal shortages, even regional money famines, and numerous ad hoc means of bypassing exchanges of actual coin, only express this intensive dependence. The organisation of rural economy, forms of administration, social relationships and property, were each pervaded by the fact of money and by the specifics of its organisation. The qualification to such claims is that we remain right at the beginnings of a true understanding of these problems. To return to Vilar (to the quotation above from p. 19), observation of money, and a small and inadequate body of research permitting such observation, indicates that money may indeed have possessed a special significance—special, at least, given the neglect of its study in the case of pre-colonial India, and given the mythology (traditional emphasis by scholars on features redolent of backwardness and archaism). What I mean by this claim is that, within the general movement synthesising the many components making up societal structure and development, money should be allowed the status of as formative a role in pre-colonial modes of production, as political forms and types of property. It indicates an institutional domain constitutive of, and central to, production relations, a cohesive force essential to political functioning (even in its military and predatory forms), and as such, by the seventeenth and eighteenth centuries, transcending the space occupied by any one of the various polities in the sub-continent. Remove money, and we would find ourselves with very different kinds of society; documentary evidence of the extent of monetary relationships is abundant. The further implication is that the problem of gold and money, as treated by Vilar, may be viewed as concerning a realm embracing even the heartland of apparently remote Asian polities.

Historians spend too little time on the problem of copper, all important for the bulk of small exchanges in which humble men were involved. The evidence for part of the 18th century Deccan (in Moharashtra, but comparatively remote from coast and European trading company), suggests that even humble labourers, rank and file troopers and husbandmen could be accustomed to handling copper and even silver coin.[7] During the seventeenth and eighteenth centuries, according to Furber [*1976: 249–251*], imports of copper into India and China were consistently high, fetched surprisingly high prices and had a vital role, first in Dutch and subsequently in English company trade in Asian waters. Vilar (98) also refers briefly to its important role in early sixteenth century Portuguese exports to Asia and in Asian currency systems. At least Furber, but not Vilar, refers to this trade as a problem 'imperfectly understood' [*1976: 249*]. Comparative neglect of copper is echoed by the generality of historians concerned with money in Asia. K N Chaudhuri's long chapter on money in his new book gives it scant space or reflection, impairing his contribution to the subject. Indeed, except for Moreland [*1923*] and Furber, it is hard to find any historian interested in the less precious moneys so far as Asia is concerned.

Given the degree of monetisation in pre-colonial India, and its role in tax gathering and in payments generally, ***and given an expanding peasant base,*** it is not hard to understand the extraordinary thirst presented by the subcontinent for coinable metals. Was it this that constituted the invisible drain into which gold and silver seemed to lose itself after import? There is little need to rely on traditional explanations like hoarding. Similarly with China. Such a thesis, with its demographic dimension, would demand that the question of prices (stable or inflationary) be assessed according to a more elaborate understanding of changes affecting Asian economies, as has indeed happened in the post-Hamilton historiography of the 16th century, European price rise.

The implication of this discussion is that of providing a positive historical basis for considering the development and growth of European overseas trade. However independently we view developments in Asia and Europe (or in different Asian regions), a certain parallelism in the chronology of growth occurred, and through the mechanisms of trade in specie and commodities a certain level of inter-dependence must surely have arisen, predating direct forms of colonial intervention. Moreover, at the start the ties of dependence may have been more on Europe's side than on India or China's.

Of course, given the state of research all this is thoroughly hypothetical and debatable, and the suggestions vague. But given available literature and knowledge, Vilar might have asked some of the questions; concerned as he is with the formation of capitalism in Europe, he might have considered whether or not such questions have serious implications for the role of gold and money in Europe's history. Nor is it a question of asking him to repeat one or another hypothesis (and certainly not mine), but rather of paying attention to the intellectual space that gives rise to them.

Be that as it may, the book remains of great value if these problems are kept in mind. Its danger—and Vilar in this respect exemplifies it—is the curious myopia of Europeanists, long-accustomed not only to see themselves at the centre of the world but even as the only world—to yawn quietly (but we would not accuse Vilar of this) whenever 'India' is mentioned. Nevertheless, Vilar has written a superb and profound introduction to the meaning of money in history—clear and precise, accessable to those without technical knowledge, and full of provocative theses and ideas. As the blurb on the cover tells us, it is indeed a 'unique and long-needed work of synthesis', even if from a non-Europeanist's viewpoint we still await a truely adequate synthesis. With respect to the latter, Vilar would serve us well if he set about writing a second edition.

NOTES

1. 'Gold' for Vilar refers to the problems of 'gold' and currency systems in the twentieth century, and his history of gold is thus a history of modern monetary media, a good deal of the book thus concerned with the problem of silver.

2. One understands Vilar's difficulty when he speaks of Asia's 'little-known economies' (92), but does he not preempt the form of the answer when he speaks of them 'as whole worlds in themselves' (92)? This is precisely what we do not know because the question of the degree of their autonomous self-development has not been asked let along investigated for the pre-colonial epoch in question. His 'manner of speaking' here is quite precisely the answer he has assumed. And further, while these economies are indeed little-known, he would have been more convincing if he had shown some awareness of that work which was available—for example, Habib's own magisterial contributions on the nature of India's pre-colonial economy [*Habib, 1963 & 1968*].

3. The pre-history of the inflation dispute takes us back to contemporary observations and speculations, although the present-day argument begins with Habib. Recent work on Asia's monetary history includes Atwell, Cartier, Chaudhuri and the various contributions of the Wisconsin Conference (see references). The subject seems indeed to have at last achieved a momentum of its own.

4. On page 200 Vilar notes a relationship between a phase of great growth in trade in the East and of a spectacular growth in the United Provinces and England, 1680–1715. A quotation and map from Chaunu, showing the fluctuations in the East, are unitegrated with the text, but refers to China's population growth in this period. These teasing, cursory remarks only accentuate the sense of a dimension missing from the text.

5. This hypothesis is based merely on impressionistic reflection on the Indian evidence. I would argue that as the eighteenth century proceeds the whole momentum of affects on independent hinterland politices produced by the many forms of European intervention elsewhere in India—and hidden within and beneath the institutional appearance of these polities—builds up to a relatively critical pitch. Again, it has yet to be studied because until recently it has not been considered a serious question.

6. I am currently formulating these views—both on the role of money in the agrarian economy (and thus the role of apparently remote European trade) and also on the general historical developments characterising these centuries (state development, etc)—in the book on which I am now working. A cursory and incipient presentation occurs in Perlin (forthcoming).

7. This is not to say that they invariably were so accustomed to handling money; this depended on the precise character, complexity and density of the social relationships of production in which peasants locally found themselves. Thus intensified lordship may in one sphere have altered previous money relationships with government into ones in kind, while in another sphere of relationships in the same locus, it may have introduced a large range of money payments (wages and the like). A single husbandman may have been affected by both changes. These are mere logical inferences, however, the actual circumstances of access being much more complex. Then, I should also add that I use the word 'accustomed' not in comparison with 20th century urbanites, but in contrast to the presuppositions about the past held by the latter, preventing proper study of the subject.

REFERENCES

Atwell, William S., 1977, 'Notes on Silver, Foreign Trade and the Late Ming Economy', *Ch'ingshih wen-t'i*, Vol. 3, No. 8, 1–33.

Carr, Raymond, 1966, *Spain 1808–1939*, Oxford.

Cartier, Michel, 1978, 'Les Importations de Metaux Monetaires en Chine', paper read at the *Collequium on History and Underdevelopment*, University of Leiden.

Chaudhuri, K. N., 1978, *The Trading World of Asia and the English East India Company, 1660–1760*, Cambridge.

Elvin, Mark, 1973, *The Pattern of the Chinese Past. A Social and Economic Interpretation*, Stabford.

Furber, Holden, 1976, *Rival Empires of Trade in the Orient 1600–1800*, Minneapolis.

Goubert, Pierre, 1960, *Beauvais et le Beauvaisis de 1600 a 1730. Contribution a l'histoire sociale de la France du XVIIe siecle*, Paris.

Habib, Irfan, 1963, *The Agrarian System of Mughal India 1556–1707*, Bombay.

Habib, Irfan, 1969, 'Potentialities of Capitalist Development in the Economy of Mughal India' *Journal of Economic History*. Vol. 29, 32–78.

Hasan, Aziza, 1969, 'The Silver Currency Output of the Mughal Empire and Prices in India during the 16th and 17th centuries', *Indian Economic and Social History Review*, Vol. VI (also *Annales E.S.C.*, 1969, 24.)

Kolm, Serge-Christophe, 1975, 'Notes sur les Lecons monetaires de l'histoire *Annales E.S.C.*, 6, 1477–1482.

Kula, Witold, 1976, *An Economic Theory of the Feudal System. Towards a Model of the Polish Economy 1500–1800*, London.

Magalhaes-Godinho, V., 1969, *L'Economie de l'Empire portugais aux XVe et XVIe siecles*, Paris.

Mauro, Francois, 1961, 'Towards an "Intercontinental Model": European Overseas Expansion between 1500–1800', *Economic History Review*, 2nd ser., XIV, 1–17.

Moreland, W. H., 1972 (1923), *From Akbar to Aurangzeb. A Study in Indian Economic History*, New Delhi.

Perkins, Dwight H., 1969, *Agricultural Development in China 1368–1968*, Edinburgh.

Perlin, Frank, forthcoming, 'The Pre-Colonial Indian State in History and Epistemology. A Reconstruction of Societal Formation in the Western Deccan from the 15th to the early 19th century', in Claessen & Skolnik, eds., *The Study of the State*, The Hague.

Perlin, Frank, forthcoming, 'Pre-Colonial South Asia and Western Penetration in 17th-19th Centuries. A problem of Epistemological Status', *Review*.

Raychaudhuri, T., 1977, (Review), *Economic History Review*, 2nd ser., XXX, 529–531.

Vilar, Pierre, 1956 (a), 'Le Temps de Quichotte', *Europe*, 34, 3–16 (also *New Left Review*, No. 68, 1971).

Vilar, Pierre, 1956 (b), 'Problems on the Formation of Capitalism', *Past and Present*, No. 10, 15–38.

Vilar, Pierre, 1962, *La Catalogne dans l'Espagne moderne*, 3 vols., Paris.

Vilar, Pierre, 1967, *Spain. A Brief History*, London etc.

Vilar, Pierre, 1973, 'Histoire marxiste, histoire en construction. Essai de dialogue avec Althusser', Annales E.S.C., 28, 165–198 (also *New Left Review*, No. 80, 1973).

Vilar, Pierre, 1974, *Or et Monnaie dans l'Histoire 1450–1920*, Paris.

Wisconsin Conference, 1977, *Comparative World History Workshop in Pre-Modern Monetary History 1200–1750 AD*, Madison, papers to be published.

11
Born with a 'Silver Spoon': The Origin of World Trade in 1571

Dennis O. Flynn and Arturo Giráldez

The birth of world trade has been described by C. R. Boxer (1969, p. 17) in the following terms: "Only after the Portuguese had worked their way down the West African coast, rounded the Cape of Good Hope, crossed the Indian Ocean and established themselves in the Spice Islands of Indonesia and on the shore of the South China Sea; only after the Spaniards had attained the same goal by way of Patagonia, the Pacific Ocean and the Philippines—then and only then was a regular and lasting maritime connection established between the four great continents."

Although Boxer does not pick a specific date for the birth of global trade, his logic leads us to choose 1571—the year the city of Manila was founded. Manila was the crucial entrepôt linking substantial, direct, and continuous trade between America and Asia for the first time in history.

For our purposes, global trade emerged when all important populated continents began to exchange products continuously—both with each other directly and indirectly via other continents—and in values sufficient to generate crucial impacts on all the trading partners. It is true that there was an important intercontinental trade before 1571, but there was no direct trade link between America and Asia, so the world market was not yet fully coherent or complete. To understand the global significance of the direct Pacific trade between America and Asia—international trade history's "missing link"—it is useful first to discuss the underlying economic forces that motivated profitable world trade in the early modern period. The singular product most responsible for the birth of world trade was silver.

The Role of Silver in Creating a World Market

More than the market for any other commodity, the silver market explains the emergence of world trade. China was the dominant buyer of silver. On the supply side, Spanish America (Mexico and Peru) erupted with unprecedented production of the white metal. Conservative official estimates indicate that Latin America alone produced about 150,000 tons of silver between 1500 and 1800 (Barrett 1990, p. 237), perhaps exceeding 80% of the entire world production over that time span (Cross 1983, p. 397). Despite America's dominance in silver production over three centuries, Japan may have been the primary exporter of silver to China in the late sixteenth and early seventeenth centuries, shipping perhaps 200 tons per year at times. Japanese silver exports, however, fell off dramatically in the second half of the seventeenth century (Innes 1980, chap. 6).[1] "The amount of Japanese silver poured into foreign trade in the heyday of Japan's overseas trade, 1615 to 1625, through Japanese, Chinese, Dutch, Portuguese and other ships, reached tremendous value, roughly estimated at 130,000–160,000 kilograms—equal to 30% or 40% of the total world silver production outside Japan. This explains why European and Asian merchants were so enthusiastic about developing trade with Japan" (Iwao 1976, p. 10). The central point is that all the great silver mines in both hemispheres sold ultimately to China.

We intentionally emphasize the role of China—and its tributary system (Hamashita 1988)—in the silver trade because the scholarly literature in general has neglected this pivotal country, certainly in terms of recognizing China as a prime causal actor. The literature on New World treasure is huge and multifaceted—sometimes focusing on sixteenth-century price inflation (the price revolution),[2] and/or the rise and fall of Spain,[3] and/or the transition from feudalism to capitalism,[4] and many other issues—but it is unified in one central respect: it focuses virtually exclusively on Europe as the fulcrum. Europe is considered the epicenter of early modern commercial activity. The sixteenth-century price revolution, for example, is mostly thought of

[1] According to the calculations made by Barrett (1990, p. 225), Japan may have produced about 30% of the world's silver in the sixteenth century and around 16% in the seventeenth century.

[2] For an overview of the main theoretical arguments of the price revolution literature, see Outhwaite 1969; Ramsey 1971; Flynn 1977, chap. 2; Flynn 1984a and 1984b.

[3] See Hamilton 1937; Elliott 1961; Kamen 1964; Gordon 1975; Kamen 1978; Israel 1981; Kamen 1981; Flynn 1982.

[4] Hamilton 1929; Keynes 1930, vol. 2; Neff 1936; Hamilton 1952; Felix 1956; Nadal 1959; Wallerstein 1974 and 1976; Brenner 1977; Hunt 1978; Flynn 1984.

in terms of European price inflation, when in reality it was a global phenomenon.[5] In a remarkably clear summary of China's participation in the global price revolution, Geiss (1979, p. 144) provides an exception to the normal Eurocentric focus: "In the late sixteenth century, however, when silver from Mexico and Japan entered the Ming empire in great quantity, the value of silver began to decline and inflation set in, for as the metal became more abundant its buying power diminished. This inflationary trend affected the values of all commodities; everything had been valued in silver and silver lost its value. Ramifications of this change touched the lives of almost everyone in the empire."

Europe's Dutch and English East India Companies are often viewed as prototypes of modern multinational corporations. The scholarly literature recognizes that huge quantities of silver flowed to Asia, but this phenomenon is considered a reflection of Europe's balance-of-trade deficit with east Asia; Europeans developed a far greater taste for Asian finery than the other way around, according to conventional wisdom, so treasure had to flow from west to east to pay for Europe's trade deficit. In short, all the key issues are normally framed in terms of European perspectives.

Acceptance of a global perspective instead of the predominant Eurocentric view outlined above yields a startlingly different view. It becomes clear that Europeans did indeed play an important role in the birth of world trade, but their role was as middlemen in the vast silver trade; they were prime movers on neither the supply side (except Spain in America) nor the demand side of the worldwide silver market. Europeans were intermediaries in the trade between the New World and China. Massive amounts of silver traversed the Atlantic. After it had reached European soil, the Portuguese in the sixteenth century and Dutch in the seventeenth century became dominant distributors of silver by a multitude of routes into Asia. Attman conservatively estimates that 150 tons of silver passed through Europe into Asia on an annual basis. "The country which reigned supreme in arranging solutions for the deficit problems in world trade was Holland, acting mainly on her own behalf in the Baltic area, the Levant and Asia, but also acting for other nations. . . . Even when other countries needed precious metals Holland acted in many cases as a clearing centre and in the final stage as an exporter of precious metals. . . . But the bulk of the precious metals required for the Asian trade came from

[5] For price inflation in China, see Cartier 1981 and Wilkinson 1980; for Japan, see Innes 1980; for Turkey, see Sahillioglu 1983; and for Russia, see Blum 1956.

Europe around the Cape" (Attman 1986, p. 6). It is worthwhile to reflect momentarily on Attman's estimate of 150 tons of silver exported each year from Europe to Asia. Attman emphasizes that his estimates include only specie shipments (i.e., bullion is excluded from consideration) and that he has studied only port records (i.e., overland trade is excluded). If Attman had included bullion shipments through ports, as well as specie and bullion shipments over land routes, his estimate of west-to-east silver flows would have been far greater. He warned readers of the conservative and partial nature of his estimates, but scholars rarely acknowledge this when citing his figures.

The Pacific leg of the China trade has not received the attention it deserves. Writing of the period 1571–1620, TePaske says that an enormous quantity of silver passed over the Pacific, especially out of Acapulco and through Manila on its way to China. "Mexican silver also flowed out to the islands in large sums, far exceeding the 500,000-peso limitation. In fact at the opening of the seventeenth century the drain of pesos from Mexico to the Orient through the Philippines was estimated at 5 million pesos [128 tons] annually, with a reported 12 million pesos [307 tons] being smuggled out in 1597" (TePaske 1981, p. 436). These are shocking figures because this alleged average of 5 million pesos (128 tons per year) over the Pacific, at the turn of the century, is only 15% less than the 150 tons minimum that Attman says Europe shipped to Asia on an annual basis. Moreover, the 12 million pesos (307 tons) in 1597 is more than double Attman's estimate of the entire European leg of the journey of silver to China. These may seem like fantastic figures to some, but Barrett (1990, p. 236) has pointed out a glaring discrepancy between Spanish American production figures and estimated exports to Europe: production seems to have exceeded exports to Europe by 5.5 million pesos (135 tons) per year. Barrett reasons that this 5.5 million pesos must have either remained in America or been exported through the Philippines.[6] It will become clear shortly that exportation of such a vast amount of silver through Manila makes sense in global terms, while its retention in the New World does not.

How has the Philippines trade been perceived by Asian scholars? Using the Blair and Robertson collection of primary sources, Chuan (1969, p. 79) estimates that at least 50 tons of silver (2 million pesos)

[6] Cross 1983, p. 420, talks of a vast smuggling trade in the sixteenth century: "Quantities of silver left the New World through the ports of Buenos Aires and Sacramento and through the Manila Galleons. At the peak of these activities, perhaps as much as 6 million pesos per year (159,000 kg), or half the output of Peru, was diverted to these channels from the Seville trade."

passed over the Pacific annually throughout the seventeenth century; silver shipments did not decline slightly after 1620 and precipitously after 1640, as previously argued by Chaunu (1960, p. 250).[7] How significant is 50 tons in relative terms? Fifty tons of silver equals the average annual exports to Asia by Portugal plus the Dutch and English East India Companies combined in the seventeenth century (Flynn and Giráldez 1994a). Fifty tons of silver was also the amount that was shipped through the entire Baltic trade (the Russian and Polish-Lithuanian markets), according to Attman (1986, p. 81). It is interesting to contrast the historiographic notoriety of the Portuguese trade, in combination with that of the East India Companies, with the situation of the Manila galleons on the Pacific side; each may have shipped approximately equal quantities of silver annually, yet the Pacific trade is rarely mentioned.

Manila had no purpose other than the trade in silver and silk. The city contained 42,000 inhabitants in the middle of the seventeenth century (Wolf 1982, p. 153), approximately the same population as Barcelona, Danzig, Marseille, and other cities with more broadly based economies (Mols 1977, pp. 42–43; deVries 1984, app. 1). Manila's population *circa* 1650 included about 15,000 Chinese, 7,350 Spaniards, and an estimated 20,124 Filipinos.[8] The Pacific route of silver to China was Spain's only avenue for entry into the lucrative Asian marketplace because the trade out of Europe in the sixteenth and seventeenth centuries was controlled first by the Portuguese and later by the Dutch. Spain's Manila galleons initiated the birth of Pacific rim trade more than 420 years ago.

Eurocentrism predisposes us to imagine that the East India Companies injected dynamism into backward Asian economies in the early modern period. Recent scholarship (Hamashita 1988, for example) suggests that the European companies simply plugged into the preexisting network of intra-Asian trade. The export of Japanese silver provides a good example of this process. As was the case in the west-to-east trade, first the Portuguese—in competition with Chinese junks and Japanese red-seal ships—and then (after 1639) the Dutch played

[7] Pierre Chaunu's conclusions were based on study of the *almojarifazgo* duties, which covered only the taxed portion of the trade. It is widely known, however, that smuggling was rampant in the Manila trade, and sources from the Asian side of the Pacific provide estimates of silver exports far in excess of Chaunu's numbers (Flynn and Giráldez 1994a and 1995).

[8] Phelan (1959, p. 178).There were thousands of Japanese in Manila, too, including Japanese Catholics fleeing persecution at home. Spaniards feared a Japanese "fifth column" within the Philippines, however, and expelled large numbers of Japanese several times (Innes 1980, p. 59).

the role of intermediaries in this crucial Sino-Japanese trade. Again within Asia's marketplace, the European role is most accurately portrayed as that of middlemen, not prime movers. Europeans were important, but potentially disposable, intermediaries who could be—and in the case of the expulsion of the Portuguese from Japan in 1637, were—replaced at the convenience of Asian trading partners.

China: The World's Silver Sink

The early modern production and distribution of silver in the western hemisphere has been studied extensively, but the world's biggest end-customer, China, is routinely eliminated from the story. This is peculiar. Nobody would think of analyzing, say, the world oil industry today without paying considerable attention to the major oil-importing industrial regions; China's dominance as an importer of silver was arguably at least as pivotal during the birth of world trade as is the industrial world's dominance as importers of oil in today's global marketplace. Godinho (1963, 1:432–65) aptly describes China as a "suction pump" *(bomba aspirante)*, a "vacuum cleaner" that attracted silver globally for centuries. Surprisingly, few scholars have continued to investigate the nature of China's metamorphosis into a seemingly bottomless silver sink; Atwell (1977, 1982, 1986, 1988) provides perhaps the most consistent exception to this bias.

The market value of silver in Ming territory was double its value elsewhere. This fact is reflected in the bimetallic ratios reported in Chuan (1969, p. 2): "From 1592 to the early seventeenth century gold was exchanged for silver in Canton at the rate of 1:5.5 to 1:7, while in Spain the exchange rate was 1:12.5 to 1:14, thus indicating that the value of silver was twice as high in China as in Spain." Divergent bimetallic ratios created tremendous prospects for profitable arbitrage trade. Economic theory predicts that gold should have flowed out of China, where it was undervalued relative to the rest of the world, in exchange for Japanese and Western silver, which was relatively overvalued in China compared with the rest of the world. This is precisely what happened from the middle of the fifteenth to the middle of the seventeenth century (Chaudhuri 1978 and 1986; Flynn 1986). It is crucial to focus on silver to understand the underlying motivation of world trade: it was the elevated value of silver inside China that created the opportunities for profit around the globe. Rather than see the west-to-east flow of silver as a reaction to Europe's trade deficit with Asia, we contend that the cause of the trade centered in China and its tributary system. Demand-side causation was of Asian origin, to which the rest of the world reacted.

How can we be confident that the arbitrage argument outlined above is superior to the traditional European trade-deficit hypothesis? The trade-deficit argument says that "money" would have to have been transshipped to Asia to cover the trade imbalance. "Money" here refers to all types of high-value coins containing internationally recognized intrinsic content, such as gold and silver. However, we have already established that gold and silver did not travel jointly into the Asian marketplace as a balancing item called "money." New World silver did indeed travel from Europe to Asia, but it crossed paths with gold coming in the opposite direction—out of Asia and into the West. Abstract "money" did not balance a trade deficit in the passive way commonly portrayed in the literature; rather, it was a specific commodity—silver—that traveled to Asia, not gold. Gold was one of the products for which silver was exchanged. The cause of this trade rests with developments endemic to the silver market, not with developments in nonsilver markets. Moreover, the exchange of silver for gold was not a Europe-versus-Asia issue in any case. Japanese silver also flowed to China in exchange for Chinese gold, which flowed into Japan, for exactly the same reasons that gold flowed to the West. In sum, Europe was not the causal center in early modern trade; moreover, the East-versus-West "trade imbalance" was not the mechanism driving world trade. There was no "trade imbalance" for which to compensate, so long as we recognize that silver itself was the key commodity distributed globally and that it was exchanged for items—mostly silk and porcelain but also gold—from the Asian mainland. Causation was located in the silver market itself, with America and Japan anchoring the supply side and China dominating the demand side.

China's metamorphosis from a paper-money system (dating from at least the eleventh century) to a silver-based economy was crucial. Overissue of paper money in China had reduced the value of this fiduciary medium to virtually nothing by the middle of the fifteenth century (Gernet 1982, p. 415). Daily commerce required a medium of exchange to replace the worthless paper money, and silver evolved as the metal of choice. Gold was too valuable for most ordinary transactions, but copper coinage was a candidate for monetary preeminence. Geiss (1979, p. 155) explains how silver defeated copper, in a passage worth quoting at length:

> The value of the coin lay in the metal, not in the mint. In that respect copper coins were hardly different from silver; each was valued as a

> piece of precious metal. While silver could, if necessary, be assayed for purity, copper coins could not. To assay a copper coin entailed its destruction. The only way to ascertain the copper content was to melt the coin, and this would defeat the purpose of coining money. But with coins of varying weight and metallic content in circulation, setting a price in copper coins became a tricky business. The rice merchant would have to specify what kind of copper coin he had in mind, each had a different value in the marketplace, and the price of the merchant's rice depended on the type of coin offered in payment. How much simpler to set the price in silver, and that in fact is what happened. Silver came to be the preferred medium of valuation and exchange.

Geiss's explanation is straightforward and compelling. Silver's gradual "conquest" of the Chinese economy may seem relatively innocuous to some at first glance—just another detail in "area studies"—but this development fundamentally altered the direction of international commerce. It also influenced the structure of power among nations throughout the world.

The Ming tried repeatedly to retard the intrusion of silver into (and from) the coastal centers of merchant power. Silver's penetration was irresistible, however, and local governments in maritime regions began specifying that taxes be paid in silver. Gradually Ming rulers abandoned their resistance to silver and implemented the Single-Whip tax system around the 1570s (Huang 1974; Liang 1970). The Single-Whip system specified two things: first, myriad existing national levies were consolidated into a single tax; second, all tax payments were to be made in the form of silver. Considering that China contained perhaps one-fourth of the earth's population by the seventeenth century, with urban centers of up to 1 million inhabitants (five to seven times greater than the largest cities in western Europe), the "silverization" of China inevitably had global ramifications.[9] China's tributary system also converted to silver, so we are talking about far more than one-quarter of the globe's population. Conversion of the world's largest economic entity to silver caused the metal's value to skyrocket in China relative to the rest of the world.

The early modern world silver trade involved structural transformations that penetrated much deeper than the mere movement of a couple of hundred thousand tons of the white metal to its most lucra-

[9] Demand for silver along China's coastal region alone must have been significant, considering that Nanjing contained more than 1 million inhabitants and Beijing around 660,000 in the late Ming period (Rodzinski 1979, p. 201).

tive market. The process yielded prodigious profits for key individuals and institutions. From mines in the Andes and Japan to the streets of China, profit was the motive force at each stage of the trade. European middlemen profited mightily from intercontinental trade and perhaps even more from the inter-Asian trade linkages, but the truly grand profiteers in the silver saga were those entities that controlled the centers of its production: imperial Spain and the Tokugawa shogunate. Conversely and ironically, the silver trade may have contributed indirectly to the overthrow of the Ming dynasty.

Silver and the Power Bases of Imperial Spain, the Tokugawa Shogunate, and Ming China

The richest silver mine in the history of the world was discovered at more than 15,000 feet altitude in the Andes in Potosí (present-day Bolivia), a one-way journey of two and one-half months via pack animal from Lima. Nothing grew at that altitude, so there was no population at the time silver was discovered in 1545. During the ensuing sixty years, Potosí's population swelled to 160,000, about equal to that of London or Paris.[10] This would be the modern-day equivalent of, say, 20 million people moving to a spot on Alaska's North Slope. Evidently something unusual was going on in Potosí.

Potosí's *cerro rico* (rich mountain) may have produced 60% of all the silver mined in the world in the second half of the sixteenth century.[11] Its veins were incredibly rich. In addition to naturally bountiful deposits, a series of new production technologies—the most famous being the mercury-amalgam "patio process"—combined to render Spanish American mines the world's lowest cost sources of silver (Jara 1966). This supply-side phenomenon was particularly fortuitous because it coincided chronologically with the extraordinary rise in the value of silver caused by the Chinese demand-side forces culminating in the Single-Whip tax reform. The combination of low supply-side production costs in Spanish America and Chinese-led demand-side elevation in silver's value in Asia generated probably the most spectacular mining boom in human history. This combination of supply-side and demand-side forces implied enormous profits.

No entity reaped greater rewards from the silver industry than the

[10] DeVries 1984, app. 1.

[11] Cross (1983, p. 404) states that "despite the decline of Potosí, beginning in the 1640s, the viceroyalty of Peru accounted for 60% of the world's silver production in the sixteenth and seventeenth centuries."

Spanish crown, which wisely allowed favored "private sector" entrepreneurs to operate New World mines, rather than attempting to do so itself.[12] Instead, the crown took a substantial fraction of mining profits through taxes. The most famous tax was the *quinto*, a 20% severance tax on gross value, but there were many indirect taxes as well.[13] According to Hamilton (1934, p. 34), 27.5% of total registered precious metals entering Seville between 1566 and 1645 belonged to the crown of Castile.[14] Revenues from overseas mines provided the fiscal foundation for the Spanish empire.

Spain was a small country of perhaps 7.5 million inhabitants by the middle of the sixteenth century, about half the population of France (Elliott 1961, p. 57). Elliott (1961, p. 62) has described Castile of 1600 as "an economy closer in many ways to that of an East European state like Poland, exporting basic raw materials and importing luxury products, than to the economies of West European states." Carande (1965, p. 340) and many others classify Spain as backward domestically, substantiating the observation that the financial foundation of the Spanish empire was based on resources outside the Iberian peninsula.[15] Mine profits were enormous, and there was no comparable profit center elsewhere, so we conclude that the New World mines supported the Spanish empire (Flynn 1982).

This view of Spain leads to surprising but inevitable conclusions. We have already established that domestic developments inside China elevated the value of silver in world markets far beyond what it could have been otherwise. The largest beneficiary of silver's high value must have been the Spanish crown, the institution that reaped enormous profits by way of its control and taxation of the low-cost New World centers of production. Thus, the silver-industry profits that financed the Spanish empire were huge because China had become the world's dominant silver customer. This implies that ultimately China was responsible for a power shift within early modern Europe. In the ab-

[12] The crown did directly control the famous mercury mines near Huancavelica on the Peruvian coast. Potosí's resurgence after 1573 was attributable to the reforms enacted by Toledo, which included successful adoption of the mercury-amalgam mining process (Cross 1983, p. 402).

[13] For discussion of the *quinto*, which frequently amounted to far less than 20% in reality, see Brading and Cross 1972.

[14] Note that the 27.5% of shipments belonging to the crown represents far more than 27.5% of the total profit. Much of the crown's revenues could be considered "profit," while huge costs had to be subtracted from private receipts before arriving at private net profit.

[15] Scholars such as Elliott (1961) and Lynch (1984, 2:i), have long emphasized both external and domestic foundations of the Spanish empire. For a more complete discussion of why we believe that the foundation of empire was external, not domestic, see Flynn and Giráldez 1995.

sence of the "silverization" of China, it is hard to imagine how Castile could have financed simultaneous wars for generations against the Ottomans in the Mediterranean; Protestant England and Holland and the French in Europe, the New World, and Asia; and against indigenous peoples in the Philippines.

Even giant China could not prop up Spain indefinitely. As tens of thousands of tons of silver accumulated on the Asian mainland, its value gradually fell there (as it had already been doing in the West and Japan) toward its cost of production. Imports eventually glutted even China's vast silver market. We know that this portrayal of silver's loss of value is accurate because by about 1635 it took about 13 ounces of silver to buy an ounce of gold in China, while a half century earlier it took 6 ounces of silver (Geiss 1979, p. 165). The value of silver also fell relative to other things, not just gold, which is to say that price inflation occurred. "In the late sixteenth century . . . when silver from Mexico and Japan entered the Ming empire in great quantity, the value of silver began to decline and inflation set in, for as the metal became more abundant, its buying power diminished. This inflationary trend affected the value of all commodities; everything had been valued in silver and silver lost its value. Ramifications of this change touched the lives of almost everyone in the empire" (Geiss 1979, p. 144). As silver lost value, more silver money was required to purchase items that had maintained their value. Price inflation is defined as the surrender of more pieces of money for a given set of items, so the descent of silver to its cost of production is what ultimately caused prices to inflate in China to about the same extent as in Europe and elsewhere (Cartier 1981, p. 464; Geiss 1979, pp. 159–64, 198; Goldstone 1991, p. 360).

The unavoidable fall in the value of silver is a crucial issue because each year as it descended closer to its cost of production in America, profit per unit of silver also shrank. Declining profits were not due to inefficient operations; rather, they were the inevitable result of the laws of demand and supply (Doherty and Flynn 1989). The existence of arbitrage profits motivated the trade, and the trade itself, in turn, led to the elimination of such profits. Faced with declining profits from its silver industry, Castile could no longer afford its vast empire. China contributed mightily to the duration of the Spanish empire, but even China's prodigious demand for silver could not prevent the eventual erosion of mine profits and therefore the decline of Spain. Spanish American silver production may have peaked in the 1590s, but large production coupled with vanishing profits per ounce of silver still implied a vanishing overall profit level by this time (Flynn 1982, p. 142).

Historians tend to focus too much attention on the quantity of silver shipments, while the participants themselves cared only about the profits associated with the trade. In the words of an executive in a standard business joke today, "Since we are losing money on each item sold, we simply must make up for it in volume!" Spain experienced multiple bankruptcies in the late sixteenth and early seventeenth centuries, during a time of record silver production, because the value of each unit of silver continued to decline. When profit per unit of a product declines to zero, multiplication of zero per-unit profit times any quantity of output must yield zero total profit. Spanish American mines were not yet yielding zero profits per unit of product, but the trend was clearly in that direction. Silver's declining value affected the crown so profoundly that interest payments on Castile's federal debt alone—"the equivalent of at least ten years' revenue" by 1623 (Parker 1979, p. 188)—eventually exceeded total crown receipts. Spain vanished as a serious Western power as its silver basis eroded, but the Iberian surge to power had been lengthy and impressive. The fact that Spain's empire owed its financial foundation to distant Ming China is a forceful reminder that much of what passes for local history in the early modern period can only be understood in terms of world history.

Since China's hunger for silver altered the balance of power in the West by transferring huge profits to the Spanish crown, it is logical to suspect an Asian power shift as a result of the inter-Asian trade in silver. The laws of supply and demand apply on all continents. As noted earlier, China's primary source of silver in the late sixteenth and early seventeenth centuries was Japan, which shipped as much as 200 tons per year at times (Innes 1980, especially chap. 6; Tashiro 1986, p. 2). Contrast this figure to the conservative average estimated by Attman (1986, p. 78) of 150 tons of silver flowing annually through Europe and into Asia in the seventeenth century. Who captured the Japanese mine profits, and what became of them?

The Tokugawa shogunate provides an interesting example of East-West comparative history because, like the Spanish crown and its American mines, the shogunate gained control over Japanese silver mines (Tashiro 1986, p. 3) and sold to China. Flynn (1991) has argued that profits from silver mines financed the defeat of hundreds of rival feudal lords (*daimyo*), thereby permitting the consolidation of Japan. The shogun was forced to align himself with the merchant class, creating an indigenous market-based economy with Asian (not Western) roots. Unlike Spain, the Tokugawa invested heavily in agricultural and urban infrastructure. Japan succeeded in withdrawing from the Chinese tributary system and even sent hundreds of thousands of troops in an unsuccessful attempt to conquer China.

Europeans were important middlemen in the Sino-Japanese silver trade, with Japan as the dominant supplier and China the end-customer. It is ironic that China's demand for Japanese silver generated the profits used by the latter to withdraw from China's tributary system. With the help of profits from its silver mines, Japan established commercial capitalism in Asia at roughly the same time that capitalism was taking root in northwestern Europe. Capitalism's Japanese track evolved independently from, and almost simultaneously with, developments in northwestern Europe. But where Japan used mining profits to establish commercial capitalism in Asia, Spain used mining profits to attack the emerging capitalist powers of northwestern Europe. It is difficult to imagine how either one of these developments could have occurred in the absence of Chinese demand for silver.

What about the impact of silver on China itself? Atwell (1977, 1982, 1986, 1988) has long argued that American silver played a critical role in the commercial and political evolution of domestic China. Basing his argument on Geiss (1979), Goldstone (1991, p. 371) has challenged Atwell's emphasis on forces external to Asia (i.e., American treasure), insisting instead that intra-Asian factors explain structural changes within Ming and Qing China. Goldstone says that domestic price inflation in late sixteenth- and early seventeenth-century China destroyed the financial basis of the Ming dynasty. Taxes formerly paid in rice had been converted to payments in a fixed quantity of silver. But over a period of a century, silver itself had lost two-thirds of its value. Even if the quantity of silver collected had increased during the late Ming, it would still be true that the purchasing power of silver taxes definitely declined. The fiscal foundation of the Ming dynasty eroded because China's tax revenues declined continuously in terms of purchasing power. Institutionalization of fixed silver taxes during an era of global price inflation (in terms of silver) may have created a fiscal crisis on the Asian mainland that led inexorably to overthrow of the Ming (accomplished by the Manchus in 1644).

It appears that the core arguments of Atwell and Goldstone are not incompatible. Along with its Japanese counterpart, American silver contributed mightily to developments inside China. On the positive side, America and Japan were instrumental in the victory of a rising merchant class that succeeded in converting China to a silver zone.

Unhappily for the Ming dynasty, however, fixing taxes in terms of silver may have created a fiscal crisis that led to the emergence of the Qing dynasty.

Summary and Conclusion

Truly global trade dates from the founding of the city of Manila in 1571, which formed the first direct and permanent trade link between America and Asia. From this date forward, all heavily populated continents traded with each other directly and indirectly in substantial volumes. Silver was the sine qua non of this global trade. Spanish America was the source of an estimated 150,000 tons of silver between 1500 and 1800, comprising perhaps 80% of world production. The second-leading source of silver was Japan, responsible for around 30% of world output in the sixteenth century and perhaps 16% in the seventeenth century. Not coincidentally, entrepôt Nagasaki was founded at virtually the same time as Manila. Silver was shipped to China, the world's dominant end-customer, regardless of whether it was produced in Asia or in the West.

Much American silver traversed the Atlantic Ocean, passing through Europe on its journey to Asia. Europe exported at least 150 tons of silver annually to Asia, a significant portion of which passed through Amsterdam. Attman has been most vocal in calling attention to the Baltic route, which carried eastward at least 50 tons per year. Traditional Mediterranean-Levantine trade routes carried vast quantities of silver too, but the Cape route was biggest of the three. Direct trade out of Acapulco, through Manila, and onward to China has been mostly ignored in the scholarly literature, despite evidence that 128 tons annually may have been shipped through Manila in the late sixteenth and early seventeenth centuries (and 307 tons in 1597, according to a single source). Recent research seems to indicate that the Manila trade did not drop off after the 1620s or the 1640s, notwithstanding the claims of Chaunu (1960). The tremendous volume of unofficial trade (smuggling) rendered official figures misleading. Chuan (1969) indicates that the Pacific leg of silver's journey carried more than 50 tons of silver per year throughout the seventeenth century, equal to the combined European shipments of Portugal and the Dutch and English East Indies Companies.

The conventional explanation of this west-to-east flow of "money" is that Europe had to send treasure to Asia because the West had to settle its trade deficit with Asia. Europeans liked Asian silks, spices, and porcelain, but Asians had not yet developed an appreciation for European wares. This conventional view is flawed for at least three reasons. First, it was not "money" or "treasure" that flowed out of Europe, but silver. Silver, not gold, was attracted to Asia; Asian gold and sometimes copper (both "money" substances) flowed in the opposite direction, into Europe. It is best to look at factors affecting the supply and demand for silver in its own right, rather than confusing the issue by aggregating this unique substance with other metals as "money," which is alleged to have played a passive, reactive role. Second, the role of Japan needs to be considered. Japan was the Asian counterpart of America (site of production) and Spain (country controlling production) combined. We prefer to focus on the supply side and the demand side of the silver industry, irrespective of which hemisphere contained centers of production, rather than to visualize global trade as an abstract East-West issue. Third, there is a basic anomaly in the treatment of America in the conventional view. Since treasure is alleged to have flowed from Europe to Asia because of a European trade deficit, then why has the Pacific trade not been explained in the same terms? We know of no one who argues that the Manila galleons carried huge quantities of treasure to Asia because of America's insatiable appetite for Asian goods, which in turn caused an American trade deficit with Asia.

Depicting precious metals as passive "money" that adapts to trade imbalances diverts attention from the central issues. Silver was produced for profit. It migrated from points of production (Japan and America) to end-customers (mostly in China). Developments within China have been largely ignored in the diverse literatures dealing with global flows of precious metals and early modern price inflation, yet China was the pivotal country. International business entities would not have shipped tens of thousands of tons of silver to China unless significant profits induced that activity. Profits from the silver trade were immense for two reasons. First, on the demand side, China's monetary and fiscal systems had substantially converted from a paper-money system to silver by the time of the Single-Whip tax reform of the 1570s. Conversion of more than one-quarter of the world's population (and its government) to silver customers contributed to the rise in the price of silver in China. Second, on the supply side, extraordinarily rich silver mines were discovered in Japan and Spanish America, and new technologies reduced production costs. Supply and demand forces created disequilibrium: silver's value in China was double its value in the rest of the world. This is what drove the silver trade—the birth of world trade—and not some abstract notion of trade deficits.

Both the Spanish empire and Tokugawa shogunate captured a substantial portion of silver profits from mines they controlled. Spain's mines financed a century of multifaceted war and empire. The shogun (and his immediate predecessors) used mine profits to finance consolidation of Japan and withdrawal from the economic domination of China. Spain nearly crushed the emerging capitalistic powers of northwestern Europe, while a market-oriented economy was established within Japan. The laws of supply and demand guaranteed that the price of silver would slowly decline to its cost of production, which is what happened. A direct effect of this process is that profit per ounce of silver was steadily squeezed out. This caused the decline of Spain. Japanese silver-mine profits plummeted too, but gold and copper production soared in Japan in the second half of the seventeenth century. The shogun had also invested heavily in improvements in infrastructure, so there was no decline of Japan in the seventeenth century.

The worldwide decline in the value of silver in the early modern period translated directly into global price inflation. When money declines in value with respect to goods, the result is called price inflation. Geiss (1979, p. 158) explains this process clearly for late Ming China: "Such massive infusions of silver diminished the purchasing power of the metal. Silver, like any commodity, lost its high value when supply exceeded the usual demand. This in turn affected the prices of almost everything in the empire, for the structure of prices was tied to the value of silver." Transportation technology permitted connection of silver markets throughout the world. The interconnection of world markets guaranteed that the fall in silver's value was global, which in turn implies that price inflation was a global phenomenon in all areas on a silver standard.

We have consciously neglected any attempt to tie the African continent into the global trade of silver. Nonetheless, it seems that the Portuguese swapped huge numbers of (mostly smuggled) African slaves directly for (mostly smuggled) New World silver via the Rio Plata in Brazil:

> During the decade 1616–1625, recorded imports for Buenos Aires (the sum of legal and confiscated goods) were 7,957,579 pesos, while exports for the same period amounted to only 360,904 pesos. The annual trade deficit, which was met with smuggled silver from Upper Peru, amounted to at least three-quarters of a million pesos per year. Between 1619 and 1623, port officials seized a total of 3,656 slaves from illegally landing vessels; their market value in Lima would have approached two million pesos. These numbers do not include those

> slaves legally imported and those which evaded port officials. The size of these figures clearly indicates a flourishing and considerable illicit trade up the Rio Plata from the 1580s until probably the 1640s. During its most successful years, no less than 1–2 million pesos (roughly 25,000 to 50,000 kg) flowed illegally from the mines of Peru out through the port of Buenos Aires. These totals equalled from 15% to 30% of the silver output of Potosí. (Cross 1983, p. 414)

Not all the slaves remained in Brazil, nor were all of them plantation laborers. Palmer (1995) has provided demographic information suggesting that between 10,000 and 20,000 Africans were domestic slaves in Mexico City in the early seventeenth century. Since, as we have argued, the Spanish enterprise in America was financed by the world silver market (as were the activities of the Portuguese traders), and since China was the dominant factor in the global silver market, then it appears that the trans-Atlantic slave trade was heavily, though indirectly, influenced by monetary and fiscal developments in Ming China. In other words, end-customer China created profitable trade in the New World, and profitable trade in America created the demand for African slaves. Clearly, a global view of early modern trade may suggest many research topics that might otherwise be overlooked.

Scholars have long been interested in the impact of Europeans on Asia (and the rest of the world). The focus has shifted in recent years, however, especially among Asian scholars who increasingly emphasize the dominant historical role of the intra-Asian marketplace. These revisionists view Europeans as having participated in a vast and sophisticated existing Asian commercial network, rather than as having introduced modernization to backward Asia. This essay is in the revisionist camp; it even suggests a reversal of causality. The economic impact of China on the West was far greater than any European influence on Asia in the early modern period. We agree with the sentiments of Moloughney and Xia (1989, p. 68), who protest that "late Ming China was not an outpost of a Seville-centered world economy."[16] Perhaps a reversal of this logic would be more accurate: Seville was an out-

---

[16] Moloughney and Xia (1989, pp. 67–68) have criticized Atwell for insisting on the importance of American silver in early modern Chinese history. The basis of the Moloughney-Xia counter-argument is empirical: the influx of American silver peaked during, rather than before, the decline of the Ming in the 1640s. This essay is consistent with both sides of this debate. American silver may have been crucial in the decline of the Ming dynasty not because of its scarcity but precisely because its abundant accumulation depressed the value of silver. Views that are contradictory from the perspective of conventional macroeconomic logic become compatible in terms of our cost-of-production model.

post of a world economy that had not one center but three (Beijing on the demand side, and America and Japan on the supply side).

The physical presence of Europeans in Asia in early modern times —and the simultaneous physical absence of Asians in the West—has understandably led scholars to pay attention mostly to the impact of the West on Asia. Superior naval firepower may explain the presence of Europeans in Asia, but the most powerful economic undercurrents ran in the opposite direction. Without the Chinese demand for silver, there would have been no finance mechanism for the Spanish empire. Without China, there would have been no century-long price revolution. Without China, the birth of world trade would have been delayed to some unknowable extent. But China did convert, both monetarily and fiscally, to silver. This fact reverberated across all continents and gave birth to world trade in 1571, providing a powerful force in shaping the modern world.

References

Attman, Artur. 1986. *American Bullion in the European World Trade, 1600–1800*. Göteborg.

Atwell, William S. 1977. "Notes on Silver, Foreign Trade, and the Late Ming Economy." *Ch'ing-shih ven-ti'i* 3/8:1–33.

———. 1982. "International Bullion Flows and the Chinese Economy circa 1530–1650." *Past and Present* 95:68–90.

———. 1986. "Some Observations on the 'Seventeenth-Century Crisis' in China and Japan." *Journal of Asian Studies* 45:223–44.

———. 1988. "Ming Observations of Ming Decline: Some Chinese Views on the 'Seventeenth Century Crisis' in Comparative Perspective." *Journal of the Royal Asiatic Society* 2:316–48.

Barrett, Ward. 1990. "World Bullion Flows, 1450–1800." In *The Rise of Merchant Empires: Long-Distance Trade in the Early Modern World, 1350–1750*, ed. James D. Tracy, pp. 224–54. Cambridge.

Blair, E. H., and J. A. Robertson. 1903–1909. *The Philippine Islands*. 55 vols. Cleveland.

Blum, Jerome. 1956. "Prices in Russia in the Sixteenth Century." *Journal of Economic History* 27:182–99.

Boxer, Charles R. 1969. *The Portuguese Seaborne Empire, 1415–1825*. New York.

Brading, D. A., and Harry E. Cross. 1972. "Colonial Silver Mining: Mexico and Peru." *The Hispanic American Historical Review* 52:545–79.

Brenner, Robert. 1977. "The Origins of Capitalist Development: A Critique of Neo-Smithian Marxism." *New Left Review* 104:25–92.

Cartier, Michel. 1981. "Les importations de métaux monétaires en Chine: Essai sur la conjoncture chinoise." *Annales E.S.C.* 36:454–66.

Chaudhuri, K. N. 1978. *The Trading World of Asia and the English East India Company, 1660–1760*. Cambridge.

———. 1986. "World Silver Flows and Monetary Factors as a Force of International Economic Integration, 1658–1758." In *The Emergence of a World Economy, 1500–1914: Papers of the IX International Congress of Economic History*, ed. W. Fisher, R. W. McInnis, and J. Schneider, 1:37–60. 2 vols. Stuttgart.

Chaunu, Pierre. 1960. *Les Philipines et le Pacifique des Ibériques (XVIe, XVIIe, XVIIIe siècles)*. Paris.

Chuan, Hang-sheng. 1969. "The Inflow of American Silver into China from the Late Ming to the Mid-Ch'ing Period." *The Journal of the Institute of Chinese Studies of the Chinese University of Hong-Kong* 2:61–75.

Cross, Harry E. 1983. "South American Bullion Production and Export, 1550–1750." In *Precious Metals in the Later Medieval and Early Modern Worlds*, ed. J. F. Richards, pp. 397–424. Durham.

DeVries, Jan. 1984. *European Urbanization, 1500–1800*. Cambridge, Mass.

Doherty, Kerry W., and D. O. Flynn. 1989. "A Microeconomic Quantity Theory of Money and the Price Revolution." In *Precious Metals, Coinage and the Changes of Monetary Structures in Latin America, Europe and Asia*, ed. Eddy H. G. van Cauwenberghe, pp. 185–208. Leuven.

Elliott, John H. 1961. "The Decline of Spain." *Past and Present* 20:52–75.

Felix, David. 1956. "Profit Inflation and Industrial Growth: The Historic Record and Contemporary Analogies." *Quarterly Journal of Economics* 70:441–63.

Flynn, Dennis O. 1977. "The Spanish Price Revolution and the Monetary Approach to the Balance of Payments." Ph.D. diss., University of Utah.

———. 1982. "Fiscal Crisis and the Decline of Spain (Castile)." *Journal of Economic History* 42:139–47.

———. 1984a. "Use and Mis-use of the Quantity Theory of Money in Early Modern Historiography." In *Munzpragung, Geldumlauf und Wechselkurse: Minting, Monetary Circulation and Exchange Rates: Akten des 8th International Economic History Congress, Section C7, Budapest, 1982*, ed. E. van Cauwenberghe and F. Irsigler, pp. 383–417. Trier.

———. 1984b. "The 'Population Thesis' View of Inflation versus Economics and History." In *Munzpragung, Geldumlauf und Wechselkurse: Minting, Monetary Circulation and Exchange Rates: Akten des 8th International Economic History Congress, Section C7, Budapest, 1982*, ed. E. van Cauwenberghe and F. Irsigler, pp. 361–82. Trier.

———. 1986. "The Microeconomics of Silver and East-West Trade in the Early Modern Period." In *The Emergence of a World Economy, 1500–1914: Papers of the IX International Congress of Economic History*, ed. W. Fisher, R. W. McInnis, and J. Schneider, 1:37–60. 2 vols. Stuttgart.

———. 1991. "Comparing the Tokugawa Shogunate with Hapsburg Spain: Two Silver-Based Empires in a Global Setting." In *The Political Economy of Merchant Empires*, ed. James D. Tracy. Cambridge.

Flynn, Dennis O., and Arturo Giráldez. 1994a. "China and the Manila Galleons." In *Japanese Industrialization and the Asian Economy*, ed. A. J. H. Latham and H. Kawakatsu. London.

———. 1995. "Arbitrage, China, and World Trade in the Early Modern Period." *Journal of the Economic and Social History of the Orient*.

———. 1994b. "Silver and the Political Economy of the Spanish Empire." Paper presented before the Society for Spanish and Portuguese Historical Studies, Chicago.

Geiss, J. P. 1979. "Peking under the Ming, 1368–1644." Ph.D. diss., Princeton University.

Gernet, Jacques. 1982. *A History of Chinese Civilization*. Cambridge.

Godinho, V. Magalhaes. 1963. *Os descobrimientos e a economia mundial*. Lisbon.

Goldstone, Jack. 1991. *Revolution and Rebellion in the Early Modern World*. Berkeley.

Hamashita, Takeshi. 1988. *The Tribute System and Modern Asia*. Tokyo.

Hamilton, Earl J. 1929. "American Treasure and the Rise of Capitalism, 1500–1700." *Economica* 9:338–57.

———. 1934. *American Treasure and the Price Revolution in Spain, 1501–1650*. Cambridge, Mass.

———. 1937. "The Decline of Spain." *Economic History Review* 8:168–79.

———. 1952. "Prices as a Factor in Business Growth." *Journal of Economic History* 12:325–49.

Huang, Ray. 1974. *Taxation and Governmental Finance in Sixteenth-Century Ming China*. Cambridge, Mass.

Hunt, Verl F. 1978. "The Rise of Feudalism in Eastern Europe: A Critical Appraisal of the Wallerstein World-System Thesis." *Science and Society* 42:43–61.

Innes, R. L. 1980. "The Door Ajar: Japan's Foreign Trade in the Seventeenth Century." Ph.D. diss., University of Michigan.

Israel, Jonathan. 1981. "Debate: The Decline of Spain: A Historical Myth." *Past and Present* 91:170–80.

Iwao, Seiichi. 1959. "Japanese Gold and Silver in the World History." In *International Symposium on History of Eastern and Western Cultural Contacts*. Tokyo.

Jara, Alvaro. 1966. "Economia minera e historia economica hispanoamericana." In *Tres ensayos sobre economia minera hispano-americana*. Santiago.

Kamen, Henry. 1964. "The Decline of Castile: The Last Crisis." *Economic History Review* 17:63–76.

———. 1978. "The Decline of Spain: A Historical Myth?" *Past and Present* 81:24–50.

———. 1981. "A Rejoinder." *Past and Present* 91:181–85.

Keynes, John Maynard. 1930. *A Treatise on Money*. 2 vols. New York.

Liang, Fang-chung. 1956. *The Single-whip Method of Taxation in China*. Cambridge, Mass.

Lynch, John. 1984. *Spain under the Habsburgs*. 2 vols. New York.

Moloughney, Brian, and Xia Weizhong. 1989. "Silver and the Fall of the Ming: A Reassessment." *Papers on Far Eastern History* (Canberra) 40:51–78.

Mols, Roger. 1977. "Population in Europe, 1500–1700." In *The Fontana Economic History of Europe*, ed. C. M. Cipolla. London.

Nadal, Oller Jorge. 1959. "La revolución de los precios en el siglo XVI." *Hispania* 19:503–29.

Nef, John U. 1934. "The Progress of Technology and the Growth of Large-Scale Industry in Great Britain, 1540–1640." *Economic History Review* 5:3–24.

Outhwaite, R. B. 1969. *Inflation in Tudor and Early Stuart England*. London.

Palmer, Colin. 1995. "From Africa to the Americas: Ethnicity in the Early Black Communities of the Americas." *Journal of World History* 6: 223–36.

Parker, Geoffrey. 1979. "War and Economic Change: The Economic Costs of the Dutch Revolt." In *Spain and the Netherlands, 1559–1659: Ten Studies*, ed. G. Parker. Glasgow.

Phelan, John L. 1959. *The Hispanization of the Philippines: Spanish Aims and Filipino Responses*. Madison.

Ramsey, Peter. 1960. "The European Economy in the Sixteenth Century." *Economic History Review* 12:456–62.

Sahillioglu, H. 1983. "The Role of International Monetary and Metal Movements in Ottoman Monetary History, 1300–1750." In *Precious Metals in the Later Medieval and Early Modern World*, ed. J. F. Richards, pp. 269–304. Durham.

Tashiro, Kazui. 1982. "Foreign Relations during the Edo Period: Sakoku Reexamined." *Journal of Japanese Studies* 8:283–306.

TePaske, John J. 1983. "New World Silver, Castile and the Philippines, 1590–1800." In *Precious Metals in the Later Medieval and Early Modern Worlds*, ed. J. F. Richards, pp. 425–46. Durham.

Wallerstein, Immanuel. 1974. *The Modern World-System: Capitalist Agriculture and the Origins of the European World-Economy in the Sixteenth Century*. New York.

Wilkinson, Endymion P. 1980. *Studies in Chinese Price History*. New York.

Wolf, Eric. 1982. *Europe and the People without History*. Berkeley.

12

Trade Between China, the Philippines and the Americas During the Sixteenth and Seventeenth Centuries

Chuan Hang-Sheng

In 1492 Columbus discovered America; in 1519 Spain subdued Mexico, and later, in 1565, went on to the Philippine Islands from Mexico. Spain's control and colonization of the Philippines greatly depended upon Mexico, and in the next two and a half centuries two or three galleons each year sailed from Acapulco to Manila to strengthen ties between the two areas. Silver flowed from America to the Philippines whereas silk was the chief cargo America imported from the Philippines*.

When the Spaniards arrived in the Philippines, they found the natural resources of the Philippines very undeveloped so that no important commodities could be exported to the Americas. Fortunately the close proximity of the Philippines to China made it possible for goods from that country to be sent to America. On the other hand, the rise of silver production and the increase in purchasing power of people in America from the 16th to the 18th centuries made for an increase in demand for silk clothing. Under these circumstances, when galleons from the Orient brought the silk which Chinese merchants had shipped to the Philippines, a large profit could be earned.

Shortly after the Spaniards reached the Philippines in 1565, the Chinese merchants shipped different Chinese goods to Manila. Most of these goods consisted of raw silk and its woven products. The silk goods received in the Philippines, except for a few goods consumed locally or sold to Japan, were then shipped by galleon for sale in Spanish America.

Aside for a few years, the galleons leaving Manila for the port of Acapulco in Mexico can be referred to as silk boats, because most of their cargo consisted of Chinese raw silk and silk goods. Before 1636 between 300 and 500 chests of silk goods were registered in each galleon, but for some ships sailing in 1636, one had more than 1,000 chests, and another as much as 1,200. The capacity of each chest, according to a galleon in 1774, was 250 pieces of pearl-coloured Cantonese taffeta and 72 pieces of

* For detailed documentation see the author, "The Chinese Silk Trade with Spanish America from the Late Ming to the Mid-Ch'ing Period", in Laurence G. Thompson, ed., *Studia Asiatica*, San Francisco, 1975, pp. 99-117; also the author's article in Chinese, "The Inflow of American Silver into China from the Late Ming to the Mid-Ch'ing Period", in *The Journal of the Institute of Chinese Studies* 2:1 (Hong Kong, Dec. 1969).

deep red thin silk, in all weighing about 250 lbs.; in addition, there were some chests that only contained foreign silk stockings, with 1,140 pairs to a chest, weighing 230 lbs. These galleons also carried Chinese raw silk with 10,000 to 12,000 bales per ship. If we consider a ship captured by the English in 1587 as an example, the cargo value of Chinese silk and other goods priced in Manila came to about one million or more pesos but when sold in Mexico, amounted to over two million pesos. Other records show that when trade flourished the total amount of silk goods annually shipped to America came to 3,000,000 or even 4,000,000 pesos.

The Chinese raw silk that arrived in Mexico was then processed and woven for shipment to Peru. According to a report in 1637, over 14,000 labourers worked in this textile industry. Most of the imported Chinese silk textiles were consumed by the Spanish colonists, their descendents, and the wealthy of mixed Spanish and Indian descent. In the 16th and 17th centuries, Peru produced more than 60 per cent of the world's silver; the export of silver had made it possible for that country to import large quantities of Chinese silk goods.

Although the Chinese trade first brought great profits to Spanish colonists, especially those in Manila, the domestic Spanish silk textile industry was later threatened by the influx of cheap Chinese silk into its South American market. By 1640 similar Chinese silk goods in the Peruvian market sold for only one-third of the price of Spanish produced silk. Spanish silk exports could not compete, and their price fell, forcing the domestic silk industry into decline. The Spanish silk manufacturers were naturally unhappy as their Share of the South American market declined.

Now let us turn to the flow of American silver into China.

In 1375 the Ming government began to issue paper money as a medium of exchange. As the quantity of notes rapidly increased, confidence in their purchasing power quickly declined. By the fifteenth century, the people had rejected paper in favor of silver which they regarded as a more reliable unit of account and medium of exchange. But the mining of silver in China failed to keep pace with demand; and the price of silver soon rose to a record high level, in fact, nearly twice that during the Sung and Yuan periods.

During the latter part of the sixteenth century, the silver output in Spanish America greatly increased, particularly from the discovery of a rich silver deposit at Potosi in Upper Peru (now Bolivia) in 1545. The annual amount of silver extracted from Potosi averaged 254,000 kg. or roughly 60 per cent of the world's output. As time passed, the silver output in Potosi began to diminish because the mine became exhausted, but in the late 17th century, Mexico became a primary world producer of silver. In Manila the Chinese merchants eagerly sought American silver, and the Spaniards quickly recognized its value in their Philippine trade. Anxious to acquire as much silver from the Spanish as possible, Chinese traders in Manila did their best to meet Spanish demands for daily necessities, silks, ammunition, etc. Beginning in the late sixteenth century, American silver began to flow continuously into China through Manila.

In 1565 when the Spaniards arrived in the Philippines, they recognized that trade with China had to be maintained with silver. In a letter to Felipe II on December 5, 1573, the viceroy of New Spain wrote: "The commerce with that land (China) must be carried on with silver, which they value above all other things." In 1609 Antonio de Morga also wrote: "The purchase price (of Chinese goods) is paid in silver and reals, for the Sangleys (Chinese traders) do not want gold, or any other articles, and will not take other things to China." Since the Spaniards understood the important role silver played in the transactions of the Sino-Filipino trade, it was natural they sent large quantities of silver to Manila each year in order to buy whatever they needed from China. The annual silver imports from Spanish America into the Philippines can be observed as follows:

Table I. Annual American Silver Imports into the Philippines, 1598–1784

Year	Amount (in pesos)	Year	Amount (in pesos)
1598	1,000,000	1729 & before	3-4,000,000
1602 & before	2,000,000	1731	2,434,121
1604	2,500,000(+)	1740	3,000,000
c. 1620	3,000,000	1746-1748	4,000,000
1633	2,000,000	1762	2,309,111(+)
1688 & before	2,000,000	1764	3,000,000(+)
1698-1699	2,070,000	1768-1773	1,500,000–2,000,000
1712 & before	2,600,000(+)	1772	2-3,000,000
1714 & before	3-4,000,000	1784	2,791,632
1723	4,000,000		

In his memorial of 1765, Francisco Leandro de Viana said that more than two hundred million pesos had arrived in the Philippines since 1565. According to the computation of De Comyn, the total amount of silver imported from Spanish America into Manila from 1571 to 1821 came to four hundred million pesos.

In order to acquire necessities for daily use, ammunition for maintaining order on the Islands, and silks for shipment back to Spanish America, the Spaniards spent most of the silver that they had brought from Mexico to Manila. The following evidence demonstrates the great flow of American silver from Manila to China.

(1) In a letter (unsigned) to Felipe II on June 17, 1586 we note: "...a large amount of silver and coin is carried thither (Manila) in exchange for them (Chinese

wares). Although a portion of this money remains in those Islands (the Philippines), all the rest is carried away by the Chinese who go thither from the mainland to sell their wares."

(2) Again in a letter from Portugual (unsigned) to Felipe II in 1590: "... if navigation is permitted from the western Indias to China, all the money and coin in the kingdom will flow thither and none will go to Hespana, because China is so large and has so much to exchange and sell that, however much coin is sent, that country will absorb it all."

(3) In his letter to Felipe II from Manila on June 28, 1597, Luis Perez Dasmarinas, governor of the Philippines, wrote: "All the money goes and passes to China, and remains there from year to year and in fact always."

(4) In his letter to Felipe II on June 24, 1598, Fray Ygnacio, archbishop of Manila, asserted: "There comes each year from Nueba Espana a million in money, contrary to the mandate of your Majesty, all of which passes on to the heathen of China."

(5) In c. 1602 a lawyer wrote: "This (silver) and all else which is carried is placed in the power of the infidels, who received it as the price of the Chinese merchandise." Also in the same year, a bishop of South America wrote: "...the Filipinas, which receive annually two million pesos of silver; all of this wealth passes into the possession of the Chinese, and is not brought to Espana..."

(6) On December 1, 1604, Felipe III decreed: "All this money (more than two millions of ducats) finally makes its way into infidel kingdoms,..."

(7) In 1630 Fray Juan de Medina, O. S. A., wrote: "For that exceedingly vast kingdom (China) abound in whatever can be desired to sustain life,... the large vessels ... and those of lesser size, are well nigh innumerable; and they sail annually to surrounding countries, laden with food and merchandise. Forty, and upwards, were wont to come to Manila alone... These vessels go to Siam, Camboja,.... They carried away all the silver in the world... Thus the kingdom of China is the most powerful in the world; and we might even call it the world's treasury, since the silver is imprisoned there, and is given an eternal prison. And if there were no more silver there than what has been taken from Mexico during sixty-six years of trade, it could make them most wealthy; and much more so, inasmuch as the Mexican silver is not the most that they get, for they take much from other quarters. They are the most greedy for and affectioned to silver of any race known. They hold it in the greatest esteem, for they withdraw the gold from their own country in order to lock up the silver therein. And when they see silver, they look at it admiringly. I am writing not from hearsay, but from the sight and experience of many years..."

(8) In his memorial of 1637,Grau y Monfalcon wrote: "...the silver is withdrawn and conveyed to the islands, whence it passes into the possession of the enemies of religion--and of this crown, Moors and pagan; and finally remains in China—which,

according to what we understand, is the general center for the silver of Europe and Asia. For as it always is current, gaining and increasing in value until it reaches that great kingdom, whence it must issue with a loss, it does not issue, and remains perpetually among those inhabitants."

(9) The amount of silver shipped to China was so great that one Spanish admiral declared in 1638 that "The Chinese king (Emperor) could use Peruvian silver to build a palace."

According to the Spanish records, the annual imports of American silver into China by way of Manila from the late sixteenth to the early nineteenth century were as follows:

Table II. Annual American Silver Imports into China by Way of Manila

Year	Amount (in pesos)
Before 1586	300,000
1586	500,000 (+)
1598 & before	800,000 – 1,000,000 (+)
1602 & before	2,000,000
1604	2,500,000 (+)
1633 & before	2,000,000
1729 & before	3-4,000,000
1815	1,550,000

De Comyn estimated that the import of silver into Manila from Spanish America, during the 250 years of exchange, 1571–1821, had amounted to 400 millions of pesos, and of this, a large share, perhaps one-half, had passed through to China.

In 1436 the Ming government allowed silver to become a legal tender, and for the next five centuries China used the silver standard. As there were few good silver mines in China, the output of silver had failed to keep pace with demand. But the fact that large quantities of American silver were imported into China must account for the successful adoption of this monetary unit in China.

13
Amsterdam, marché mondial des métaux précieux au XVIIe et XVIIIe siècles

J.-G. Van Dillen

Vers la fin du moyen âge, le commerce s'est développé en Europe avec une grande rapidité. Le système des échanges en argent, ce qu'on appelle en allemand *Geldwirtschaft*, se répandit de plus en plus et, par conséquent, le besoin des métaux précieux se fit de plus en plus sentir. Ce phénomène explique le succès de la doctrine dite « mercantiliste ». Quelles qu'en soient les erreurs scientifiques, je suis convaincu que les circonstances historiques l'expliquent et la justifient dans une forte mesure. Aussi est-il très important de bien étudier la production et le commerce des métaux précieux ; c'est à juste titre que, dans la nouvelle édition de son livre intitulé *Der moderne Kapitalismus*[1], M. Werner Sombart commence par cette étude le volume consacré au commerce international pour la période qui s'étend du XVIe au XVIIIe siècle. C'est le trafic des métaux précieux, dit-il, qui en forme, pour ainsi dire, le pivot. Aussi convient-il d'attirer l'attention des érudits sur le commerce des métaux précieux, tant en barres qu'en espèces, à cette époque, et, en particulier, sur le rôle important joué par le marché d'Amsterdam.

1. Il importe de signaler la nouvelle édition de cet ouvrage. L'auteur l'a entièrement remanié, et il faut dire que, s'il a de grands défauts, il possède aussi de grandes qualités.

Je n'insisterai pas, parce que les faits sont bien connus, sur le grand changement qui s'opéra au cours du XVIe siècle dans la production de l'or et de l'argent. On sait que c'est surtout à partir de 1545, année où furent découvertes les mines de Potosi, que le flot d'argent ne cessa de couler d'Amérique en Europe. L'Espagne n'a pas pu garder pour elle cette quantité d'argent qui grossissait toujours et, malgré la défense d'exportation, l'argent commença de se répandre hors de ses frontières. C'est la France qui en fut d'abord inondée. L'abondance de l'argent, monnayé et en barres, y causa un renchérissement de la vie dont les contemporains, à l'exception de Bodin, ne virent pas bien les causes ; c'est ce qu'a montré avec une clarté parfaite M. Henri Hauser dans le chapitre intitulé « Controverse sur les monnaies, 1566-1578 », de son livre si intéressant : *Travailleurs et marchands dans l'ancienne France.*

Cependant, le flot d'argent se déversa aussi vers les Pays-Bas. Ce fut naturellement à Anvers, le marché mondial le plus célèbre de ce temps, qu'il fit son apparition. Les Fugger, les fameux banquiers d'Augsbourg, dirigèrent sur cette place l'argent provenant de l'Amérique, que Charles-Quint et Philippe II leur cédèrent en remboursement de leurs emprunts incessants. Mais, en 1585, le duc de Parme s'empara d'Anvers et les Hollandais fermèrent les bouches de l'Escaut. La prospérité de la ville en souffrit gravement. Par contre, les provinces septentrionales prospérèrent de plus en plus. C'est surtout vers la fin du XVIe siècle que le commerce hollandais se développa d'une façon merveilleuse. Peu à peu la ville d'Amsterdam l'emporta sur ses rivales dans la lutte commerciale qui s'engagea alors entre les villes de la Hollande et de la Zélande ; elle finit par devenir le grand centre du commerce international.

On se demandera peut-être d'où venaient aux Hollandais l'or et l'argent indispensables à leurs transactions commerciales. La réponse est facile : l'or leur vint en assez grande quantité d'Afrique, de la région de la Guinée qu'on appelle justement la Côte d'Or. Quant à l'argent, c'est l'Espagne qui le leur fournissait ; en effet, même pendant la guerre de Quatre-vingts ans, le commerce entre l'Espagne et la Hollande n'a jamais cessé entièrement, parce que l'Espagne ne pouvait se passer des marchandises apportées par les navires hollandais. Ce trafic, il est vrai, ne se faisait pas sans difficultés. L'auteur inconnu d'un rapport au roi d'Espagne, rédigé en français et datant du début du XVIIe siècle[1], le décrit ainsi :

Le traficq et navigation que les dictz rebelles ont eu par le passé en Espagne a été sous dessoubz la couverture de France, d'Angleterre et d'Allemagne, et, par icelle navigation d'Espagne que les rebelles y ont eu depuis vingt-deux ans en ça, ont remporté à leurs villes et provinces force argent et or pour retour en fourmages, bleds, beurre, harens, toutes sortes de manufactures, chair, bière, bré, tere, cire et

1. Il a été édité par le professeur P. J. Blok, de Leyde.

autres marchandises de Oostlande[1] ; et, par ce moyen, ont acquis encore de plus grands trésors qu'ilz ne pouvoient par leur pêcherie et leur navigation d'Oostlande ; et avec grandes dissimulations et tromperies donnent à entendre estre de Dannemarcq, Oostlande et Norvègue et sujetz à l'Empire par fausses certiffications et passeports contrefaicts...

J'ai pu constater, en outre, d'après des documents trouvés aux archives des États-Généraux de La Haye, que les commerçants hollandais se procuraient souvent l'argent espagnol-américain, qu'ils recherchaient si ardemment, par l'intermédiaire de commerçants de Hambourg.

Dès que la paix avec l'Espagne eut été conclue (1648), la navigation hollandaise vers Cadix, ville où se faisait le commerce d'Amérique, prit tout son essor. Désormais, pendant la deuxième moitié du XVIIe siècle, Amsterdam fut sans contredit le marché principal pour le commerce international des métaux précieux. Chaque année, à l'automne, une flotte d'argent, composée de trente à cinquante voiles, escortée de vaisseaux de guerre, venant de Cadix, arrivait à Amsterdam ; c'était pour la ville, comme pour la Bourse, un événement important. Selon une évaluation digne de foi, vers la fin du XVIIe siècle la moitié environ de la production des mines d'argent de l'Amérique était débarquée sur le marché d'Amsterdam. Toutefois, tout cet argent n'appartenait pas aux commerçants hollandais. Des Espagnols envoyaient de l'argent à Amsterdam pour leur propre compte ; en outre, des négociants des Pays-Bas méridionaux et même du nord de la France, par exemple ceux de Lille, faisaient venir de l'argent par la voie d'Amsterdam et par l'intermédiaire de la flotte d'argent hollandaise ; ils y étaient incités par la sûreté et par le bon marché du fret. On estime que la moitié seulement de l'argent parvenant au port d'Amsterdam restait entre les mains des commerçants hollandais.

En tout cas, il est prouvé qu'à Cadix le commerce hollandais dépassait alors de beaucoup celui des autres nations, notamment celui des Anglais et des Français. Un écrivain contemporain bien connu, Pierre-Daniel Huet, évêque d'Avranches, dans ses *Mémoires sur le commerce des Hollandois*, se plaint amèrement de la prépondérance des Hollandais dans le commerce avec l'Espagne. Nous avons aussi le témoignage de l'économiste anglais Sir Josuah Child, président de la Compagnie anglaise des Indes orientales ; dans la préface de son livre *New Discourse of Trade* (1690), il constate également la supériorité des Hollandais dans le commerce de Cadix, et il l'attribue principalement au faible taux de l'intérêt en Hollande, qui permet aux commerçants d'accumuler à Cadix de grands « stocks » de marchandises longtemps avant l'arrivée de la flotte et des galions d'Amérique.

Un économiste hollandais, M. Pierson, a fait observer cependant que le faible taux ne suffit pas pour expliquer ce phénomène ; c'est la richesse capi-

1. C'est-à-dire de la Baltique.

talisée des commerçants hollandais qui leur a permis d'accumuler des « stocks » et aussi d'accorder de longs crédits ; le faible taux de l'intérêt fut uniquement un *symptôme* de cette richesse. C'est cette richesse en capitaux qui explique le fait étonnant de la prépondérance des Hollandais dans le commerce international de cette époque. Les commerçants d'Amsterdam ne se bornaient pas à acheter de l'argent à Cadix, ils s'appliquaient aussi à faire le trafic de l'Amérique — interdit aux étrangers — et à cet effet ils employaient l'intermédiaire de leurs commissionnaires espagnols. Ils confient des marchandises — écrit Pierre Huet — « secrettement à des Espagnols dont la probité est connue, qui les vont vendre aux Indes sous leurs noms et en rapportent ensuite le prix à celui qui les leur a confiées ». Huet ajoute : « La cour de Madrid n'ignore pas ces intelligences, mais elle les dissimule par politique. » Cette pratique ne tarda pas à être imitée par les commerçants anglais et français.

D'ailleurs, en dehors du commerce de Cadix, les Hollandais faisaient, dans l'Amérique espagnole, un commerce dit interlope. « Les Hollandais » — écrit encore Huet — « ont même trouvé le moyen d'y trafiquer secrètement, ou pour mieux dire, directement par le moyen de l'isle de Curaçao, qui n'est pas fort éloignée de Carthagène. »

C'est donc l'abondance relative des métaux précieux qui a rendu possible le développement du commerce hollandais, et cela pour deux raisons. En premier lieu, l'argent en espèces était indispensable pour le commerce aux Indes, parce que les Portugais avaient habitué les indigènes à être payés en réaux espagnols. Dans la suite, on a toujours exporté aux Indes des quantités considérables d'or et d'argent, parce que, au XVIIe et au XVIIIe siècle, les métaux précieux étaient les principaux objets d'échange que l'Europe pouvait offrir à l'Orient. Ils s'y exportaient sous forme soit de monnaies, soit en barres. Puis aux réaux espagnols vinrent s'ajouter les ducatons hollandais qu'on y envoyait en masse. En second lieu, le commerce hollandais dans les diverses parties de l'Europe exigeait l'exportation d'or et d'argent, la balance du commerce avec les autres pays étant souvent déficitaire. Un fait très remarquable est que cette exportation se faisait en grande partie en monnaies hollandaises. Aussi une grande partie de l'or et de l'argent importés en Hollande y était-elle monnayée en *negotie-penningen*, c'est-à-dire en monnaies de négoce, qui n'étaient pas destinées à la circulation intérieure, mais uniquement ou principalement à l'exportation. Au XVIIe siècle, il n'y eut pas moins de quatorze hôtels des monnaies dans la République des Provinces-Unies. La monnaie qu'on y frappait était en général de bon aloi et de bon poids ; on s'explique donc aisément que l'argent hollandais ait été fort recherché dans tous les pays qui ne disposaient pas eux-mêmes de bons moyens d'échange. Ainsi les *leeuwendaalders* s'en allaient-ils au Levant et en Asie Mineure. Les *rijkrsdaalders* hollandais étaient fort recherchés dans la Baltique, surtout en Pologne, tandis que les ducats d'or allaient en masse en Russie. La Compagnie des Indes orientales exportait les « ducatons » en grande quantité aux Indes et en Chine. La Compagnie anglaise des Indes orientales, à partir de 1670 et au cours du XVIIIe siècle, achetait annuellement à Amsterdam des ducatons, pour une valeur de sept à huit millions de florins, destinés aux Indes.

Cette grande exportation d'espèces a duré pendant toute l'existence de la République. Encore vers la fin du XVIIIe siècle, en Pologne, les paysans ne voulaient être payés, dans le commerce des grains, qu'en ducats d'or hollandais. Il est vrai que l'argent monnayé a été aussi exporté en d'autres pays ; mais, à l'exception des réaux ou piastres espagnols, je ne crois pas qu'il y ait eu à cette époque, en Europe et en Orient, une monnaie plus recherchée que celle de la Hollande. Si j'insiste tant sur ce fait, c'est que M. Sombart, dans l'ouvrage que je viens de mentionner, l'a passé complètement sous silence.

Contrairement à la doctrine mercantiliste et aussi à la pratique usitée dans presque tous les autres pays, l'exportation de la monnaie n'a jamais été interdite dans la République. Quant aux métaux précieux en barres, l'exportation en fut proscrite dans la première motié du XVIIe siècle, mais dès l'année 1647 elle fut pratiquement autorisée. Sur ce point, il y a toujours eu une lutte, souvent acharnée, entre les directeurs généraux de la Monnaie, opposés résolument à l'exportation des métaux précieux en barres, et, d'autre part, les commerçants d'Amsterdam, partisans du libre-échange. Dans ce cas, ainsi d'ailleurs qu'il en était presque toujours dans la République, c'est l'intérêt du commerce qui l'emporta. Les idées mercantilistes n'ont donc exercé que très peu d'influence en Hollande.

Cette liberté d'exportation de l'argent monnayé et des lingots a beaucoup contribué à la grande stabilité du cours du change à Amsterdam, de sorte que le commerce des lettres de change n'y a pas été moins actif que celui des métaux précieux.

Vers la fin du XVIIe siècle, Amsterdam était certainement le premier marché du monde, tant pour les marchandises que pour les traites et pour les métaux précieux. Voici ce qu'en dit François d'Usson de Bonrepaus, ambassadeur de France, dans un rapport au roi datant de l'année 1699, rapport qui se trouve aux Archives des Affaires étrangères à Paris :

> ... Amsterdam est sans contredit de toutes les villes des Provinces-Unies la plus considérable par sa grandeur, par ses richesses et par l'estendue de son commerce. Il y a même peu de villes en Europe qui peuvent l'égaler à ces deux derniers esgards ; son commerce s'estend dans les deux parties du monde et ses richesses sont si grandes qu'elle a fourny pendant la guerre jusqu'à cinquante millions par an, et même davantage.

Au XVIIIe siècle, les rôles changèrent peu à peu ; le commerce anglais l'emporte à Cadix comme ailleurs. Le traité de commerce conclu en 1713 entre l'Angleterre et l'Espagne ouvre même, dans une certaine mesure, les colonies espagnoles au commerce anglais. Par conséquent, le rôle joué par l'Angleterre dans le commerce de l'argent est beaucoup plus considérable qu'au siècle précédent. D'ailleurs, dès le commencement du siècle, l'or devient plus abondant, à cause de la meilleure exploitation des mines du Brésil. Le Brésil étant une colonie portugaise, le commerce anglais s'empare de l'or brésilien, parce que le Portugal se trouve, depuis le traité négocié en 1703 par Methuen, dans la dépendance politique et économique de l'Angleterre. Si l'on considère, en outre, l'accroissement de la navigation anglaise vers la Côte d'Or de Guinée, on s'explique aisément l'importance grandissante du commerce anglais des métaux précieux dès le début du XVIIIe siècle. Néanmoins, même alors, et c'est un fait que Sombart n'a pas remarqué, l'activité hollandaise ne s'était pas ralentie, le marché d'Amsterdam continua d'exercer une forte attraction sur les métaux précieux. Le commerce avec Cadix fut toujours considérable et la navigation à la Côte d'Or de Guinée garda toute son importance. Il y avait, en outre, des relations commerciales directes entre la Hollande et le Portugal. Les Anglais eux-mêmes furent obligés d'envoyer une partie de leur or et de leur argent au marché d'Amsterdam. Cette ville resta donc pendant la plus grande partie du siècle le plus grand marché mondial, soit pour les marchandises, soit pour les traites et pour le crédit. C'est surtout le crédit, largement accordé par les riches banquiers d'Amsterdam, acceptant des lettres de change tirées sur eux de tous les coins de l'Europe, qui lui conserva la plus grande partie de son ancienne importance internationale. C'était donc à Amsterdam qu'on pouvait faire les meilleurs prix pour les métaux précieux.

Du reste, il ne faut pas oublier sur ce point le rôle joué par le numéraire hollandais comme article d'exportation. Assurément, la guinée anglaise et le louis d'or français avaient pris une grande importance. Mais le ducat d'or hollandais garda et même accrut son ancienne popularité dans une grande partie du monde. Il s'exportait en masse vers la Baltique, en Pologne, en Russie, au Levant et en Orient. Comme au siècle précédent, les hôtels des monnaies furent, si l'on peut dire, une industrie d'exportation, la seule industrie d'ailleurs qui fût encore florissante en Hollande à cette époque.

La célèbre Banque d'Amsterdam contribua beaucoup aussi à soutenir le commerce des métaux précieux et des espèces. Fondée au début du XVIIe siècle, à l'exemple de celle de Venise, pour remédier aux désordres monétaires, elle était à l'origine simplement une banque de virements et d'échange de monnaie ; elle fut, pour ainsi dire, le caissier général des négociants. Voici comment un auteur français (dans un rapport de l'année 1775) explique cette fonction :

... Tous les paiemens se font par un simple transport en vertu de l'assignation du propriétaire, de sorte que celui qui est créancier sur les livres de la Banque cesse de l'être dès qu'il a assigné en faveur d'un autre. Celui-ci y est couché pour créancier en sa place et ainsi consécutivement, les parties ne faisant que changer de nom, sans qu'il se fasse un payement réel en espèces.

Les virements à part, la Banque faisait, au moins au XVIIe siècle, le trafic des métaux précieux en barres et en espèces, en fournissant des matériaux aux hôtels de monnaie et en procurant des monnaies de négoce aux commerçants qui en avaient besoin. Pourtant elle n'a jamais fait d'avances sur titres, ni escompté de lettres de change, ni émis de billets. Ce n'était donc pas une banque de crédit , et elle ne ressemblait point à la Banque d'Angleterre. Mais, vers la fin du XVIIe siècle, ses attributions furent élargies. Elle fit, désormais, des avances sur ses livres, en argent de banque, aux commerçants qui déposaient chez elle de l'or ou de l'argent, en espèces et en barres. Sur son compte en banque, le dépositaire d'argent pouvait disposer immédiatement d'une somme presque équivalente à la valeur de son dépôt, tout en se réservant le droit de retirer ce dépôt au moment où cela lui conviendrait, par exemple en cas d'une hausse dans le prix des métaux ou des espèces. La Banque faisait ces avances au taux très modique d'un quart pour cent pour l'argent et d'un demi pour cent pour l'or, tous les six mois. Elle délivrait aux déposants des récépissés, c'est-à-dire des billets de dépôt, qui se vendaient couramment en Bourse. Ainsi un négociant pouvait, avec un capital modeste, faire de grandes affaires dans le commerce des métaux précieux et des espèces.

Au début du XIXe siècle (en 1802), quand commença la décadence de la Banque, voici comment le consul de France à Amsterdam expose cette manière d'opérer [2] :

Les métaux précieux arrivent du nouveau monde en Europe non par un écoulement constant et régulier, mais à des intervalles plus ou moins considérables et qui ressemblent mieux à la marche irrégulière d'un torrent qu'au cours paisible d'une rivière. A la fin surtout d'une guerre maritime qui a retenu dans les colonies espagnoles et portugaises les trésors provenant de leurs mines, l'Europe se trouve tout à coup inondée d'or et d'argent dans une proportion qui excède le besoin qu'on a de ces signes, et qui les aviliroit si on les versoit aussitôt et tout à coup dans la circulation. Que faisoient en ce cas les Amsterdamois? Ils déposoient ces métaux en lingots à la banque, où ils étoient à leur disposition moyennant une rétribution très modique, et ils les en retiroient peu à peu pour les envoyer dans les différens pays à mesure que la hausse du change en indiquoit le besoin. De cette manière ces signes utiles, dont l'affluence trop rapide eût fait excessivement monter le prix de toutes les choses, au grand préjudice des personnes qui n'ont que des revenus fixes et bornés, se distribuoient insensiblement par une multitude de canaux, vivifioient

1. Pourtant, il convient de faire exception pour les sommes souvent considérables que la Banque avançait régulièrement à la Compagnie des Indes orientales, ce qui a, plus tard, causé sa perte.

2. Voy. ma publication de documents : *De Wisselbanken van Amsterdam, Middelburg, Delft en Rotterdam*, 2 vol. La Haye, Nijhoff, 1925.

l'industrie, facilitoient et animoient les échanges. Ainsi la Banque d'Amsterdam ne bornoit pas ses bons effets à l'utilité particulière des négocians de cette ville, l'Europe entière lui devoit plus de stabilité dans les prix, d'équilibre dans les échanges, une proportion plus constante entre les deux métaux qui font concurremment l'office de monnoie, et l'on peut dire que, si cette banque ne se rétablissoit pas, il manqueroit un rouage essentiel à la grande machine du commerce et de l'économie politique des nations civilisées.

Cette conclusion est un peu exagérée, j'en conviens. Le monde a très bien su se passer de la Banque d'Amsterdam ; mais, en tout cas, il résulte de ce document, comme de tant d'autres, que, pendant tout le cours du XVIIIe siècle, le commerce des métaux précieux et des espèces à Amsterdam fut d'une grande importance et que la Bourse fut un des centres financiers les plus célèbres du monde. De grands emprunts s'y faisaient, tandis que le libre échange des métaux précieux contribuait beaucoup à la stabilité des cours du change. La guerre de Sept ans marque l'apogée de la Bourse. Cette guerre terminée, le déclin commence et il s'accélère pendant le dernier quart du siècle. Il existait d'ailleurs, disons-le tout de suite, d'autres centres pour le commerce international, aussi riches en capitaux et aussi bien, et peut-être même mieux outillés, notamment le marché de Londres. Au XVIIIe siècle, l'Angleterre disposait d'une grande industrie d'exportation qui manquait justement à la Hollande.

Ainsi, au dernier quart du siècle, Amsterdam, en tant que marché mondial, est en pleine décadence. Puis vinrent les désastres accumulés par les guerres de la Révolution et de l'Empire. L'impérialisme de Napoléon a été néfaste au commerce hollandais. Après Waterloo, Amsterdam reste, pour bien des années, une ville morte ; les capitaux avaient diminué et les anciennes relations commerciales étaient perdues. Le commerce des métaux précieux, comme celui des marchandises, avait abandonné la ville.

C'est seulement vers la fin du siècle passé que la navigation, le commerce et la Bourse d'Amsterdam ont reconquis une partie au moins de leur ancienne activité. Assurément, Amsterdam n'est plus, comme au XVIIe et au XVIIIe siècle, le centre du commerce mondial ; elle a été dépassée par d'autres centres beaucoup plus importants et elle doit se contenter d'un rôle plus modeste qu'autrefois. Cependant, ce rôle n'est pas sans grandeur et, si tous les signes ne nous trompent pas, il deviendra encore plus important, une fois la crise actuelle passée.

14

World Silver Flows and Monetary Factors as a Force of International Economic Integration, 1658–1758

K.N. Chaudhuri

The centuries spanning the discovery of America (1492) and the termination of the Napoleonic Wars (1814) have exercised a remarkable fascination for the historians of Europe. Western societies, their way of life and thinking, experienced in those years an internal qualitative expansion which also contained a geographical or spatial expression. Individual societies have always undergone in the recorded history of mankind phases of expansion and contraction. The outward movements which found their most concrete and wondrous symbol in the geographical discoveries of that more-than exciting decade at the end of the fifteenth century were not an entirely new phenomenon in social locomotion. The fundamental meaning and significance of those changes lie elsewhere: in the creation of structural discontinuities or transients. Before we address ourselves to this large and controversial subject, we should perhaps raise a more practical question. What are the visible and material identities of these mysterious and often invisible forces of change which past and present historians have so phrenetically tried to analyse? The answer would probably point in the direction of four factors and their dispersion in the physical and mental domains of time. First of all the discovery and the diffusion of enormous quantities of precious metals in America (and to a lesser extent in Japan) during the sixteenth century provided at both theoretical and practical level the means for the creation of an alternative system of economic production and distribution, based on the general acceptance of a quantitative index of value. The social entitlement to goods and services began to move from a physical plane to a more abstract one. In the Classical period, as Michel Foucault has reminded us, there was no political economy because "in the order of knowledge" production did not exist. In the seventeenth and eighteenth centuries (and he might have added in the sixteenth too) there did exist as so many partial objects, the notions of value, price, trade, circulation, income, interest.[1] It is doubtful whether any of these abstract notions could have gained the ascendancy that they did, without a wider usage of money created by American treasure within and outside the European epicentre. The operation of a price system administered through an international mechanism of money and currency was the "sign language" of a new mental world in human civilisation, which has evolved and gained increasing refinement from the early sixteenth century to the present times.

The second material expression of the expanding change was the invention and the continued development of the "round" ship, reaching its most advanced form in the Spanish galleons and the Portuguese carreiras, floating fortresses and floating warehouses. The carriage of goods by sea and the control of wide oceanic space were revolutionised by the addition of the gun-deck and gun-ports; vessels of large size could be adequately defended without sacrificing cargo-storage. As a mobile fortress and warehouse the galleon also made it possible to build and maintain territorial bases far from home on the principle of enclaves exposed to the powers of neighbours whose goodwill was questionable. The political and economic conquest of the Atlantic, the Indian Ocean, and the Pacific removed the last tyranny of the great oceans. The third element in our set is a historical development which followed on from the second: the combination of warfare by sea and land, backed up by the artillery, muskets, and new deployment of the cavalry and infantry. The extension of the Mediterranean style of warfare by land and sea to the Atlantic and the Indian Ocean was as revolutionary in its political and economic implications as the fourth and final factor in our quartet, the discovery of the steam engine leading to the Industrial Revolution of the late eighteenth century. The Chinese civilisation by 1313 had mastered and applied the technological principle of converting rotary motion into longitudinal motion by the use of crank and piston-rod, which was the mirror image of the movement of a steam engine. But Chinese industrial production remained non-mechanised as did that of India.[2] One of the possible explanations for the rapid application of machinery in the West may be found in the changing ratio between geographical space and population brought about by the discovery of the New World. In any case, there is no doubt that the prime motivation for the first Industrial Revolution came from finding alternative ways of saving or increasing energy previously provided by human labour in another continent (Indian cotton textile industry) and by animals at home. The railways completed the transport revolution set in motion by the galleon.

This description and review of the four material expressions of European expansion is prompted by a belief that the last three contributed to a significant extent to the economic integration of different and hitherto self-contained parts of the world and that none of them could have been effective without a substantial increase in financial liquidity. After all it is almost axiomatic that ships and steam engines cannot be built or commissioned by feudal lords in the service of a monarch sanctified by the coronation ceremony. Why were the directors of the English East India Company so afraid in the late seventeenth century of the rival Dutch Company seizing the financial control of the world pepper trade? The logic was spelled out clearly in 1687. "If the present mis-understandings between the two nations should ferment to an open war, it would be thought by the vulgar but a war for pepper which they think to be a slight thing, because each family spends but a little of it. But at the bottom it will prove a war for the dominion of the British as well as the Indian seas. Because if ever they come to be sole masters of that commodity, as they are already of nutmetgs, mace, cloves, and cinnamon, the sole profit of that one commodity, pepper being of general use, will be more to them than all the rest and in probability sufficient to defray the constant charge of a great navy in Europe."[3] The reasoning contained in the above passage assumes three inter-linked mechanisms in an ongoing economic system: a network of international supply and distribution from individual producers to consumers, a pattern of family expenditures that enables a part of the money income to be spent on marginal items, and finally, capital accumulation and investment which creates further fixed capital and means for the international control of resources (in this case, ships). Two sets of questions and problems immediately follow. Was the increased mining and flow of New World gold and silver in their monetary form a necessary condition for the emergence and operation of the commercial capitalism of seventeenth century Amsterdam and London? And in what way was the economic decisions of, let us say, the pepper farmers in Indonesia and the Malabar influenced by financial events taking place in a narrow stretch of land reclaimed from the cold wind-swept sandbanks of the North Sea? Did the city on the Zuiderzee derive its own economic destiny from the blazing fortunes of another southern metropolis on the slow-moving Guadalquivir? In other word, would the historical course of commercial and industrial capitalism in Europe, America, and Asia have followed an identical path without the "sign language" written by New World treasure?

Through these rhetorical questions we are able to identify and illuminate the linkages, the movements, and the communication channels that did exist between various component parts of human society structured in time, space, physical and mental domains. But what we cannot do is to measure precisely the relative weight of each linkage in its functional role and assert positively that if a particular linkage failed or was simply absent in the operating system, an adequate substitute would have been found and applied. There is no difficulty in agreeing with Fernand Braudel that capitalism or capitalistic activities had appeared very early in the Near East, India, China or Europe, perhaps as soon as division of labour appeared in these societies, and it is of course a necessary condition of capitalism that there should be a system of measuring values. In the words of Braudel, "Far in advance, there were signs announcing the coming of capitalism: the rise of the towns and of trade, the emergence of a labour market, the increasing density of society, the spread of the use of money, the rise in output, the expansion of long-distance trade or to put it another way the international market. When, in the first century AD, India seized or at any rate penetrated the islands of the East Indies; when Rome held an area even greater than the Mediterranean in her power; when China invented paper money in the ninth century; when the West reconquered the Mediterranean between the eleventh and thirteenth century; when a world market began to take shape in the sixteenth century, the "biography of capital" was starting to be written in one form or another."[4]

It is clear that money and prices, whether measured in units of commodities such as a basket of rice, a roll of silk, a moulded block of salt or in units of precious metals, gold and silver, copper and rare cowrie shells, had served the economic purposes of societies in old Eurasia for a long time.[5] But from the sixteenth century onwards commercial capitalism and the emerging international order, which we should like to describe as a process of economic (and political) integration, assumed a new reality and a new meaning in the conceptual vocabulary of the collective mind. It is important to remember that neither the "reality" nor the "meaning" of the concept of international "integration" remained of the same order in its structural relationships or quantitative intensity. Failure to take account of these abstract and concrete thresholds often leads to absurd historical statements which would describe, as for example, the economies of India and China as forever being commercialised from a state of imperfect monetisation. The huge influx of gold and silver from the New World to Europe from the sixteenth century to the nineteenth is seen by many European historians as one of the fundamental determinants of economic expenditures. But the outflow of the same precious metals to the Middle East, India, and China in the paths of an ancient trans-continental trade, we are told, owed its explanation to a totally different reasoning, an eastern psychology which assigned a higher value to stored wealth than to current material consumptions. The point has been made many times by myself and others that the absorption of gold and silver by the Asian economies in the early modern period had little to do with a "hoarding" social mentality but was grounded on an international pattern of economic specialisation, on payments mechanism, and socially-determined demand which had existed for at least a millennium.[6]

What then was "new" about international economic integration? The question raises problems of both periodisation and structural relationships. The concept of integration is defined in this essay in a three-dimensional plane. Firstly, in the amplitude of temporal fluctuations the deviations from an acceptable norm should steadily approach this value, indicating that the controlling mechanisms in the system are functioning properly or that they are being improved. If the standard deviation in grain prices for certain years steadily assumes lower values, we can say that the structure of cereal production is approaching temporal integration. Secondly, the activities arranged in order of time between two or more points in space should display respectively sychronous movements, a sign of closer relationships between the points. This is taken as geographical integration. Thirdly, if the "output" generated by different structural systems show increasing joint dependence on one another, we treat this phenomenon as evidence of a structural integration. Yet one other basic proposition of this paper is that gold and silver in their intrinsic or "surrogate" forms acted as a crucial instrument of exchange that was not merely economic but also social and political. The capacity to undertake and increase international trade was as much influenced by the monetary role of precious metals as the ability of state systems to wage wars and extract an economic surplus from agrarian production oriented towards satisfying household demand. Needless to say that no historian howsoever pain-staking he is in his methodology can provide a rigorous quantitative measurement of integration in its three above definitions. But there can be little harm in keeping the theoretical points in mind when we try through broken quantitative and impressionistic evidence to isolate the larger effects and consequences of world-wide bullion flows in the seventeenth and eighteenth centuries. If the inter-national economic integration in our period of study increased according to any of the above criteria, there were other factors in this development than the circulation of New World treasure, just as in a central-heating system there are operating parts other than copper pipes distributing the heated water.

By the middle of the seventeenth century, the economic role of American treasure in the network of European and world trade began to assume different forms. Adam Smith in his role as the confident historian had no difficulty in asserting that between 1630 and 1640 or about 1636 precisely, the effect of the discovery of silver mines in America in raising the ratio between units of silver and commodities had passed its peak. This fact was proven by the relative stability of grain prices expressed in silver.[7] Remarkably, in Hamilton's calculations of treasure imports into Castille from America the decade 1631-40 recorded a fall of more than a third from the level of the previous ten years (see Table 1), although the really significant decline did not come about till the 1650s. Contemporary observers, whether merchants engaged in long-distance trade involving the use of precious metals or governement officials concerned with currency and financial matters showed little interest in estimating the long-term effects of American bullion imports on either prices or levels of private and public expenditures. They were however acutely aware of its short-term effects on all aspects of domestic and international economy and certainly of the accelerating velocity of bullion circulation along the existing channels of international trade. As early as 1553 Thomas Gresham noted the financial changes taking place in the money markets of Antwerp, which acted as a key centre of control in Iberian trade with northern and central Europe, and he especially singled out the strange displacement of gold by the coined silver of Spain in the city's bourse.[8] A few years later Padre Thomas de Mercado pointed out that Spain's Atlantic trade had the effect of raising the prices of all kinds of cloth and other export goods at home. He concluded his descriptions of the activities of Spanish traders in the wholesale markets of Seville with the remark:"a este tono succede cada dia en Sevilla en los precios, ansi merceria que viene de Flandres, como enlos panos de Segovia, y Toledo, en el vino, y azeyte que se coge en esse Axarase."[9] By 1622 both Bacon and Duarte Gomes could reflect on the immense territorial expansion in European activities which the floods of American treasure had fuelled and the Spanish writer accurately traced the path taken by the New World bullion in its eastward migration to India and China from about 1588.[10]

The century which followed the conclusion of peace between Spain and Holland (1648) was important for several structural changes in the distribution and movements of American treasure. The general belief that the production of American silver and its rate of transfer to the Casa de Contratacion in Seville slowed down in the second half of the seventeenth century has been challenged by the research of Michel Morineau, who has taken his data from the Dutch commercial reports. If we are to believe the latter sources, the magnitude of imports remained substantial until the turn of the century, though from 1717 the general level appeared to have declined .[11] The Dutch gazettes help to correct the impression given by the official Spanish records of a steep decline in the receipts of New World treasure in Seville. The figures compiled by Lutgardo Garcia Fuentes from the records of the Casa de Contratación for 1650-1700, however, are so much smaller than the Dutch estimates that it is impossible to reconcile them rationally. Research conducted by myself in the Archivo de Indias, Seville, confirms the accuracy of the figures given in Garcia Fuentes' work.[12] Furthermore, John TePaske, Herbert Klein, and Hernandez Palomo have reconstructed in detail the accounts of receipts and expenditures in the Indies and their tables also demonstrate the fall in European remissions.[13] There is no doubt that Spanish accounting methods at home and in the Indies, in spite of their obsession with arithmetic accuracy, leave much to be desired when compared with the impeccable doubtle-entry system followed by the Dutch and English financial institutions. But the problem in this case is to explain a difference by a factor of ten. It is possible that the real volume of Spanish American bullion, both silver and gold, did not substantially decline during the second half of the seventeenth century and that the value of private imports were significantly under-stated in the official records.

Table 1
Import of gold and silver from Spanish America (in kg fine)

Période	Silver	Gold
1503-1510	.....	4 965
1511-1520	.....	9 153
1521-1530	148	4 889
1531-1540	86 193	14 466
1541-1550	177 573	24 957
1551-1560	303 121	42 620
1561-1570	942 858	11 530
1571-1580	1 118 591	9 429
1581-1590	2 103 027	12 101
1591-1600	2 707 626	19 451
1601-1610	2 213 631	11 764
1611-1620	2 192 255	8 855
1621-1630	2 145 339	3 889
1631-1640	1 396 759	1 240
1641-1650	1 056 430	1 549
1651-1660	443 256	469
1503-1660	16 886 807	181 327

After E.J. Hamilton, American Tresure and the price revolution in Spain 1501-1650 (Cambridge, Mass., 1934), p. 42, the figures have been converted into quantities.

Table 2
The export of gold and silver to Asia by the English East India Company 1660-1760
(in kg fine)

Période	Silver	Gold
1660-1665	40,145	1,074,47
1666-1670	22,910	1,673,66
1671-1675	49,828	3,669,50
1676-1680	179,252	5,156,62
1681-1685	240,952	6,931,61
1686-1690	30,567	879,18
1691-1695	7,687	221,14
1696-1700	131,511	491,22
1701-1705	166,885	-
1706-1710	173,833	141,11
1711-1715	167,503	145,79
1716-1720	250,851	-
1721-1725	289,349	-
1726-1730	261,401	-
1731-1735	260,102	-
1736-1740	260,378	-
1741-1745	257,882	-
1746-1750	366,289	-
1751-1755	398,041	-
1756-1760	193,458	-

From K.N. Chaudhuri, The Trading World of Asia and the English East India Company 1660-1760 (Cambridge, 1978).

What light do the qualitative records throw on the availability and distribution of precious metals in Europe during this period? First of all, it is necessary to take account of three basic historical developments associated with the bullion trade. The diffusion of American treasure throughout the European financial centres had originally followed an Italian, German, and Flemish axis. From the 1650s the distribution network began to move towards Amsterdam and London. In gradually dismantling the old restrictions on the re-export of bullion, the financial authorities in Holland and England bowed to pressure from their merchant communities and recognised the new role of the national money markets in intra- and extra-European trade.[14] For the outflow of American silver towards Asia also began to shift in quantitative terms from the Mediterranean-Middle Eastern route to the Cape passage. The Dutch and English East India Companies were of course the main commercial instruments of the redistribution in favour of the Indian Ocean. The expansion in European imports of Asian goods, so striking in the 1680s, was largely financed by increasing bullion exports, although the export of commodities also increased in volume and value.

The monetary implications of these regional and quantitative shifts were quite clear. Previously, precious metals in bullion and money had travelled from the Mediterranean to the Red Sea and further east in a series of short-distance stages with a large number of national currencies exchanging one against the other.[15] In this system of Eurasian exchange, passing through many intermediate and semi-independent economic zones, the monetary effects created by the abundance or shortage of precious metals were not immediately transmitted over any great distance and were generally a protracted process in time. When the Cape route became the dominant one in the West-to-East trade, it shortened at once the journey-time taken by gold and silver to reach their Asian destinations and established nearly an unbroken line of economic communication between not only Europe and Asia but also between America and Asia. The institution of direct commercial voyages between Acapulco and Manila had the same effects so far as the movement of treasure was concerned. Indian and Chinese bullion dealers were aware within six to eight months, of the variations in the fineness and weight of the Spanish real of eight, which became the principal trading currency throughout the world. At the same time, European domestic currencies suffered from a chronic erosion of intrinsic value, resulting in successive devaluations. The outflow of substantial quantities of American treasure to the Baltic areas and to the leading commercial nations of Asia coincided with a marked tendency in Europe to adopt token currency and to increase financial liquidity by a greater usage of bank deposits, bills of exchange, commercial bonds, and government securities. India and China however remained faithful to monetary conventions with an intrinsic standard of 99-100 per cent pure gold and silver.

The available historical evidence seems to suggest with varying degrees of strength the possibility for four somewhat contradictory developments in the trade and finance of Spanish-America, Europe, and Asia in the second half of the seventeenth century and the early eighteenth. The production of gold and silver in the central and south American mines fell in the 1640s and 1650s, with a slow recovery taking place later. In terms of actual figures, the import of American treasure into Spain (accepting Morineau's estimates) was not insignificant. But as we shall see, the bullion trade experienced in Europe a period of acute crisis in the last quarter of the seventeenth century which lasted well into the first two decades of the next. The Dutch and English trade with Asia recorded a rapid increase in volume and Europe was flooded with eastern goods, most of which were paid for with Spanish-American treasure. Finally, European monetary system after suffering from repeated devaluations and destabilising movements of currency between different countries began to look for other alternatives than precious metals alone for sustaining its financial requirement. Asian monetary practices in contrast retained their adherence to the metallic standard of currency, though historians have underestimated the role of traditional banking instruments in the economies of the Middle East, India, and China.

Between 1660 and 1685, the correspondence between the English East India Company's Committee of Treasure (responsible for buying the required quantity of bullion destined for Asia) and the leading bankers in London, Amsterdam, and Puerto de Santa Maria suggest little indications of any kind of crisis in the supply of American treasure. In 1670 however Edward Backwell, the leading goldsmith-banker in London at the time, failed to supply the Company with the contracted amount of silver because of the non-arrival of the Spanish plate-fleet. The tables of both Morineau and Garcia Fuentes show that there was indeed a break in treasure imports in Seville during 1668-9.[16] We may note the immediate contemporary connection beween the inflow of treasure in the Casa de Contratación and the financial liquidity of at least one substantial European banker. Edward Backwell asked for permission in 1670 to export gold, which gives us a clue to a point not hitherto emphasised in modern historical works. We know that both the Dutch and English East India Companies exported large quantities of gold to the East Indies between 1660-1685 and that the gold was supplied in London by resident Jewish bankers such as Jeronimo Miranda, Solomon del Medina, and Alphonso and Gomez Rodriguez. Originally, it was assumed by myself that this develpment was largely due to monetary conditions in Asia, the fact that European trade was moving in quantitative terms towards the gold-currency areas of southern India and that the bimetallic ratio in the Mughal empire was turning against silver.[17] I did not ask myself at the time where the gold came from in Europe but felt that gold production was increasing in West Africa. It was a great surprise to discover from the records of the Casa de Contratación that the annual plate fleet was importing gold on a surprisingly large scale in the third quarter of the century. To take only one example, the records of the "compradores de plata y oro" provide the following "Resumen" in 1675:

> "Cargo y datas Generales de la Plata y Oro en pasta que han recevido los quatro Compradores asaber de la Flota de 1671, de la 1672, de los Galeones del 1673, y Flota del 1674 ..."[18]

71

	Cargo		Data		Demasias	
(in marcos)	Plata	Oro	Plata	Oro	Plata	Oro
Ybarburu y Co.	255,549	244,574	277,027	473,385	21,478	228,811
Ochoa y Co.	85,916	94,712	108,759	254,700	22,843	159,988
Gainza y Co.	29,392	33,428	39,428	89,150	10,036	55,484
Morales y Co.	137,212	149,931	137,493	306.900	281	156,969
	508,069	522,883	562,708	1124,135	54,638	601,252
(in kg)	233,712	240,526	258,846	517,102	25,134	276,576

The balance between silver and gold in the annual importation of Spanish treasure in the second half of the seventeenth century needs to be further investigated in a systematic way. But the influx of gold on such dimensions in the 1670s must have had a considerable impact on the availability of silver, its market and mint price, and the contemporary monetary disease, the problem of clipping and debasement. It is possible that the outflow of silver in the direction of Asia and the Baltic areas exercised strong pressures on current stocks in western Europe, so that when the real silver famine began in the 1690s continuing through the Wars of the Spanish Succession, the disruption in supplies immediately forced up the market price of silver and removed silver coins from circulation. We know that in India the bimetallic ratio had been steadily rising, a sign of the increasing cheapness of silver, and that gold was used mainly for savings and making large payments. But in the mid-1670s the supplies of silver through the Red Sea failed and Indian bullion markets experienced a sudden rise in the price of silver and depreciation of gold. Is it possible to believe that what was happening in the Indian banking and financial world in these years was connected with the possible fall in silver stocks in Spain and Europe generally? If true, it points to a most striking demonstration of both temporal and structural integration in at least monetary connections between Europe and the western Indian Ocean.

In any case, there is ample qualitative historical evidence which show that merchants, financiers, and even the general public in India and China were perfectly acquainted with the fact that the Mexican and Peruvian mines upheld the monetary liquidity of the entire commercial world. In 1635 one of the officials of the English East India Company commented after returning from a trading voyage to Poruguese Macao in China that the best-selling commodity there was silver, particualrly the peso of eight reals. So great was the Chinese demand for silver that merchants tenaciously followed its routes of supplies as if silver was their life blood.[18] The amount of silver carried to China by the junk traders of the Philippines, both Chinese and "barbarian", served as a sensitive indicator to mandarins and bankers of the possible variations in the supplies of silver.[20] The metal was imported to Manila not only in the annual Acapulco galleons but also in the Indian ships which exported both silver and fine cotton textiles from Bengal and Coromandel. In the 1690s and again after 1710 India-Manila trade was seriously disrupted by the crisis in American silver supplies as was also the trade to China.[21] After the War of the Spanish Succession, as European trade with Asia recovered in value and volume, the export of silver also stabilised in terms of annual variations. Bengal perhaps received the largest quantities of the treasure imported by Europeans and the Nawab's finance ministers kept a sharp watch on the level of the inflow and the extent of coinage at the government mints.[22]

The extent to which the silver trade had become integrated between Mexico City, Madrid, and the Indian money markets can be verified from the historical records. Shroffs and mint officials in Surat and Cassimbazar who purchased Mexican and Peruvian silver from the Dutch and English Companies kept themselves informed of the currency changes by regular metallurgical tests as well as financial news. In 1728 Philip V attempted to reform the Spanish American currency by a series of measures, the result of which was a reduction in the fineness of silver contained in a peso of eight reals (93.06 per cent to 91.67). He also installed new and modern machinery in the Mexican mints for coinage. The new technical improvements did not work quite satisfactorily for a time and there was a considerable variation in the weight and the reduction of silver to the right alloy mixture. The complaints from silver merchants in Spain led to a detailed investigation by both the Casa de Moneda in the Mexico City and that of Madrid. In 1736 the Madrid mint carried out detailed assays and established the measure of variations in a random sample. At about the same time (1736-7), a similar but totally independent assay was carried out in the Surat mint and the results were passed on to the officials of the English East India Company. A comparison between the Spanish tables and the Indian ones reveals almost complete statistical similarities.[23] Five years later Fatehchand Jagatshet, the leading banker and the manager of the Bengal mint, was still complaining of the deteriorating standards of pillar dollars.[24] As the Indian bankers and merchants had close business associates and even personal friends among European officials of the East India Companies and kept in touch with them by correspondence, they would certianly have been informed of the monetary events taking place in Séville and Amsterdam. The remittance of European funds left in India with these bankers and the calculation of exchange rates were an immediate incentive for the exchange of financial news.

Our historical sources, whether we look for them in Seville, Mexico City, Vera Cruz, in London, Amsterdam, or Istanbul, or in Mocha, Surat, and Canton, speak with the same voice of the power of Spanish-American treasure and the peso of eight reales to penetrate the commercial economies of the leading trading nations in the world and to destabilise their national monetary systems. Quite effectively the silver peso of 28.75 grams weight with 0.931 fine metal carrying the stamp of the Assayadores of imperial Spain became the international monetary standard. In the seventeenth century the Ottoman government closed its independent mints with the exception of those in Istanbul and Cairo and the fiscal authorities gave up collecting or accounting for taxes in gold and adopted the peso or the Lion Dollar as their money of account.[26] At the epicentre of the financial tremours, in Madrid itself, Don Geronimo de Uztariz spoke in the early eighteenth century how the peso of eight reales had fled the Spanish dominion to enrich the coffers of the enemies of the Holy Church, leaving only debased coins of brass to circulate in the mother country.[27] From the Isthmus of Panama to the straits of Bab al-Mandeb and Hormuz, no government was able to declare monetary independence from the piece of eight. There were however two great exceptions in the Indian Ocean. The Mughal and the Ming empires with perhaps the two strongest commercial economies in the world at the time used Spanish money for international payments but retained intact their domestic monetary conventions. The reasons were different in each case but quite unique otherwise. The Mughal government in Delhi insisted that its silver rupees (11.6 grams weight, 0.98.5 fine) carrying the name of the reigning emperor and an inscription from the Quran should be the sole legal tender in the empire; foreign imported coins were immediately put in the furnace of the mints. The Celestial court on the other hand did not care in what form silver circulated in China, as its monetary value was derived from bullion weight and not from the metrology of coinage. Chinese traders, merchants, and bankers, having to make payments in silver, used a pair of precision scales and metal cutters. The stamp of the bullion dealers authenticating a bar of silver or a Mexican dollar was sufficient proof of its financial acceptability. The administrative strength of the Indian and the Chinese empire is illustrated nowhere better than in the independence of the rupee and the "sycee" silver. Does this mean that the Mughal or the Ming economies were not "integrated" in a proper sense with the international economic system? That will be a hasty conclusion.

We have so far hinted at the existence of several contradictions in the historical working and development of world monetary and economic system during the century 1650-1750. When examining the problem or the phenomenon of "integration" it is vital to distinguish between the purely monetary mechanism and the larger, economic, social, and political environment and to look at the exact location of the "gates" which connected them together. In the Mediterranean and the Atlantic economies, the internal currency arrangements were destabilised during the century 1650-1750 mainly because there were two separate markets for metallic money, one for the peso of eight reals and the other for the national mints. When the price of silver diverged between these markets, the dominant peso market always won and the same rationale also applied to most of the Near and Middle East. With contiguous geographical areas and short sea voyages, it was impossible for frontier officials to supervise strictly the movements of treasure and equally difficult to restrain mercantile activities. In India and China on the other hand treasure mostly arrived by sea and had to pass through a strict customs check. Although there was an apparently large smuggling trade in gold, Indian merchants were prevented by the political power of the state from using metallic money in any other form than the government-approved currency and many of them were deeply involved in collaborating with the mint officials in maintaining a semi-monopoly in the supply of money. The collection of huge money sums in land taxes alone ensured the cooperation of the bankers in India and China with the state authorities on monetary matters.[28] But neither India nor China could have kept their commercial economy and public expenditures at a high level without an adequate supply of metallic money. The quantitative equation between Spanish-American treasure and national currencies was an essential factor in the operation of the Mughal and Ming-Ch'ing imperial economy. The gradual commutation of the land tax into silver units which occurred in both the empires from the middle of the sixteenth century may be an indication of the greater availability of metallic currency. It is worth considering that the ability of the Indian provincial governors in the early part of the eighteenth century to intercept a large part of the silver import and raise standing armies was joined to the Maratha military expeditions to all parts of India in search of financial indemnities as two significant factors in the decline of the Mughal empire. Was the rising level of warfare in the Indian subcontinent during the eighteenth century comparable to what happened in Europe in the seventeenth with a time-lag and did the imperial unity of India pay the price for the subcontinent's integration with the world commercial economy?

I have argued in another place that one of the effects of the injection monetary liquidity into the European economies in the form of American treasure was to divert part of the aggregate demand in the direction of Asia.[29] The increase in the volume of Asian imports reflected this fundamental inner reasoning. However, it is evident that without an international comparison of money prices such a development could not have taken place and more important, unless European consumers were able to maintain or even increase their money income, the growth in world trade would have slowed down or even come to an end. Was the "crisis" of the seventeenth century associated, in any way, with the declining availability of New World treasure? We do not know. Prices of manufactured products in England at least tended to stabilise during the period 1650-1750, though currency inflation disturbed from time to time the long-term profile. What is clear is that when the commercial and the banking world was faced with the shortage of bullion, alternative forms of money were invented. European economy did not return to the "feudal" mode of production. The essential relationships between money, prices, income, and expenditures continued in both production and distribution.[30]

We can now draw together the various threads of arguments presented so far. The price mechanism in international trade was the fundamental operating factor in the distribution of goods and it operated in two ways. Prices fluctuated in an upward or downward direction according to the costs of production. Higher wages and an increase in the cost of raw material were often mentioned in India as reasons for the rising prices of industrial manufactures.[31] An expansion in demand ahead of supplies also put pressure on prices. When foreign merchants encountered rising prices, expressed in units of silver or gold, they were presented with a problem common to the past and present, that of selecting alternative sources of supply. The measurement of the prices of internationally traded goods was an absolutely indispensable condition of profitable operations. It seems reasonable to suggest that the redistribution of American treasure to the countries of Western Europe created a new source of consumer demand. The imports purchased in the Near East. India, Indonesia, and China were financed by exporting silver (and occasionally gold) which could be produced at a lower cost relative to goods. In this sense, the Western economies definitely enjoyed a comparative advantage in the creation of purchasing-power. The industrial producers in Asia also enjoyed a considerable advantage in the cost of export goods.

The difference in the money prices between Asia and Europe enabled the European East India Companies to market their imports at a mark-up varying from two to four hundred per cent. The economic "gates" for integrating Asian prices with those prevailing in Europe were there and remained resolutely open.[31] The continuing production of silver in the American mines kept the westward flow of commodities unchanged until the beginning of the nineteenth century. Yet another development in this period was the rise of plantation agriculture in the New World and the associated forced migration of labour, the Atlantic slave trade with west Africa. The final sequence of international division of labour between 1550 and 1750 may be drawn up in the following scheme. American treasure helped to finance Spain's balance of indebtedness. The cost advantage enjoyed by the Genoese bankers, and later by the Dutch and English in entrepot trade accumulated capital in the hands of Europe's most efficient entrepreneurs. This capital helped to finance the imports of Indian cotton textiles, Asian spices, and Chinese silks and tea. Indian textiles and cowries were in turn exchanged for African slaves who produced new agricultural commodities in the New World for consumption in the Old. The circuit seems to have ended as Barbados sugar sweetened Chinese tea in porcelain cups copied from the products of Ching-te-chen kilns in imperial China.

A postscript to the above comments and conclusions on the role of precious metals in world trade may be added in view of the discussion that took place on several papers presented at the International Workshop at Pommersfelden, Bamberg, in June 1985. The central question before any historian of the monetary developments during the period from 1500 to 1750 is the continuing and accelerating outflow of silver and occasionally gold from the West to the East. While the phenomenon itself was observed, commented upon, and sometimes explained at the time, no modern scholar has wholly succeeded in finding a theoretical reason for this astonishingly long period of what can be called a process of monetary adjustment. The paper presented by Dennis Flynn using stock demand supply and dynamic adjustment analysis is most interesting in suggesting a possible and satisfactory answer to the historic puzzle. The mining activities in the New World, while large in terms of absolute figures, must be seen against a background of huge silver stocks already being held in Europe and Asia in the middle of the sixteenth century. The annual additions to stocks obviously reacted on the demand for silver in a slow fashion and if the existing technological possibilities are taken into account the changes in demand and supply were more likely to produce changes in the structure of international trade rather than on production functions. Dennis Flynn's paper has re-emphasised the problem I raised in my book The Trading World of Asia (1978) that the monetary situation in Asia was split up into two major geographical areas, the Indian subcontinent and China and that the adjustment mechanisms within the sub-regions of these two wider areas combined with the inability of the Western countries to generate an industrial or technological Revolution much before 1800 kept silver mining in the New World profitable and the Asian economic producers able to undercut European competitors.

The theoretical reasoning suggested by Flynn and myself helps to put in true perspective Kristof Glamann's criticism of my earlier works. In an essay written in highly academic German (1981), Glamann warned, "Against this complex of explanation, a series of objections can be raised. Firstly, with regard to comparative differences in the level of wages and prices, it can be said that they presuppose that the factors which are being compared are identical and comparable. Here lies the main rub. The major part of the imported Asiatic goods, pepper and other spices, tea, coffee, sugar, cotton were often agricultural and without any European counterpart. Hence, comparisons are meaningless.."[32] Glamann's accusing finger is of course pointing more towards David Ricardo rather than me. The inventor of the theory of comparative costs took the unfortunate example of England and Portugal as two trading partners and two commodities, wine and cloth, to illustrate his famous law. Did Ricardo not know that Portuguese wine had no direct counterpart in England? In which English counties can the conditions of the Douro be replicated to make the English Factory in Oporto obsolete if and when English wine-makers became efficient? Glamann has forgotten how the theory of comparative cost works. When production costs measured in money prices become unequal between different trading areas, the adjustment process seldom leads to direct substitutions of imports, although in the case of English cotton industry it did so. More often the changes occur along the entire range of economic production with different sets of cost. In other words, if the Western nations found themselves importing more pepper, spices, tea, and coffee than they were able to pay for with their export of goods alone, then they must produce silver, gold or foreign currency which the Asian merchants are willing to accept. Silver mining in the New World became uneconomic in cost terms from time to time, but Glamann and other historians who share his views must find an explanation why production continued in Central and South America for more than three hundred years.

Glamann's point about the imperfect nature of Asian market for precious metals has been mentioned by F.S. Gaastra to illuminate the Dutch policy on the export of precious metals. It is worth pointing out that the notion of a "perfect market" is strictly a theoretical technique and should not be used to describe the actual reality as "imperfect". Glamann has drawn on his experience of the historical material from the archives of the VOC, particularly the Company's bullion and copper trade, to criticise my conclusions on the operations of the bimetallic ratios in Eurasian silver-gold flows. "The ratio explanation too does not withstand closer examination", he writes in his article (1981). "If for example, one takes 17th century India, one notices that the demand for precious metals differs widely from region to region. The Coromandel coast had gold standard, Bengal double silver-copper standard ... Recently undertaken but not yet published study by a young American historian Joseph J. Brennig on the role of the trading place of Surat in the west coast in the export by the English East India Company of precious metals shows that the bimetallic ratio in the years 1665 to 1670 was above the European level,that it was 1:16 wheras the ratio in Europe was 1:14. In gold terms, this meant a loss if one sent silver from London to Surat, therefore, the people of the English Company tried at that time in Surat to pay in other commodities than silver." This evidence and the conclusion to be drawn from it appear so astonishing that Hermann Kellenbenz summarising all the papers in the volume also referred to the point.[33] Since Glamann wrote his essay, Brennig's study has been published in a volume edited by J.F. Richards (1983). Brennig cites various figures for the higher gold-silver ratio in India for the 1660s but does not give references to the original sources from which these data are taken. His graph of the ratios in India and Europe has a reference to the article by Braudel and Spooner but no citations for the Indian ratio.[34] Why did Brennig and Glamann fail to cite references to the original sources? In any case my article in the Economic History Review (1968) pointed out for the first time how gold-silver ratio had risen in India in those years?[35] I was able to show in that article how large quantities of gold were exported to India by the English East Company in preference to silver and my conclusions are fully supported by references to the historical documents. Glamann is certainly aware of the existence of the study, as he has utilised it in his other writings.[36] Moreover, the bimetallic variations are fully discussed in my later study (1978). To use one's own research findings to criticise the conclusions is an academic novelty which I hope will not become fashionable. To Glamann my use of economic theory, particularly concepts such as "liquidity preferences", "transactions demands", "income and investment levels", "flow mechanisms", "multipliers", and "elasticities" may be no more than Modeworten, fashionable jargon. Economists will be astonished to hear this and all I can say is that the use of economic theory has not precluded me from examining the relevant historical facts which support or contradict that theory.

Reference Notes

1 Michel Foucault, The Order of Things: an archeology of the human sciences, London, 1970,155.

2 Joseph Needham, The Grand Titration: science and society in East and West, London, 1979, 96, M. Elvin, The Pattern of the Chinese Past, Standford, 1973, 312-3, and "The High Level Equilibrium Trap," in W.E. Willmott, ed., Economic Organisation in Chinese Society, Standford, 1972, K. N. Chauhuri, The Trading World of Asia and the English East India Company 1660-1760, Cambridge, 1978, 272-4.

3 India Office Records, East India Company, Despatch Book, 3 August 1687, vol. 91,321.

4 Fernand Braudel, Civilization and Capitalism 15-18th Century, Volume III, The Perspective of the World, London, 1984, 620.

5 K. N. Chaudhuri, Trade and Civilisation in the Indian Ocean: an economic history from the rise of Islam to 1750, Cambridge, 1985, Chapters 1,10,11.

6 K. N. Chaudhuri, The English East India Company, the study of an early joint-stock company 1600-1640, London, 1965, The Trading World of Asia, "Some Reflections on the World Trade of the XVIIth and XVIIIth Century: A Reply," Journal of European Economic History , vo. 7, 1978, 223-229, "Circuits monetaires internationaux, prix compares et specialisation economique 1500-1750," in John Day,ed., Etudes d'Histoire Monetaire XIIe-XIXe Siecles, Lille, 1984, Om Prakash,, "Bullion for Goods: international trade and the economy of early eighteenth century Bengal," Indian Economic and Social History Review, vol. 12, 1976, 159-87.

7 Adam Smith, The Wealth of Nations, 1776, reprinted from the 6th edition 1905, vol.1, Book I, Chapter XI, 201-2.

8 Gresham to the Council in London, 20 december 1553, Public Record Office, London State Papers 69/2, quoted by F.C. Spooner, The International Economy and Monetary Movements in France, 1493-1725, Cambridge, Mass., 1972-23.

9 Thomas de Mercado, Summa de Tratos y Contratos, Sevilla, 1571, fol. 95.

10 Sir Francis Bacon, "Advertisment touching an Holy Warre" (1622), Works, vii, 20 quoted by J.H. Elliott, The Old World and the New 1492-1650, Cambridge, 1970, 87, Duarte Gomes, Discursos sobre los comercios de las dos Indias, 1622, Lisboa, 1943, 49.

11 Michel Morineau, Incroyables gazettes et fabuleux metaux, les retours des tresors americains d'apres les gazettes hollandaises (XVIe-XVIIIe siecles), Cambridge, 1985, Antonio Garcia-Baquero Gonzalez, Cadiz y El Atlantico (1717-1778) (El comercio colonial español bajo el monopolio qaditano), 2 vols., Sevilla, 1976.

12 For example, in a volume entitled "Relaciones de oro plata y efectos que vinieron de India los años 1666 as 1698", Archivo de Indias, Sevilla, Contratación Legajo 4926, document dated 1 November 1674, the following figures are given for the value of treasure imported by the fleet of Don Pedro Corvete from Nueva España: (in maravedis, one peso of eight reals equalled 272 maravedis:)Su Magestad, 99 015 072, Santa Cruzado 53 601 408, Donativo 130 016, Particulares, 212 859 312, total 365 605 808, which is exactly the figure given by Garcia Fuentes in his El Comercio Español con America 1650-1700, Sevilla, 1980. Morineau on the other hand gives the value for this year for the same flota as 12 182 566 pesos, see Morineau, Incroyable gazettes, 234.

81

13 John TePaske "New World Silver, Castile, and the Far East (1590-1758)", Paper presented at Comparative World History Workshop, Madison, University of Wisconsin, 1977, John TePaske and Herbert S. Klein, "The seventeenth-century crisis in New Spain: myth or reality,"Past and Present, No. 90, 1981, 116-35, John TePaske and Jose y Mari Luz Hernandez Palomo, La Real Hacienda de Nueva España: la real caja de Mexico (1575-1816), Mexico, 1976.

14 J.G. van Dillen, History of the Principal Public Banks, The Hague, 1934, 93, K.N. Chaudhuri, The Trading World of Asia, 169, A. Attman, Dutch Enterprise in the World Bullion Trade 1550-1800, Goteborg, 1983.

15 S.D. Goitein, A Mediterranean Society, vol. 1, Berkeley and Los Angeles, 1967, E. Ashtor, Studies on the Levantine Trade in the Middle Ages, London, 1978, John Day, "The great bullion famine of the fifteenth century,"Past and Present, No. 79, 1978, 3-54.

16 K.N. Chaudhuri, The Trading World of Asia, 169, Morineau, Incroyables gazettes, 233, Garcia Fuentes, El Comercio Español, 388.

17 K.N. Chaudhuri, "Treasure and Trade Balances: the East India Company's export trade 1660-1720," Economic History Review, 2nd series, vol. XXI, 1968, 487, and The Trading World of Asia.

18 Archivo de India, Sevilla, Contratación Legajo 4954, 13 December 1675. There is another "Resumen" in the same file but the figures differ slightly the cited table. The full names of the bullion firms were as follows: Lorenzo de Ybarburu y Compania, Juan Ochoa, Juan Cruz Gainza, and Don Joseph de Morales. For the imports of gold see also Garcia Fuentes, El Comercio Espaniol, 40001. One marco = 8 onzas = 230 grams = 8.4 pesos = 67 reales. See Humberto F. Burzio, Dicconarios de la Moneda Hispaniamericana, 2 vols., Santiago de Chile, 1956-8.

19 Henry Bornford at Surat to the Company, 29 April 1636, The English Factories in India 1643-1636, ed. William Foster, Oxford, 1911, 228-9.

20 See Chao Ch'uan-ch'eng, "A ship's voyage from Luzon to China in the eighteenth century," in E-Tu Zen Sun and John de Francis, eds., Chinese Social History: translations of selected studies, Washington, D.C., 1956.

21 See K.N. Chaudhuri, The Trading World of Asia, Chapter 8.

22 See K.N. Chaudhuri, The Trading World of Asia, Chapter 8.

23 Archivo de Indias, Sevilla, Audiencia de Mexico, Casa de Moneda de Mexico, Madrid, 26 de Agosto de 1736, Legajo 751, India Office Records, London, East India Company, Bombay Public Proceedings, 17 February 1737, vol. 9, Memoranda of the Committee of Correspondence, vol. 101, 1736-42.

24 K.N. Chaudhuri, The Trading World of Asia, p.584, n.35.

25 Examples of such correspondence can be found in the private papers of Sir Robert Cowan, governor of Bombay, Public Record Office of Northern Ireland, Belfast, and the correspondence of Robert Adams, India Office Records, Home Miscellaneous Series, vol. 36, and the private papers of Thomas Hall, the East India Company's shipping agent, Public Record Office, London, Chancery Masters' Exhibits, C.103, No. 130-33.

82

26 Halil Sahillioglu, "The role of international monetary and metal movements in Ottoman Empire monetary history 1300-1700," paper presented at Comparative World History Workshop, Madison, 1977, Omer Lutfi Barkan, "The price revolution of the sixteenth century- a turning point in the economic history of the Near East," International Journal of Middle Eastern Studies, vol. 6, 1975, 3-28, Christopher E. Challis, "Spanish bullion and monetary inflation in England in the later sixteenth century,"Journal of European Economic History, vol. 4, 1975, 381-92, Pierre Vilar, A History of Gold and Money 1450-1920, London, 1976.

27 Don Geronymi de Uztariz, Theorica y Practica de Comercio y de Marina, 1742. ed. Gabriel Franco, Madrid, 1968, 7-8.

28 For Braudel's inimitable descrption of the process of exchange see, Civilisation and Capitalism, vol.III, The Perspective of the World, 500, W.S. Atwell, "International bullion flows and the Chinese economy circa 1530-1650," Past and Present, No. 95, 1982, 68-90.

29 K.N. Chaudhuri, The Trading World of Asia, Chapter 1.

30 For a somewhat different interpretation of bullion's role in the late seventeenth century see, Immanuel Wallerstein, The Modern World System II, New York, 1980, 110-12. For a theoretical discussion of the problems see also, Krzysztof Pomian, "The secular evolution of the concept of cycles," Review, vol. II, 1979, 563-646, originally published as "Cicli" in Enciclopedia Einaudi, vol. II, Torino, 1977.

31 K.N. Chaudhuri, The Trading World of Asia, 265.

32 Kristof Glamann, "Edelmetallströme und Welthandel vom 16. bis 18. Jahrhundert. Einige Gesichtspunkte", in Herman Kellenbenz, ed., Precious Metals in the Age of Expansion, Stuttgart, 1981.

33 Glamann, Ibid.

34 Joseph J. Brennig, "Silver in seventeenth-century Surat: Monetary circulation and the price revolution in Mughal India", in J.F. Richards, ed., Precious Metals in the later Medieval and Early Modern Worlds, Durham, North Carolina, 1983.

35 K.N. Chaudhuri, "Treasure and Trade Balances: the East India Company's Export trade, 1660-1720", Economic History Review, 2nd ser., vol.21, 1968, 480,502.

36 Kristof Glamann, "The Changing Patterns of Trade", Chapter IV, in E.E. Rich and C.H. Wilson, eds., Cambridge Economic History of Europe, volume V, Cambridge, 1977, see p. 259 and p. 646.

15

The Disintegration of the Ottoman Monetary System During the Seventeenth Century

Pamuk Şevket

Monetary history remains one of the least studied areas in the historiography of the Ottoman Empire and more generally of the Middle East. In the Ottoman case, despite the availability of considerable material, archival and otherwise, the existing literature is rather dated and incomplete. It is also ridden with gaps, factual errors, and problems of interpretation. The numismatics literature, however, may be useful in illuminating problem areas for which there is little evidence, but it is limited in scope, and the interests and emphases of numismatists are often quite different from the concerns of economic and social historians.

As a result, we know little today about the basic monetary patterns, such as the degree of market orientation and the use of money and credit in rural and urban areas during different historical periods. Our knowledge about the policies of the state for ensuring a steady supply of money is similarly limited. For the most part, economic and social historians of the Ottoman Empire are still unable to make sense of the most basic of monetary magnitudes involving prices, wages, and

* Earlier versions of this paper were presented at the Near Eastern Studies Department of Princeton University, History Department of Ohio State University, and the Sixth International Conference on the Economic and Social History of the Ottoman Empire held at Aix-en-Provence in July 1992. The author would like to thank the participants of these meetings for helpful comments and criticism.

wealth. Intertemporal comparisons of these magnitudes—which are the most basic prerequisites for studying the long duree and comparing the Ottoman case with others—are still virtually impossible. To cite a small but important example, it is not clear today how the 10 akçe daily wage of a mason in 1550 compares with a wage of 120 akçes or 1 guruş for the same kind of work in 1780. Similarly, we do not yet know when real wages and the standards of living in Europe began to move ahead of those in the Middle East. There is no doubt that a better understanding of monetary history will broaden our insights into the economic and social history of the region. It is also quite probable that new research in this area will necessitate the revision of many of the currently held views and generalizations about the economy, society, and state in the Ottoman Empire.[1]

The study of monetary history also underlines the importance of long-distance linkages, of trade and specie flows between different regions of the world in the early modern era. It reminds us once again that the Ottoman Empire was located at a rather vulnerable intersection of long-distance trade routes and bullion flows. During the sixteenth and seventeenth centuries, large amounts of gold and silver arrived from the Americas. A large part of these precious metals were then used by Europeans to pay for imports of spices, silk, and other wares from southeast Asia. As the Middle East was often a transit zone for these intercontinental bullion flows, the monetary system of the empire came under immense pressure.

Ever since Earl J. Hamilton reformulated the monetarist argument with evidence gathered from Spanish archives more than half a century ago, historians have been debating the causal linkage between the arrival of large amounts of gold and silver from the Americas and the rise in prices in the Old World during the sixteenth and early seventeenth centuries.[2] Most prominent amongst the defenders of this hypothesis in recent decades have been economic historians of monetarist persuasion and the adherents of the Annales School.[3] New evidence recently compiled by Morineau has shown, however, that Hamilton's data were incomplete and that the volume of specie flows into Europe continued to increase during the seventeenth century even after prices had started their descent. His findings now cast serious doubt about the validity of the causal linkage between bullion inflows and inflation.[4]

Despite the critical importance of these debates for understanding Ottoman economy, their impact on Ottoman historiography has been very limited. Two decades ago, Ömer Lütfi Barkan provided the most notable attempt to insert the Ottoman case into the debate when he linked the Ottoman price increases until the 1580s to trade and specie inflows from Europe using a theoretical framework very similar to that of the Annales School.[5] Despite the availability of archival data, virtually no empirical work has been undertaken since that time on money and prices to examine these arguments in greater detail in the Ottoman context. Similarly, alternative explanations of the so-called Price Revolution such as demographic change and its consequences have not yet received serious attention from Ottoman historians.

This article focuses not on the Price Revolution but on a monetary phenomenon that took place somewhat later. In contrast to the economic expansion of the sixteenth century, it is well known that the seventeenth century was a period of monetary disorders as well as economic crises in many parts of Europe. Starting somewhere towards the middle of the century and lasting for at least half a century in the areas close to Istanbul and considerably longer in many of the provinces, Ottoman mint output declined and Ottoman coinage disappeared from local markets. In their place arrived European coinage, gold, silver, and, increasingly, copper. Foreign coins always circulated in Ottoman domains side by side with Ottoman coinage from as early as the fourteenth century. What happened during the seventeenth century, however, was qualitatively different.

While government documents and court records continued to refer to akçe and, to a lesser extent, to para, these Ottoman monetary units were in fact reduced to units of account. Actual payments both in the capital city and the provinces were made in European coins. It would be appropriate, therefore, to talk about a disintegration of the Ottoman monetary system and view the monetary developments of the period from this new perspective. This paper will examine the reasons behind the disappearance of the akçe and the ascent of European coinage. To set the stage, the next section will provide a brief outline of the Ottoman monetary system before the seventeenth century.[6]

THE OTTOMAN MONETARY SYSTEM: AN OUTLINE

Until the sixteenth century, the Ottoman domains covering most of Anatolia and the Balkans had a unified monetary system based on the silver akçe (which dated back to the earliest days) and the gold sultani (first minted in the 1470s). At the bottom of the hierarchy were copper coins called *mangır* or *pul*, with nominal values and used for small daily transactions. For the most part, the mints were kept open for the coinage of both silver and gold subject to payments to the state. Foreign coins circulated extensively and without any form of government intervention. The government often announced the official rates at which gold and silver coins, Ottoman as well as foreign, would be accepted as payment. Similarly, in local markets, Ottoman and foreign coin changed hands on the basis of their market rates of exchange.

It is perhaps worth noting that, contrary to the prevailing view, the use of money was not limited to segments of the urban population but was quite widely spread during the sixteenth century. Recent research on the rural economy has shown that through their participation in markets and because of state taxation of a wide range of economic activities, large segments of the rural population came to use coinage, especially the small denominations of akçe and the mangır.[7] A close examination of Ottoman provincial codes of the sixteenth century, whose primary purpose was to establish the sources of fiscal revenue for the state, similarly point to an economy with strong urban-rural linkages, considerable market orientation, and frequent collections of small amounts of taxes in money from artisans and merchants as well as from nomads and sedentary peasants.[8]

As the emerging Ottoman state territorially expanded to become a full-fledged empire, however, the relatively simple monetary system based on the akçe and the sultani became increasingly more complex. The newly conquered territories, each of which was subject to different economic forces, already had well-established currency systems of their own. In most cases, the Ottoman governments did not attempt to change them as they wished to avoid economic disruption and possibly popular unrest. Although the coinage minted in these territories began to bear the name of the sultan, Ottoman authorities in Istanbul could not fully control the evolving monetary structures, each of which remained distinct from the Istanbul-based akçe/sultani system.

While this pragmatic attitude differed from the practices of other contemporary states, both European and Asian, which extended over considerably smaller territory, it is consistent with Ottoman administrative practices in other fields. With respect to land-tenure systems, for example, the Ottoman central administration did not attempt to impose the timar regime in all conquered territories. In many of the more distant areas administered more loosely from the capital, such as eastern Anatolia, Iraq, Egypt, Hungary, or the Maghreb, the Ottomans were content to collect taxes, but changed the existing land regime either to a limited extent or not at all. As a result, in monetary as well as other matters, there emerged, inside the empire, different zones with varying degrees of administrative control. At the core were the areas most closely administered by the capital, with institutions most closely resembling those in the Istanbul region. With increasing distance from the capital, these institutions and administrative practices reflected the power balances between the capital and the local forces in the provinces.

During the sixteenth and seventeenth centuries, the Balkans, together with western and central Anatolia, remained as the core regions of the akçe/sultani system described earlier.[9] In outlying Wallachia, Moldavia, and Hungary, however, Austrian, Polish, Hungarian, and German coins were used more widely than Ottoman coins. In the Crimea, coins were minted in the name of local khans even though Istanbul exerted some influence in monetary affairs, especially since the prices of agricultural imports from the Crimea played an important role in the economic life of Istanbul.

When the Ottomans conquered Egypt, they retained the standard small silver coin called *medin, nisf*, or *nisf fidda*, which dated back to the early fifteenth century. Over time, this unit began to be called *para* and remained the basic silver standard in Egypt until the end of the eighteenth century.[10] During the sixteenth century, the medin also circulated in Arabia and Yemen. Other Ottoman coins were also minted in Yemen from the 1520s to the middle of the seventeenth century, but they do not appear to have been significant economically. In Syria, which remained as transitional monetary zone between Egypt and Anatolia, akçes circulated together with medins.

The areas neighboring Iran, from Eastern Anatolia to Iraq, were especially sensitive for the Ottoman government. In this region, the Ottoman mint produced a coin called *dirham* by the numismatists and *shahi* by the Ottoman documents and the local population. Its weight and silver content were similar to the shahis of Persia.[11] Political factors may have contributed to this strategy of having a separate monetary zone next to the Safavid Empire. Perhaps more importantly, trade with the East exhibited large deficits during the sixteenth century, and the outflow of specie occurred mostly through this region. The minting of the shahi might, therefore, be interpreted as part of unsuccessful Ottoman attempts to control the outflow of specie to the East.[12] Finally, in the Maghreb, the political and economic ties to Istanbul remained rather weak. Even though local coins carried the name of the Ottoman ruler, Istanbul exerted little influence over the evolution of the monetary system in Algeria, Tunisia, and around Tripoli. European coins circulated extensively in this region.

THE SEVENTEENTH CENTURY

The sixteenth century until the 1580s was a period of relative fiscal and monetary stability coupled with demographic and economic expansion in the Ottoman Empire. The territorial conquests generated additional fiscal revenues for the state especially from Egypt. Along with increases in population, land under cultivation and internal as well as long-distance trade expanded. With growing commercialization, economic ties between the countryside and urban areas became stronger. The increasing monetary needs of this economy were apparently met by the increased availability of gold, primarily from Egypt, and silver arriving from the Americas by way of Europe. The silver content of the akçe and the gold sultani remained relatively stable between the 1480s and the 1580s.[13]

It is now believed that these favorable trends were reversed starting in the 1580s. It appears that the long-term increases in population and economic activity came to a halt towards the end of the century. Later, combined with social and political upheavals, population and economic activity may have declined during the seventeenth century.[14] Equally important for the monetary history of the period, Ottoman state finances began to experience serious difficulties late in the sixteenth century. The inability to conquer new territories, long and protracted wars against Iran and the Habsburgs, the adverse consequences of world-wide inflation on state revenues all played their part in intensifying fiscal difficulties. The onset of the Celali rebellions during the last decade of the century exacerbated budgetary problems even further.[15]

At the same time, the arrival of large amounts of silver from the New World lowered the relative price of that metal, leading to the closure of silver mines in Serbia and Bosnia that had served as the principal sources of specie for Ottoman coins until the sixteenth century.[16] When fiscal pressures began to intensify, therefore, the state could not fall back on the earlier sources of silver to maintain steady supplies of coinage.

As a result, the period from the massive debasement of 1585 or 1586 until the 1640s was one of exceptional instability for the akçe. Its silver content fluctuated sharply and often. In addition, since the government was not always successful in collecting the earlier coinage or adjusting the official rates of exchange after each debasement, the clipped versions of the old akçes often circulated together with the new versions, leading to further confusion. Not surprisingly, the counterfeiting of silver coins flourished in this environment. Each time these problems reached crisis proportions, with adverse effects for the economy, the government attempted to establish a new standard for the akçe. These operations, called *tashih-i sikke* (correction of coinage), were carried out in 1600, 1618, 1624, and 1640.[17]

This half a century of debasements turned the akçe into an impractical coin. With its weight and silver content reduced from 0.7 grams until the 1580s to less than 0.3 grams, the akçe became a very small and very thin coin. Similarly, its purchasing power became too small for many daily transactions. In response, the Ottoman government eventually began to mint a new coin called para, which carried three times as much silver as the akçe and which was based on the monetary unit in circulation in Egypt and parts of Syria. It appears that the first minting of para in Istanbul was undertaken during the reign of Murad IV (1623–1640), possibly in 1640, but its volume remained very small.[18]

This protracted period of instability led to a serious loss of confidence in the Ottoman currency with important long-term consequences. The frequent fluctuations and decline in the silver content of the coins coupled with the disappearance of the better-quality coins from circulation made it increasingly risky to hold akçes. There emerged, as a result, greater demand for the more stable European coinage. Measured in terms of their silver content, the European coins began to circulate at a premium over their Ottoman counterparts.[19]

It would be misleading, however, to examine the monetary difficulties in the Ottoman Empire solely in terms of the fiscal problems being experienced by the state. In many ways, the state finances and the Ottoman economy were coming under the influence of powerful monetary forces operating on a world scale. Large amounts of gold and, especially, silver had arrived from the Americas since the early part of the sixteenth century. The Middle East, however, was often a transit zone for these intercontinental flows. While the Ottoman government welcomed the arrival of large amounts of silver and silver coinage from the West, it could not prevent their outflow towards Iran and India, as the empire continued to run trade deficits towards the East. Hence, the sharp contrast between the experiences of the Ottoman Empire and Safavid Iran, on the one hand, and Mughal India, on the other. While the Ottoman and Safavid economies struggled with monetary instability, the monetary system in Mughal India benefitted from trade surpluses during the seventeenth century.

Despite the recent findings by Morineau indicating that the volume of American silver arriving in Europe continued to increase, the seventeenth century was a period of great monetary turbulence and apparent scarcity of specie in many parts of Europe.[20] With the disappearance of silver from circulation, the government came to rely more on copper coinage. As Frank Spooner pointed out so aptly, however, copper currency acted like yeast in everyday transactions, driving silver further away. By the middle of the century, copper coinage was being used heavily in parts of Europe. In Spain, Sweden, and Russia, the great majority of daily transactions were being undertaken with copper coinage. Even in France, the government was forced to limit the circulation of copper in order to contain the damage.[21]

As it became increasingly more difficult for the Ottoman government to locate supplies of silver and control the quality of coinage produced in the provinces, it began to close down the mints. The volume of silver and gold coin production at the Istanbul mint declined significantly during the 1640s. The important exception was the *tecdid-i sikke* (renewal of coinage) operations associated with the accession to the throne of Mehmed IV in 1648. Minting activity declined even further after the mid-1650s. It appears that for the next three decades, until the mid-1680s, the limited volume of gold and silver coins produced at Istanbul was used primarily by the sultan and his retinue in ceremonial occasions.[22]

THE MINTS

There is very little information about the activities of the provincial mints during this period. In sharp contrast to the large numbers of active mints in the earlier period, available archival evidence points out to limited activity in a small number of mints during these decades. In orders sent to provincial mints, the government expressed its reluctance to maintain their operations in view of the poor quality of coinage being produced. It might be added that the low level of activity and the poor quality of coinage at the provincial mints were related to their inability to locate or attract supplies of silver.[23]

While the mint records are mostly unavailable, numismatic evidence can be employed to shed additional light on the operations of Ottoman mints during this period. Since most Ottoman coins carry the name of the sultan, the year of his accession, and the location of the mint, it is possible to follow, on the basis of coins available in numismatic collections and catalogues, the decline in the numbers of provincial mints active during each reign.

According to numismatic catalogues, the number of active mints producing silver akçes in the Ottoman Empire peaked late in the sixteenth century. During the twenty-one year reign of Murad III (1574–1595), the akçe was minted in at least 38 locations around the empire, most of them in Anatolia and the Balkans. The number of mints producing akçes remained at the same level, close to 40, during the eight-year reign of Mehmed III (1595–1603). More than a quarter century later, during the seventeen-year reign of Murad IV (1623–1648), close to 30 mints produced akçes.[24]

The number of active mints declined either during the latter part of the reign of Murad IV or during the 1640s. For during the reign of the next sultan, Ibrahim I (1640–1648), numismatic catalogues can point to no more than four locations producing the akçe. After this shift, the number of mints producing silver coinage remained low until the 1680s. For the long reign of Mehmed IV (1648–1687), the same catalogues can currently list no more than seven locations for the akçe and the larger para.[25]

Clearly, these numismatic lists of locations are not final, and there might be a few, but not many, additions in the future. Even with some additions, however, this emerging pattern of fewer active mints beginning some time in the second quarter of the century should remain intact. In addition, it has already been established from archival evidence that the mint in Istanbul did not increase its output to compensate for the decline in the provinces. On the contrary, its volume remained sharply lower until the 1680s. Combining the two sources of evidence, we can state with confidence that there occurred a substantial decline in the production of Ottoman coinage during these decades.

It should be noted that for reasons that are yet to be fully established, minting of copper coinage also remained limited during this period.[26] This is quite intriguing, since many states in Europe and elsewhere relied on copper coinage in the face of scarcities in silver.[27] One explanation for the absence or limited use of copper coinage is the small size of the akçe. It has been argued that since the nominal value of the mangir had always been in fractions of the akçe, the decline in the value and purchasing power of the akçe might have limited the usefulness of any coin even smaller in value.[28]

EUROPEAN COINAGE AND THEIR DEBASED VERSIONS

When the Ottoman government could not or did not meet the economy's demand for money, this need was met increasingly by European coins. Although foreign coins had always circulated in Ottoman lands since the fourteenth century, they played a qualitatively different role during the seventeenth and early eighteenth centuries. As Ottoman coinage disappeared, the akçe was reduced to little more than a unit of account, an invisible unit in which monetary magnitudes were cited and the values of actual coins were measured.[29] Foreign coins in fact became the leading forms of actual money in the akçe region, from the Balkans and Istanbul to Anatolia and Syria. Local court records, the observations of well-known travellers such as Tavernier and Chardin, and, more recently, the studies by economic and social historians on Ottoman provinces provide ample evidence in this respect.[30]

One of the more prominent silver coins in circulation from the Balkans to Egypt was the Dutch thaler. It was called *esedi guruş* or *aslanlı guruş*, as it had the inscription of two lions.[31] Even more important were the Spanish "pieces of eight" (*reales de a ocho*) called the *riyal guruş*, which was, in fact the most widely used coin of the world economy during the sixteenth and seventeenth centuries. There were others such as the rix-thaler *(kara guruş)* and the Polish isolette, or *zolota*, which was later imitated by Dutch and English merchants. The Venetian gold ducat remained the most important gold coin.[32] The Ottoman government did not attempt to restrict the circulation of these coins. In fact, the government regularly accepted, and sometimes even demanded, payment in European coinage.

The debased versions of some of the leading Europeans coins, the Venetian ducat, the Dutch thaler, and the Polish zolota all made their appearance in Ottoman markets at one time or another during the seventeenth century. In addition, towards the middle of the century, as shortage of coinage intensified, counterfeits or debased versions of small European coins began to arrive literally in shiploads. The inflow of these small coins reached their peak, not surprisingly, between 1656 and 1669, during the long war with Venice for Crete, when the Ottoman government was preoccupied with military and political problems, and mint output had declined to very low levels if not stopped altogether.[33]

CONCLUSION

As the series of demanding wars began to wind down towards the end of the century, the Ottoman government tried hard to establish a new monetary system around a new and larger unit. Not surprisingly, this new unit was called the Ottoman guruş. The first Ottoman guruşes, containing roughly the same amount of silver as their European counterparts and with the standard (1 guruş = 120 akçes), were minted in 1690.[34] Because of fiscal and economic difficulties, however, it took a long time for the new unit to establish itself in the provinces. In many parts of the empire, the Ottoman guruş did not become the leading currency until the second half of the eighteenth century.

The monetary history of the sixteenth and seventeenth centuries raises interesting questions regarding the power of the Ottoman state as well as its policies and attitudes towards money, and, more generally, the economy. The inability of the state to impose a single monetary unit for the empire after the territorial conquests of the sixteenth century provides strong evidence regarding the limitations of its power. Similarly, the inability of the state to maintain the akçe even in the core areas of the empire during the seventeenth century provides additional evidence in the same direction. It is well known, of course, that the fiscal, administrative, and political powers of the Ottoman state showed considerable decline during the seventeenth century.

Secondly and in the face of these limitations in its power, Ottoman state policy in monetary affairs appears remarkable for its pragmatism. From the fourteenth century, the Ottoman governments were quite aware of the vulnerability of the economy and state finances to recurring shortages of specie and the difficulties associated with relieving these shortages. As a result, they chose to encourage the circulation of foreign coinage rather than demand that they be surrendered to authorities. The experiences of other contemporary states suggest that the prohibition of foreign coins would have been difficult to enforce, would have driven these coins underground, and, equally importantly, it might have deprived the state treasury of an important source of bullion. The Ottomans' willingness, after the conquests of the sixteenth century, to allow the circulation of medin, shahi, and other local coinage rather than attempt to impose the akçe in the new territories needs to be interpreted as another important example of the same attitude. Finally, given its inability to provide a steady output of coins during the seventeenth century, the Ottoman government appears to have accepted the circulation of not only European coins but even its debased or counterfeit versions. This pragmatism may help explain why the Ottoman archives are curiously silent about the debased coins while European observers paid a good deal of attention to the same phenomenon. For the economy, the circulation of debased coins was better than having no coins. Similarly, the state continued to receive tax revenues and make payments with the debased coins. The state finances would have been even worse off without them.

All of this need not imply that the decline and disappearance of the akçe did not pose serious challenges to the Ottoman administration. Without control over the currency, its control over the economy certainly diminished considerably. In addition, without its own currency, the state could not use debasement as a means of obtaining fiscal revenue in times of difficulty. It is for these reasons that the Ottoman state attempted to establish a new monetary system as soon as its military and fiscal problems subsided.

Finally, let us turn to the connection between coinage and sovereignty. It is well known that since antiquity the issuing of coinage with the ruler's name has been considered an important symbol of sovereignty in the Mediterranean basin. Being born into the same region, Islamic states embraced this principle from the earliest period. Yet, it also needs to be pointed out that ever since the Classical Age in Greece, issuing of a limited amount of coinage for symbolic purposes has been viewed as being quite distinct from maintaining a stable currency. The latter requires more power, and, for states with modest resources, it often came into conflict with other goals, most notably, monetary, economic, and fiscal stability.[35] In other words, the linkage between coinage and sovereignty may well have been an important abstract principle for the Ottomans. When that principle came into conflict with some of their more practical concerns, however, the Ottomans chose to remain pragmatic.

BIBLIOGRAPHY

Barkan, Ömer Lütfi. *Zirai Ekonominin Mali ve Hukuki Temelleri, Kanunnameler, Cilt 1.* Istanbul, 1942.

———. "The Price Revolution of the Sixteenth Century: A Turning Point in the Economic History of the Near East." *International Journal of Middle East Studies* 6 (1975):3–28.

Braudel, Fernand, and Spooner, Frank. "Prices in Europe from 1450 to 1750." In *The Cambridge Economic History of Europe*, edited by E.E. Rich and C.H. Wilson, vol. 4. Cambridge: Cambridge University Press, 1967.

Chardin, Sir John. *Voyages du Chevalier Chardin en Perse et aux Indes Orientales*. London: Moses Pitt, 1686.

Çizakça, Murat. "Osmanli Ekonomisinde Akçe Tağşişinin Sebebleri Üzerinde Kısa bir Inceleme," *Boğaziçi University Journal, Administrative Sciences and Economics* 4–5 (1976–77):21–27.

Erder, Leila, and Faroqhi, Suraiya. "Population Rise and Fall in Anatolia, 1550–1620." *Middle Eastern Studies* 15 (1979):322–45.

Erüreten, Metin. "Osmanlı Akçeleri Darp Yerleri," *The Turkish Numismatic Society Bülten*, no. 17 (1985), pp. 12–21.

Establet, Colette, and Pascal, Jean-Paul. "Damascene Probate Inventories of the 17th and 18th Centuries: Some Preliminary Approaches and Results." *International Journal of Middle East Studies* 24 (1992):373–93.

Faroqhi, Suraiya. "Sixteenth Century Periodic Markets in Various Anatolian Sancaks." *Journal of the Economic and Social History of the Orient* 22 (1979):32–80.

———. "A Great Foundation in Difficulties: Or Some Evidence on Economic Contraction in the Ottoman Empire of the Mid-Seventeenth Century." In *Mélanges Professeur Robert Mantran*, edited by A. Temimi, pp. 109–21. Zaghouan, CEROMDI, 1988.

Hasluck, F. W. "The Levantine Coinage." *Numismatic Chronicle, Fifth Series* 1 (1921):339–91.

Inalcık, Halil. "Osmanlılar' da Raiyyet Rüsumu." *Belleten* 23 (1959):575–608.

———. "Impact of the Annales School on Ottoman Studies and New Findings." *Review*, Fernand Braudel Center 1 (1978):69–96.

———. "Military and Fiscal Transformation in the Ottoman Empire, 1600–1700." *Archivum Ottomanicum* 6 (1980):283–337.

———. "Osmanlı Idare, Sosyal ve Ekonomik Tarihiyle Ilgili Belgeler: Bursa Kadı Sicillerinden Seçmeler." *Türk Tarih Kurumu, Belgeler*, vol. 10, no. 14 (1981):1–91.

Jennings, Ronald C. "Loans and Credit in Early 17th-Century Ottoman Judicial Records." *Journal of the Economic and Social History of the Orient* 16 (1973):168–216.

Kafadar, Cemal. "Les Troubles Monetaires de la Fin du XVIe Siècle et la Prise de Conscience Ottoman du Declin." *Annales, Économies, Societés, Civilisations*, no. 2 (1991), pp. 381–400.

Kütükoğlu, Mübahat S. *Osmanlılarda Narh Müessesesi ve 1640 Tarihli Narh Defteri*. Istanbul: Enderun Kitabevi, 1983.

Mantran, Robert. *Istanbul dans la seconde Moitié du XVIIe Siècle*. Paris: Librairie Adrien Maisonneuve, 1962.

———. "Politique, Économie et Monnaie dans l'Empire Ottoman au XVIIème Siècle." In *Social and Economic History of Turkey (1071–1920), Papers Presented to the First International Congress on the Social and Economic History of Turkey*, edited by O. Okyar and H. Inalcik, pp. 123–25. Ankara: Hacettepe University, 1980.

Martin, Thomas R. *Sovereignty and Coinage in Classical Greece*. Princeton, New Jersey: Princeton University Press, 1985.

Masson, Paul. *Le Commerce Français dans le Levant au XVIIe Siècle*. Paris: Hachette, 1896.

Masters, Bruce. *The Origins of Western Economic Dominance in the Middle East: Mercantilism and the Islamic Economy in Aleppo, 1660–1750*. New York: New York University Press, 1988.

Morineau, Michel. *Incroyables gazettes et fabuleux métaux: Les retours des tresors Americains d'après les gazettes Hollandaises (XVIe–XCIII[e] siècles)*. New York and Paris: Cambridge University Press and Editions de la Maison des Sciences de l'Homme, 1985.

Pamuk, Şevket. "Money in the Ottoman Empire, 1326 to 1914." In *The Ottoman Empire: Its Economy and Society: 1300–1914*," edited by H. Inalcık and D. Quataert. Cambridge University Press, forthcoming, 1994.

———. "17. Yüzyılda Avrupa'da Basılan Kalp Paralar Osmanlı Piyasalarında Niçin Kabul Görüyordu?" *Toplym ve Ekonomi*. Istanbul, no. 2 (1991):55–62.

Richards, J.F., ed. *Precious Metals in the Later Medieval and Early Modern Worlds*. Durham, North Carolina: Carolina Academic Press, 1983.

———. (ed.) *The Imperial Monetary System of Mughal India*. Delhi: Oxford University Press, 1987.

Sahillioğlu, Halil. *Kuruluştan XVII. Asrın Sonlarına Kadar Osmanlı Para Tarihi Üzerine Bir Deneme*. Ph.D. dissertation, University of Istanbul, 1958. Available in the Library of the Faculty of Economics.

———. *Bir Asırlık Osmanlı Para Tarihi*. Thesis for docent University of Istanbul, 1965. Available in the Library of the Faculty of Economics.

———. "Bakır Para Üzerine bir Enflasyon Denemesi (H. 1099–1103/M. 1687–1691)." *The Turkish Numismatic Society Bülten*, no. 10 (1982):7–40.

———. "The Role of International Monetary and Metal Movements in Ottoman Monetary History." In *Precious Metals in the Later Medieval and Early Modern Worlds*, edited by Richards, pp. 269–304, 1983.

Schaendlinger, Anton C. *Osmanische Numismatik*. Braunschweig: Klinkhardt & Biermann, 1973.

Spooner, Frank C. *The International Economy and Monetary Movements in France*. Cambridge, Mass.: Harvard University Press, 1972.

Spufford, Peter. *Money and its Use in Medieval Europe*. Cambridge: Cambridge University Press, 1988.

Sundhaussen, Holm. "Die 'Preisrevolution' im Osmanischen Reich während der zweiten Halfte des 16. Jahrhunderts." *Südost-Forschungen* 42 (1983):169–81.

Tavernier, J. B. *Nouvelle relations de l'interieur du Serrail du Grand Seigneur*. Paris: G. Clouzier, 1675.

NOTES

1. The most important piece on Ottoman monetary history to appear in English in recent decades is Sahillioğlu (1983). A more comprehensive survey is attempted in Pamuk (1994). That the monetary history of India's Mughal Empire has received much attention in recent years is evidenced by the important collection by Richards (1987). The relative strength and stability of the Mughal monetary system is one important reason for this contrast.

2. Hamilton (1929).

3. See for example, Braudel and Spooner (1967) and Spooner (1972).

4. Morineau (1985). For another recent and important volume providing a global perspective on monetary developments during the sixteenth and seventeenth centuries, see Richards (1983).

5. Barkan (1975). Amongst other notable contributions to the debate are Çizakça (1976–77); Inalcık (1978); Sundhaussen (1983); and Kafadar (1991).

6. In recent decades, Mantran (1962, 1980) and Sahillioğlu (1965) have also drawn attention to the extraordinary aspects of these monetary developments for Ottoman history.

7. Faroqhi (1978, 1979); Jennings (1973); Inalcık (1981).

8. A compilation of these provincial codes is available from Barkan (1942).

9. For details see Pamuk (1994).

10. For the Egyptian medin and para, the most important source is Raymond (1974).

11. Sahillioğlu (1983).

12. Geographically, a good deal of overlap existed between the three units, akçe, medin (or para), and the shaki. For example, mints in the cities of Amid (Diyarbakır) and Aleppo struck all three of these coins, and a number of other mints struck two of them during the second half of the sixteenth century. Schaendlinger (1973), pp. 99–105.

13. Pamuk (1994) and also Sahillioğlu (1983).

14. See, for example, Erder and Faroqhi (1979); Inalcık (1980); and Faroqhi (1988).

15. Inalcık (1980).

16. Sahillioğlu (1983).

17. Sahillioğlu (1983); Kütükoğlu (1983); Kafadar (1991).

18. Schaendlinger (1973), p. 110.

19. Pamuk (1994).

20. Morineau (1985).

21. Spooner (1972).

22. Sahillioğlu (1965), pp. 18–36.

23. Sahillioğlu (1965), pp. 36–37, provides evidence of minting activity for Bagdad, Damascus, Aleppo, and Belgrade. This is remarkably consistent with the numismatic evidence cited later.

24. These were the mints at Istanbul, Diyarbakır, Damascus, and Cairo. Schaendlinger (1973), pp. 102–106, and Erüreten (1985), pp. 15–18.

25. The mints cited were in Istanbul, Belgrade, Noveberde, Diyarbakır, Damascus, Aleppo, and Cairo. Schaendlinger (1973), pp. 106–1,13, and Erüreten (1985), pp. 18–19.

26. The big Ottoman experiment with copper coinage came between 1687 and 1691, when the coins were produced in very large volume strictly to raise fiscal revenue. Sahillioğlu (1982) provides a detailed account of this episode.

27. Spooner (1972), chapter 1.

28. The decline in the value of the akçe might have also brought the costs of production of the mangır quite close to, if not above, their nominal values. Production and distribution of mangir might have then become less profitable or not profitable at

all for the private individuals who usually purchased the local rights to issue mangir from the government. Another possibility that needs to be examined is whether supplies of copper were available. The other alternative, of course, is that because of a variety of reasons, fiscal, political, and administrative, the Ottoman government was simply unable to conduct any meaningful policy in monetary affairs.

29. For discussions of the term "unit of account," see Braudel and Spooner (1967), pp. 378–83, and Spufford (1988), Appendix II.

30. Tavernier (1675) and Chardin (1686). In recent times, Robert Mantran was one of the first to point out the decline of the akçe in local markets. His account is especially striking since it describes the conditions in the capital city. Mantran (1962), pp. 233–85. With the appearance, during the last decade, of new studies on the economic and social history of the provincial cities that make extensive use of local court records, it is now possible to get a geographically more detailed account of the disappearance of the akçe and the rise of European coinage. See for example, Masters (1988) and Establet and Pascal (1992).

31. Most European silver coins were called *guruş*, which was the local adaptation of *groschen*, a diminutive for gross or grosso, a term used for silver coins in many European countries going as far back as the thirteenth century. The *esedi guruş* was called *ebukelb* in the Arab provinces, apparently because the lions were taken for dogs.

32. Mantran (1962), pp. 233-71; Sahillioğlu (1965), pp. 38-53; Masson (1896), pp. 492-94.

33. Sahillioğlu (1965), pp. 45-49. Contemporary accounts by Tavernier (1675) and Chardin (1686) provide detailed descriptions of this phenomenon. The most detailed numismatic treatment is given by Hasluck (1921). For a recent study of this phenomenon, see Pamuk (1991).

34. Sahillioğlu (1965), pp. 90-122; Sahillioğlu (1983); and Pamuk (1994).

35. Martin (1985).

16

Imports of Treasure and Surat's Trade in the Seventeenth Century

V.B. Gupta

It is well known that the domestic production of gold and silver in India has always been negligible. As a result, the availability of these metals to meet sub-continental demand has usually been dependent on imports[2]. It is also well-known that the sub-continent received a part of these metals from Europe either directly or via the Middle East. Bullion also flowed into India from China, Taiwan, Japan and Africa. A very diverse group of people and organizations contributed to the inflow of gold and silver into this country in the early modern period. The list of importers of precious metals into India thus included the European companies, the European merchants and factors on their private account, the Mughal emperors, princes and princesses, nobles and above all the Asian merchants, both big and small.

The primary object of this note is to analyse the inflow of bullion into Gujarat from the Middle East during the seventeenth century. This note is based largely on the Dutch archival sources. It has also drawn heavily on a Ph.D. dissertation by Hans Walther van Santen[3].

---

[1] I am grateful to Om Prakash and Sanjay Subrahmanyam for comments on an earlier version of this paper. Opinions expressed here are, however, those of the author alone.

[2] Om Prakash, *Foreign Merchants and Indian Mints in the Seventeenth and Early Eighteenth Century*, in: J.F.Richards (ed.), The Imperial Monetary System of Mughal India, Delhi, 1987, p. 171; I. Habib, *The Monetary System and Prices*, in: T. Raychaudhuri and I. Habib (eds.), The Cambridge Economic History of India, Delhi, 1984 (reprint), vol. 1, p. 361.

[3] H.W. van Santen, *De Verenigde Oost Indische Compagnie in Gujarat en Hindustan, 1620-1660*, unpublished Ph.D. dissertation, Leiden University, 1982, pp. 74–78.

In the course of discussion, I have attempted to show that there was no discernible trend in the inflow of precious metals into Gujarat during the seventeenth century. It was only during the first decade of the eighteenth century that one notes some signs of expansion in the imports of bullion. However, this hypothesis of post-1700 expansion remains as yet tentative.

No precise estimate of the inflow or net inflow of bullion into Gujarat for the period can be made. In the first place, at times the information is available in such diverse units that no comparable figures can be compiled for different years. For instance, one often notes import figures in terms of bags of *abbasis*, bundles of Spanish *rials*, chest of rix dollars, canisters of gold and silver and so on; it is therefore not possible to make precise calculations unless information is available as to how many *abbasis* were contained in a bag or how many *rials* there were in a bundle, chest or canister. The problem is more serious since we find that there was no standard rate of conversion between a chest or bag and the number of pieces[4]. In addition to this, one may find cases when a bag or chest had mixture of currencies of different values[5].

A second problem relates to incomplete coverage. Information on the amounts brought by the Indian merchants through the European companies' vessels in the 17th century is available only rarely. We cannot, however, ignore the value of the bullion brought by these merchants on freight in the companies' ships because the occasional evidence that does exist shows that the amounts brought in by the Indian merchants on the companies' vessels could exceed the amounts brought in their own. For instance, in 1645, according to Dutch sources, Gujarati ships from Persia brought gold and silver worth a little more than 200,000 *rupees* into Surat[6]. But at the same time it was

---

[4] For instance, an entry in the Dutch records of 1676 shows that one chest contained 4,000 *rupees* and another entry in the diary kept at the Surat factory dated 1.10.1692 shows that one chest had as much as 40,000 *rupees*. Batavia - Holland (hereafter B.XVII, that means Batavia to XVII Directors) 30.9.1676, VOC 1315, f. 69; Diary maintained at Surat, VOC 1476, f. 525.

[5] In 1681, for instance, a Muslim merchant's ship brought into Surat 7 canisters containing gold and silver coins and the *Salamatras* of Mulla Abdul Ghafur brought 77 bags containing a mix of *rials*, ducats and *abbasis* from Jeddah; Shipping list 1681-1682, VOC 1379, ff. 2555-2564.

[6] H.W. van Santen, *o.c.*, p. 75.

reported that annually about 20 *lakh rupees* worth of gold and silver was imported by these merchants into Surat from Persia[7]. If these two reports are taken at their face value, it implies that the remaining 18 *lakh rupees* worth of bullion (some nine times of what was brought by the Asian traders in their own vessels) was brought on the European ships as freight. Finally, the data relate only to the arrivals at Surat. Ships bringing precious metals to other ports of Gujarat are not included. This may, however, not be a major problem during the second half of the seventeenth century as Surat by that time had emerged as the major port of Gujarat and thus attracted most of the long distance shipping destined for the region as a whole.

Keeping these reservations in mind, let us first examine the broad trends in the import of bullion into Gujarat in the period. A glance at Table 1 would clearly indicate that the Indian merchants imported a large amount of gold and silver in minted and unminted form during the seventeenth century into Surat from Mocha, Basra and Persia. The fact that Surat was largely dependent on these sources of precious metals is evidenced by frequent reports in the Dutch records that whenever the annual fleets from the Red Sea and the Persian Gulf failed to bring large amounts of treasure, the Surat money market experienced a considerable shortage of liquid funds, that this often could cause a significant rise in the rate of interest. The data further reveal that during the 17th century the value of imports of bullion from the Red Sea fluctuated widely, at times being as low as 5 *lakh rupees*, but on other instances attaining a level as high as 50 *lakh rupees*. It was only during the early 18th century that the value rose to 60 *lakh rupees* (viz. 6 million) or more. In contrast to the relatively plentiful information on the Red Sea-Surat trade, import figures in respect of the Persian Gulf are available only for a few years, and are difficult to interpret.

Table 1 also clearly reveals that until about the beginning of the 18th century no clear cut expansion is visible in bullion imports into Surat from the Middle East. In this connection, it is important to note that no particular year should be taken in isolation. A conspicuously large import in a particular year could be the result of too little inflow in the preceding year or years. In 1687, for example, the Dutch factor

[7] *Ibid.*

Table 1

Estmates of Import of Gold and Silver into Surat by the Indian Merchants' Ships from the Red Sea, Persia and Basra (in hundred thousand *rupees*)

year	from the Red Sea	from Persia and Basra
1582	12	N.A.
1622	41[a]	N.A.
1624	25	N.A.
1628	12	N.A.
1630	20	N.A.
1638	5	N.A.
1641	35	N.A.
1643	7	Negligible[b]
1644	13	N.A.
1645	8	2
1646	11	2
1647	16	1.7
1649	18-24	
1651	8.5	7.5
1653	20	N.A.
1654	21	N.A.
1657	N.A.	20-25
1666	N.A.	Negligible[c]
1676	28	N.A.
1679	50	N.A.
1680	38-43	N.A.
1681-1682	5[d]	-
1682-1683	N.A.[e]	4.7[f]
1685	50	N.A.
1686-1687	15[g]	-
1692	16.4[h]	-
1708	62	-
1710	60	-

Source: M.N. PEARSON, *Merchants and Rulers of Gujarat*, p. 101. W. P. COOLHAAS (ed.), *Pietervan den Broecke in Azie*, vol. 2, pp. 273, 275, 296, 297 and 341. S.B. 8.10.1630, VOC 110, f. 202^{v}; VOC 1166, ff. 797-803. G.M. II, dated 19.1.1654, p. 729, dated 26.1.1655, p. 800. XVII. B. 16.5.1676, VOC 320, the document is not foliated. S.B. 7.9.1680, VOC 1361, ff. 572-575. For 1681-1682, 1682-1683 and 1686-1687, see the arrival of ships at Surat, VOC 1379, ff. 2555-2564, VOC 1383, ff. 630-632 and VOC 1414, ff. 1539-1546.

The information for the years 1638 to 1646, 1647, 1649, 1657, 1679, 1685, 1708 and 1710 is from H.W. VAN SANTEN, *De VerenigdeOost-Indische Compagnie in Gujarat en Hindustan 1620-1660*, p. 76.

[a] The value of precious metals brought from Jeddah on 1.11.1622, is not available in the source, so not included. W.P. COOLHAAS (ed.), *Van den Broecke in Azië*, vol. 2, p. 278.

[b] In 1643, Indian merchants import of bullion was worth only *rupees* 1,906. See Om Prakash's Presidential Address at Bodh Gaya, p. 11.

[c] Indian ships from Persia reportedly brought just 93,000 *rupees* worth *abbasis* S.B. 6.6.1666, VOC 1259, f. 3233.

[d] See Table 2, n. 4. The figures include import both from Red Sea and the Persian Gulf.

[e] Information is given in boxes.

[f] In addition these merchants brought 186 bags containing cash, 92 bags of *abbasis*, 15 bundles of gold ducats and 32 packets of gold *rupees*.

[g] See Table 2, n. 7. The figures include import from both the Red Sea and the Persian Gulf.

[h] Refers to the ships arrived in January, May and September 1692.

from Surat reported that there was a great scarcity of money as very little gold and silver had flowed into the town during the past two to three years[8]. The point which I wish to emphasize here is that the data given in the Table 1 as well as the additional evidence that one finds in the Dutch archives is not sufficient to conclude with any degree of certainty that there was a significant expansion in the import of bullion by the Indian merchants into Surat from the Middle East.

In the first place, the value of the import of bullion during the first half of the seventeenth century was as much as, if not more than 4 million *rupees*. In 1622, for instance, the Indian merchants imported bullion from the Red Sea worth more than 41 *lakh rupees*. Similarly,

[8] S.B. 12.5.1687, VOC 1434, ff. 330-330v.

in 1624, Pieter van den Broecke, the Dutch director at Surat, recorded in his journal the arrival of two Gujarat ships from the Red Sea[9]. One of these had on board gold, silver and merchandise worth 25 *lakh rupees* and the other was also reported to be richly laden with treasure. Although the value of treasure brought by the second ship is not stated, the phrase »richly laden« is suggestive enough. These two ships thus had brought bullion worth well over 25 *lakh rupees* from the Red Sea alone. An earlier Dutch report from Surat in 1620 clearly indicates that large quantities of bullion were flowing into the town at that time[10]. In 1628 also, bullion worth 12 *lakh rupees* was brought by one imperial ship alone while another Muslim ship arrived at Surat from Mocha on September 6th of the same year[11]. Unfortunately, the sources are silent on the value of bullion brought by the latter vessel. Then in 1630, Gujarati ships from Mocha and Jeddah reportedly brought gold and silver worth 20 *lakh rupees*[12]. In 1641, the value of these imports from the Red Sea was as high as 35 *lakh rupees*[13]. On the other hand, we find many instances when the import of bullion from the Red Sea during the period following 1650 was no more than 20 *lakh rupees*. In 1651, for instance, it was worth only 8.5 *lakh rupees*[14]. During 1653 and 1654, these merchants imported cash specie worth 20 and 21 *lakh rupees* respectively[15].

It has been suggested by the American historian Joseph J. Brennig that during the second half of the seventeenth century, much larger amounts of gold and silver flowed into Surat in comparison to the quantities imported before 1650[16]. He compares the value of bullion imported by the Asian merchants into Surat during 1643-1644 with the value given in a Dutch memoir referring to the end of the seven-

---

9 W.P. COOLHAAS (ed.), *Pieter van den Broecke in Azie*, The Hague, 1962, vol. 2, pp. 275 and 296-297.

10 OM PRAKASH (ed.), *The Dutch Factories in India 1617-1623*, New Delhi, 1984, p. 138.

11 W.P. COOLHAAS, *o.c.*, p. 341.

12 S.B. 8.10.1630, VOC 1100, f. 202.

13 H.W. VAN SANTEN, *o.c.*, Table 5, p. 76.

14 See Table 1.

15 W.P. COOLHAAS (ed.), *Generale Missiven van Gouverneurs-Generaal en Raden aan Heren XVII der VOC*, The Hague, 1964, vol. 2, p. 729, letter dated 19.1.1654; also p. 800, letter dated 26.1.1655.

16 J.J. BRENNIG, *Silver in Seventeenth Century Surat: Monetary Circulation and the Price Revolution in Mughal India*, in: J.F. RICHARDS (ed.), Precious Metals in the Later Medieval and Early Modern Worlds, Durham, 1983, pp. 480-481.

teenth century[17]. The question arises of the methodological soundness of a comparison between the value of imports at the end of the century with only one year of the first half of the seventeenth century. We have already noted that in certain years prior to 1650, bullion imports into Surat by Indian merchants were as high as 40 *lakh rupees* or more. Brennig has also argued for another significant change in the pattern of bullion flows into and out of Surat in the course of the 17th century, suggesting that imports of silver during the years 1690 to 1720 were so great that the VOC occasionally found it possible to export silver from Surat to its other factories. In this connection, some points need clarification. In the first place, the Dutch Company's export of silver from Surat to their other factories was much larger from the mid 1660s to 1680s in comparison to the amounts sent during the 1690s. In fact, the Surat factory sent cash specie to other factories in very small amounts during the last decade of the century. On the contrary, as Brennig himself has noted, six Dutch ships from Amsterdam imported 214,512 *rials* into Surat during 1703–1704[18]. In fact, the amount of silver *rupees* sent by the Dutch from Surat to other factories depended on many considerations, such as: a) The needs of other factories that could not be met by Holland/Batavia; b) The size of the surplus generated by the Dutch at Surat from the sale of their merchandise; c) The amount that the Surat factory could raise in the local money market and d) The relative cost of borrowing at Surat in comparison to other establishments.

It is obvious therefore that large transfers of silver *rupees* by the Dutch from Surat are not necessarily closely correlated with bullion flows into Surat. One more point needs to be emphasized. Sometimes the amount of »capital« imported by the Indian merchants is confused with the amount of bullion imported. The word »capital« generally included both goods and bullion. Thus a letter from Surat in 1688 reported that the merchants from Mocha, Jeddah, Basra, Persia etc. usually brought a capital of nearly 60 *lakh rupees*[19]. If goods and bullion were imported in the ratio of 1:2, it would mean that bullion worth 40 *lakh rupees* was the normal value of imports of Indian merchants during the late 1680s.

---

[17] Brennig cites K.A. 1689, *Memoir ter order van de Hoog Edele gesorenge Heer Joan van Hoorn*, p. 164.

[18] J.J. Brennig, *o.c.*, p. 481.

[19] S.B. 17.5.1688, VOC 1454, f. 1317v.

That commodities other than bullion had a certain importance in the import trade of Surat does not seem to be unrealistic. An idea about the composition of bullion and goods in the imports from the Middle East can be formed from the following example. In 1691, 16 Indian merchants' ships from Basra, Jeddah, Muscat, Persia etc. reportedly brought into Surat jewels, gold, silver and other goods worth 61.5 *lakh rupees*, comprising the items in Table 2[20].

We have seen that at least some past authors favour the hypothesis that there was an expansion in the inflow of bullion into Surat over the course of the 17th century. If this were indeed so, to what causes might one attribute such a phenomenon? H.W. van Santen suggests that the »monopolization« of freight trade from Surat to west Asia by Shah Jahan and Aurangzeb could have played some role in this regard[21]. In 1654, Gerardo Pelgrom, the Dutch director at Surat had also argued that the Mughal Emperor Shah Jahan had »monopolized« freight trade with the aim of encouraging import of bullion as there were no gold and silver mines in the country[22]. However, to the extent that Shah Jahan in the 1650s monopolised the freight trade, his activities must in part have displaced the Indian merchants' and European companies' shipping to the Red Sea and the Persian Gulf. The policy was modified only in 1662 when his son Aurangzeb (now the Mughal emperor) instructed his officers to dispose off five out of his seven ships with the aim of sending only two ships to Mocha and Jeddah[23].

In fact, the so-called monopolization of freight trade by Shah Jahan in the early 1650s was not a great innovation. In 1651 for instance, out of the total import of bullion worth 852,493 *rupees* from the Red Sea, as much as 748,394 *rupees*, forming approximately 88 per cent of the total, was brought by the imperial ships[24]. Similarly in the case of the import from the Persian Gulf, the Mughal emperor's vessels brought in the same year 780,000 *abbasis* out of the total import of 978,950

[20] Arrival of ships at Surat, 1691, VOC 1529, ff. 403-404.

[21] H.W. VAN SANTEN, *o.c.*, p. 75.

[22] Van Santen cites G. Pelgrom's »Remonstrantie« to J. Maetsuyker, 24.5.1655, VOC 1209, f. 485.

[23] S.XVII (that means Surat to XVII Directors), 14.11.1662, VOC 1239, f. 1242.

[24] Arrival of ships at Surat, February 1651 to 27th December 1651, VOC 1188, ff. 566-572. The Emperor's ships and the prince's ship brought 328,899 *rials* of eight in the total of 376,470 *rials*. Similarly, out of the total ducats of 7,597, the imperial ships brought 7,527.

Table 2

List of Items Shipped by 16 Indian *Increhants*, 1691

item	quantities
"Real" Pearls	15 bags and 24 bundles worth Rs. 29,610
"False" Pearls	1 packet
Venetian ducats	33,766 and 9 bags
Gold	6 bundles
Silver	1 bundle
Spanish *reals*	9 chests and 1639 bags
Persian *abbasis*	21,600
"Tack" Coral	3 packets
Tortoise shells	24 packets
Ivory	131 packets
Black coral	48 packets
Almonds	758 packets
Coffee	248 packets
Red dyestuff (low quality)	137 packets
Red dyestuff	36 packets
Tobacco	16 packets
Sulphur	11 packets
"Rasmalle"	19 vats
Saffron	1 packet
Persian Wine	351 packets
Dates	276 packets
Myrrh	79 packets
Horses	62
and any more items	

abbasis[25]. In this case also nearly 80 per cent of the total bullion was brought by the royal ships. The imperial involvement in the shipping to the Red Sea can be traced back to the year 1582, when one of Akbar's vessels brought from the Red Sea a rich treasure worth 12 *lakh rupees*[26]. The Mughal emperors sent one, two, or more ships to

[25] *Ibid.*

[26] M.N. Pearson, *Merchants and Rulers in Gujarat: The Response to the Portuguese in the Sixteenth Century*, New Delhi, 1976, p. 101.

Mocha almost every year. These imperial ships carried hundreds of pilgrims, each with a few packets of textiles, indigo and other merchandise, and who on their return brought quantities of gold and silver in minted and unminted form. In 1621 Prince Khurram's (the future emperor Shah Jahan) ships arrived at Surat with a large amount of precious metal from Mocha[27]. Again in 1622, the »Salamati«, an imperial ship, brought 200,000 ducats from the same port. Similarly in 1628, royal ships from the Red Sea reported to have brought gold and silver worth 12 *lakh rupees*[28]. In 1637 the Mughal emperor instructed his officials that the merchants should be permitted to make purchases at Taj Ganj (near Agra) only after cotton textiles worth 60,000 to 70,000 *rupees* had already been bought for his ships destined for Mocha[29]. Again in 1638, cotton stuffs and indigo worth 95,000 *rupees* were purchased at Ahmedabad and Baroda for the royal ships[30]. In 1643-1644, Shah Jahan's ships from Jeddah brought into Surat 10,700 ducats and 113,428 *rials*[31]. Similary in 1647, the »Jang Johar« of Shah Jahan brought 474 *tolas* gold, 29 *sers* silver, 138,974 *rials* of eight, and 15,617 venetian ducats, and during the same year, the »Shahbi« of the Mughal prince had on board 104,479 *rials* of eight, 160 *tolas* gold, 4 *sers* silver and 8,202 venetians[32]. Imperial involvement in maritime trade declined somewhat in the early 1660s when Aurangzeb instructed his officials to reduce the size of the imperial fleet. The interest, however, remained all through the seventeenth century. This is testified to, *inter alia*, by a Dutch letter of 1679 from Surat. This letter reported that as usual 3 vessels of the Mughal emperor had been despatched for Mocha with a good amount of freight goods[33]. In 1681, 3 royal ships from Mocha were believed to have brought 22 chests of *rials* and then in 1684, the »Ganj-i-sawai« of the Mughal brought 28 chests of gold and silver[34].

---

[27] W.P. Coolhaas (ed.), *Pieter van den Broecke in Azie*, vol. 2, p. 264.

[28] W.P. Coolhaas, *o.c.*, pp. 275 and 341.

[29] H.W. van Santen, *o.c.*, pp. 70-71.

[30] H.W. van Santen, *o.c.*, p. 71.

[31] J.J.Brennig, *o.c.*, p. 480.

[32] List of ships arriving at and departing from Surat, 1647, VOC 1166, ff. 800-803.

[33] S.B. 5.4.1679, VOC 1349, f. 1637. These three vessels reportedly had on board merchandise worth 300,000 *rupees*.

[34] List of ships that arrived at Surat 1681-1682, VOC 1379, ff. 2555- 2564; list of ship arrivals, 1684-1685, VOC 1414, ff. 1539-1546.

How diverse the group of people engaged in the maritime trade between Surat and the Red Sea was can be judged from the following details. An unfortunate Mughal ship, the »Shahi« that ran aground on its way from Surat to Mocha in 1637, had on board the goods of the following[35]:

- *shahbandar* of Surat, Haji	*rupees* 100,000
- *mutasaddi* Hakim	70,000
- Mirza Muhammad	20,000
- the VOC's *sarraf*, Nan	30,000
- the EIC's *sarraf*, Tapi Das	20,000

Some have argued that as Surat had a much bigger marine in the second half of the seventeenth century, a larger number of Asian ships were engaged in the trade between Surat and the Middle East and this resulted in a larger amount of inflow of bullion into Surat. There is absolutely no doubt that the number of ships on these routes had increased in this period. For instance, a Dutch factor reported in 1661 that the Company had granted as many as 13 passes for Mocha to the merchants of Surat and 3 passes for the Mughal emperor's vessels[36]. In 1665 the Company granted 21 passes for the Red Sea and 23 passes in the following year[37]. However, there is no direct correspondence between the number of ships engaged and the value of bullion imported. Om Prakash has rightly shown that in 1651 and 1654, while the number of ships arriving from the Persian Gulf was considerably larger than those from the Red Sea, the value of import of bullion was greater from the latter[38]. Secondly, the size of the vessels engaged in the trade also play their role in the quantities imported. Pearson has suggested that by the middle of the seventeenth century the large sized ships had lost their superiority and thus one finds that much of the Gujarat-Mocha trade was handled in the

[35] W.P. COOLHAAS (ed.), *Generale Missiven*, vol. 1, p. 622.

[36] S.XVII, 29.1.1661, VOC 1232, f. 399v.

[37] S.XVII. 5.1.1666, VOC 1259, ff. 3252-3253; S.XVII. 21.2.1667, VOC 1264, f. 629v.

[38] OM PRAKASH, *Some Aspects of the Trade of Mughal India*. Presidential Address to the Indian History Congress, Medieval Section, 42nd Session, Bodh Gaya, 1981, p. 11, n. 15. The number of ships arriving from the Red Sea was 8 and that from the Persian Gulf was 14 in 1651. The corresponding figures for 1654 were 12 and 17. According to his estimates the value of the precious metals imported from the Persian Gulf during 1651 and 1654 was 575,904 and 1,173,300 *rupees* respectively while that from the Red Sea 822,322 and 1,931,620 *rupees*.

medium sized ships of about 200 tons[39]. Moreover, it was not always the case that all the vessels from the Red Sea and the Persian Gulf had bullion on board. This is clear from the information available on the import of bullion for the year 1682–1683[40]. In all twelve Indian merchant ships came from Persia and Basra. Four of these ships had only goods, and the remaining eight had goods and bullion. These eight vessels carried on board *rials*, *abbasis* and gold worth 469,453 *rupees* plus 186 bags containing cash, 92 bags of *abbasis*, 75 bundles gold and ducats and 32 packets of gold *rupees*. During this year, seven ships came from Mocha and Jeddah. One of these had no bullion on board. The other six brought cash, ducats, *rials*, gold and silver.

Finally, we should not ignore the fact that the Indian merchants' shipping was often affected by wars and piracy. In such cases, these merchants had to change their areas of operation. In 1678, for instance, it was reported that Indian merchants' demand for passes for the south-east Asian ports had increased because their movement to the Middle East had declined[41]. Similarly in 1695, between October and December, ten ships arrived at Surat, but of these, not even one came from the Red Sea or the Persian Gulf[42].

All in all, it appears difficult to conclude from available evidence that in the course of the seventeenth century any significant expansion in the import of bullion by the Indian merchants into Surat had taken place. In fact what appears an expansion to some may have been no more than a revival of Indian merchants' trade to the Red Sea which had almost come to a standstill during the wars between the Turks and the Arabs around late 1620's to late 1630's[43]. This is evident from the fact that in 1641 the import of bullion from the Red Sea registered an increase of nearly seven times over the value imported in 1638 i.e. from 500,000 to 3,500,000 *rupees*[44]. Another point that needs to be emphasized is that the Mocha market, the most important entrepot in the Red Sea, was a highly unpredictable centre of trade. A

---

39 M.N. Pearson, *Introduction*, in: A. Das Gupta and M.N. Pearson (eds.), *India and the Indian Ocean, 1500–1800*, Calcutta, 1987, p. 15.

40 List of ships arriving at Surat, 1682–1683, VOC 1383, ff. 630–632.

41 B.XVII. 15.2.1678, VOC 1323, f. 575v.

42 List of ships arriving at Surat, October–December 1695, dated 18.12.1695, VOC 1582, ff. 117–119.

43 W.P. Coolhaas (ed.), *Pieter van den Broecke in Azie*, vol. 2, p. 333.

44 H.W. van Santen, *o.c.*, p. 76.

Dutch report on Mocha in 1683 made a specific point that Mocha market was subject to violent fluctuations[45]. There is ample evidence in the Dutch documents that confirms that such uncertainties characterised this centre of trade[46]. It was perhaps in this sense that Niels Steensgaard has described Asian markets as non-transparent markets. It may therefore not be safe to use figures of high value of import of certain years in the late seventeenth century to establish any definite trend.

We may now briefly examine the question of where the Indian merchants imported a larger amount of bullion - from the Red Sea or the Persian Gulf. If we take the case of bullion brought in their own vessels, the picture is unambiguous. Om Prakash has shown, on the basis of comparable data available for the years 1643, 1645, 1647, 1651 and 1654, that the value of import of bullion by the Indian merchants into Surat from the Red Sea had been greater than that from the Persian Gulf[47]. Whether the Red Sea could maintain its lead over the Persian Gulf is a big question. An import of 20-25 *lakh rupees* worth of bullion from the Persian Gulf in 1657 was very close to the value imported from the Red Sea between 1649 and 1654[48]. The Indian merchants' imports from the Persian Gulf were, as has already been noted, much larger on the European ships than in their own. This was so basically for two reasons. First, soon after the fall of Hormuz in 1622, the Indian merchants generally feared an attack by Portuguese vessels on the Surat-Persia route. Secondly, a much larger number of European companies' and European merchants' vessels were found on the Surat-Hormuz run than on the Surat-Mocha run. Both the English and the Dutch had their trading establishments at Surat and Gombroon, whereas voyages to Mocha by the European companies' vessels were made only occasionally. There is enough evidence in the

---

45 Report by factors Cloek and Spender on Mocha, 28.3.1683, VOC 1406, ff. 1114-1115.

46 For instance, while in 1665 and 1666, Gujarat's merchants reportedly obtained extremely low profits on their merchandise at Mocha, in 1667, these merchants were reported to have sold their goods at high prices and earned substantial rates of profit in Mocha, Aden etc. Similarly while in 1680, these merchants earned as much as 100 per cent on their goods at Mocha, in 1684 they could not even cover their costs. S.B. 11.2.1666, VOC 1259, ff. 3113-3115; S.XVII 21.2.1667, VOC 1264, ff. 630-630v; S.XVII. 12.11.1667, VOC 1261, f. 751; S.B. 14.9.1680, VOC 1361, f. 571; S.B. 20.5.1684, VOC 1406, f. 955v.

47 Om Prakash, *o.c.*, p. 11.

48 H.W. van Santen, *o.c.*, Table 5, p. 76.

Dutch records to establish that both the VOC and the EIC earned annually a substantial amount by carrying Indian merchants' goods on freight between Surat and Persia.

This brings us to the part played by the European companies in the import of bullion into Gujarat. We know that in the composition of the imports of the north European companies into the East, in terms of value, bullion imports far exceeded the import of goods. Until about the mid-1660s the EIC and the VOC, and thereafter the EIC and the French Company, brought precious metals into Surat. It is suggested sometimes that the amount of bullion imported by the European companies during the seventeenth century was not large enough to make a difference in the overall picture[49]. In this connection we must note that the Europeans' import of bullion into Gujarat was not simply from Europe but also from the Middle East, China, Japan and Taiwan. Any judgement on the relative share of the Europeans in the total imports of precious metal must take into consideration the fact that the value of bullion imported by them depended upon a number of factors and thus varied enormously from year to year. The VOC for instance, imported bullion into Surat worth between 350,000 *rupees* and a little more than 800,000 *rupees* during the period 1628-1629 and 1640-1641 and between 180,000 to 950,000 between 1641-1642 to 1659-1660[50]. But during the period following 1667, the VOC hardly imported any bullion into Surat until at least the end of the seventeenth century. Instead, the Dutch exported cash specie from Surat to other establishments is Asia for the greater part of the latter half of the seventeenth century. The value of the EIC import of bullion also varied from year to year. In 1619, the English Company brought into Surat 240,000 *rials* (or 500,000 *rupees*) from England and a large amount of gold and silver for Mocha[51]. In 1629, the EIC imported into Asia bullion worth 16 *lakh rupees* but only 1.3 *lakh rupees* worth in 1631 and then 9.2 *lakh rupees* worth in 1633[52].

---

[49] A. Das Gupta, *Indian Merchants and the Indian Ocean*, in: T. Raychaudhuri and I. Habib (eds.), *The Cambridge Economic History of India*, vol. 1, p. 417.

[50] H.W. van Santen, *o.c.*, Table 2, p. 37. His table, however, misses an import of Japanese silver worth 321,648 *rupees* from Taiwan, mentioned in VOC 1188, f. 575.

[51] Om Prakash (ed.), *The Dutch Factories in India*, p. 123.

[52] K.N. Chaudhuri, *The English East India Company: The Study of an Early Joint Stock Company, 1600-1640*, London, 1965, p. 115.

These fluctuating imports into Asia would be reflected in the amounts imported to Surat. Fortunately we have comparable figures of import of bullion into Surat by the VOC, the EIC and Asian merchants for 1643-1644, 1647, 1651, 1681-1682 and 1686-1687.

A glance at Table 2 would reveal that while the relative share of the VOC and the EIC changed from year to year, the European import of bullion into Surat clearly had a lead over the imports by the Indian merchants at least during 1643-1644, and 1651. The picture for the years 1681-1682 and 1686-1687 is not so clear because the figures noted in the table are incomplete. On the face of it, while in 1681-1682 the European import of bullion far exceeds the Indian import, during 1686-1687 the Indian merchants' import had a lead over the European imports. These figures thus clearly suggest that the European companies and European merchants import of precious metal into Surat was not at all insignificant. Not only this, the EIC import of bullion showed an increasing trend at least until 1681-1682, when the Company imported more than 24 *lakh rupees* worth bullion into Surat. According to another estimate, the English Company's import of bullion into Surat increased from 298,764 *rupees* per annum during 1658-1667 first to 398,017 *rupees* during 1668-1677 and then to 1,367,404 *rupees* per annum during 1678-1681[53]. The English Company imported bullion worth *rupees* 1,150,000 during 1683[54]. The import of bullion by the French Company was also not insignificant. In Feb. 1671, for instance, two ships on behalf of the French Company brought from France into Surat bullion worth 361,657 *rupees* (50,000 *rials*, 15 ponds gold and 5035 ponds unminted silver)[55]. Later in December 1671, seven French ships brought into Surat bullion worth 565,124 *rupees*[56]. Similary during 1681-1682, the French Company imported into Surat bullion worth *rupees* 379,791[57].

It is thus clear to us that the European imports of bullion into Surat were not inconsiderable during the seventeenth century. It has also been established that the European companies provided freight

[53] Bal Krishna, *Commercial Relations between India and England, 1601-1757*, London, 1924, p. 299.

[54] S.XVII. 31.3.1683, VOC 1388, f. 2092.

[55] S.XVII. 21.2.1671, VOC 1284, f. 2209.

[56] S.B. 12.12.1671, VOC 1279, f. 872.

[57] List of ships arriving at Surat during 1681-1682, VOC 1379, ff. 2555-2564.

Table 3

Import of Bullion into Surat, 1643-44, 1647, 1651, 1681-82 and 1686-87
(in *rupees*)

year	Indian merchants			Europeans			
	in their own ships	on freight	total	EIC	VOC	french	total
1643-1644	819,779	108,180	999,959	569,604	773,293	-	1,342,897
1647	966,628	165,194	1,131,822	875,727	180,833	-	1,056,560
1651	1,604,378[a]	N.A.	1,604,378	1,081,433[b]	794,300[c]	-	1,875,643
1681-1682	507,488[d]	N.A.	507,488	2,423,611[e]	nil	379,791[f]	2,803,402
1686-1687	1,499,981	N.A.	1,499,981	777,373[g]	nil	426,405	1,203,778

Source: 1643-1644, J.J. BRENNIG, *Silver in Seventeenth Century Surat*, Table 1; For VOC, H.W. VAN SANTEN, *De VOC in Gujarat*, p. 37, Table 2. 1647, Arrival of ships at Surat, 1647, VOC 1166, ff. 797-803. 1651, Arrival of ships at Surat, February 1651 to December 1651, VOC 1188, ff. 566-572. 1681-1682, Arrival of ships at Surat, 1681-1682, VOC 1379, ff. 2555-2564. 1686-1687, Arrival of ships at Surat, 1686-1687, VOC 1414, ff. 1539-1546.

[a] From the Red Sea worth 852,493 *rupees* and from the Persian Gulf and Basra worth 751,885 *rupees*.

[b] This includes 583,333 *rupees* brought by the English factors in their private account.

[c] 472,652 *rupees* from Persia and Basra and 321,648 *rupees* from Taiwan.

[d] In addition, 198 bags *abbasis*, 25 boxes and 1 *canaster* (perhaps tin box) *rials*, 508 bags *abbasis* and *rials*, 92 bundles of ducats, 4 bundle of gold *rupees*, 19 *canasters* and 1 chest gold and silver, 77 bags containing *rials*, ducats and *abbasis*.

[e] In addition, many bags of *abbasis* and *rials* from Persia.

[f] In addition, many bags of *rials* and *abbasis*.

[g] In addition, 166 bags and 19 bundles *rials* and 84 *canasters* plus 1 chest rix dollars.

[h] This includes imports from Europe, Persia, Mocha and Siam.

service to the Indian merchants to bring, especially from the Persian Gulf, a large amount of precious metal in this period. The value of Indian merchants' import of bullion from the Middle East fluctuated violently throughout the seventeenth century and thus it cannot be demonstrated that the second half of the seventeenth century witnessed a distinct expansion in bullion imports over the first half of the century. And finally, in the relative contribution of the Persian Gulf and the Red Sea, while the latter had a lead until about 1650s, no such conclusion can be drawn for the later period as not much is a-vailable in the documents on the amounts brought as freight from the Persian Gulf.

Index

Please note page numbers which appear in italics are references to tables or illustrations.